Acknowledgments

First and foremost, we thank the Commission and its co-chairs, William Reilly and Senator Bob Graham, for supporting our investigation throughout. We also thank the Commission's executive director, Richard Lazarus, for his extraordinary leadership.

Richard Sears, the Commission's senior science and engineering advisor, was an integral part of our team from the very beginning. We relied heavily on his considerable technical expertise as well as his deep understanding of industry business practices and culture. On the rare instances in which Richard could not answer a question himself, he summoned support from the broad network of industry professionals he created over his 33 years at Shell.

Many individuals involved in the Macondo incident spoke with us voluntarily. They included rig crew members, cementers, mudloggers, equipment suppliers, and shore-based engineers. Notably, many members of BP's Macondo well team met repeatedly with us to explain the chain of events that led to the blowout. BP also released a report of its own nonprivileged investigation of the blowout, and then provided us access to supporting documents. While BP's report differs from ours in scope, purpose, and conclusions, it aided our efforts, and we commend BP for undertaking it. Transocean graciously invited us to visit one of its deepwater drilling rigs to appreciate firsthand the scale and complexity of its operations. Halliburton provided samples of cement materials for testing. And regulators from the former Minerals Management Service spoke candidly about their roles and capabilities.

To support the work of our own retained experts, experts from every corner of the drilling industry met with us, sometimes at great length, to help us understand both the basic science and engineering of offshore drilling and the possible causes of the blowout. Chevron went even further by providing cement testing services and the assistance of a uniquely qualified group of cement testing experts. We also extend special thanks to Darryl Bourgoyne, Steve Lewis, Erik Nelson, and Dr. John Smith.

Megan O'Leary, William Lane, Devin Price, Vincent Yuen, Christina Crankshaw, Sammy Joe Osborne, Corinne Bilyeu, and the rest of the staff at TrialGraphix created the sophisticated graphics and animations that we have used to explain the story of the Macondo blowout, and produced the Web and print versions of this Report. Their results speak for themselves. In addition, Paul Ortiz of Cisco Systems assisted in preparing our public hearing presentation.

Finally, we thank most of all our own very dedicated staff: counsels J. Jackson Eaton, Brent C. Harris, Jon Izak Monger, and Saritha Komatireddy Tice, our paralegal Joseph Hernandez, and our assistant Michelle Farmer. They showed once again how much a small group of talented and hardworking individuals can accomplish. We and the Commission owe them great thanks.

Fred H. Bartlit, Jr., Chief Counsel

Sambhav N. Sankar
Deputy Chief Counsel

Sean C. Grimsley
Deputy Chief Counsel

Table of Contents

For sale by the Superintendent of Documents, U.S. Government Printing Office
Internet: bookstore.gpo.gov Phone: toll free (866) 512-1800; DC area (202) 512-1800
Fax: (202) 512-2104 Mail: Stop IDCC, Washington, DC 20402-0001

ISBN 978-0-16-087963-0

Figures and Tables

Figures

Chapter 4.4| Foamed Cement Stability

Chapter 4.5| Temporary Abandonment

Chapter 4.6| Negative Pressure Test

Foreword

When President Obama created the National Commission on the BP Deepwater Horizon Oil Spill and Offshore Drilling he instructed the Commission to "examine the relevant facts and circumstances concerning the root causes of the *Deepwater Horizon* oil disaster." This Chief Counsel's Report is the result of an extraordinary effort to carry out that mission. It is essential reading for anyone looking to extract the lessons of Macondo and avoid a repeat of that terrible tragedy.

Notwithstanding demanding time constraints, the Commission's Chief Counsel Fred Bartlit and his investigative team have provided the most comprehensive, coherent, and detailed account of the events leading up to the blowout and explosion. The wealth of material presented here offers new details and documentation in support of the Commission's final report, *Deep Water: The Gulf Oil Disaster and the Future of Offshore Drilling*, released on January 11, 2011.

What the investigation makes clear, above all else, is that management failures, not mechanical failings, were the ultimate source of the disaster. In clear, precise, and unflinching detail, this Report lays out the confusion, lack of communication, disorganization, and inattention to crucial safety issues and test results that led to the deaths of 11 men and the largest offshore oil spill in our nation's history.

The Chief Counsel's efforts were integral to the Commission's deliberations and findings. For that reason, this report is an important companion to the full report of the Commission. It stands on its own as well—a durable contribution to our understanding of the importance of responsible management systems and state-of-the-art practices, and the dire consequences when they fail.

Senator Bob Graham
Commission Co-Chair

William K. Reilly
Commission Co-Chair

Executive Summary of Findings

The Macondo blowout happened because a number of separate risk factors, oversights, and outright mistakes combined to overwhelm the safeguards meant to prevent such an event. The Chief Counsel's team identified a number of technical risk factors in the design, execution, and testing of the Macondo well. The team was also able to trace all of these failures back to an overarching failure of management. Better management of personnel, risk, and communications by BP and its contractors would almost certainly have prevented the blowout. The Macondo disaster was not inevitable.

Technical Findings

The root technical cause of the blowout is now clear: The cement that BP and Halliburton pumped to the bottom of the well did not seal off hydrocarbons in the formation. While we may never know for certain the exact reason why the cement failed, several factors increased the risk of cement failure at Macondo. They include the following: First, drilling complications forced engineers to plan a "finesse" cement job that called for, among other things, a low overall volume of cement. Second, the cement slurry itself was poorly designed—some of Halliburton's own internal tests showed that the design was unstable, and subsequent testing by the Chief Counsel's team raised further concerns. Third, BP's temporary abandonment procedures—finalized only at the last minute—called for rig personnel to severely "underbalance" the well before installing any additional barriers to back up the cement job.

BP missed a key opportunity to recognize the cement failure during the negative pressure test that its well site leaders and Transocean personnel conducted on April 20. The test clearly showed that hydrocarbons were leaking into the well, but BP's well site leaders misinterpreted the result. It appears they did so in part because they accepted a facially implausible theory suggested by certain experienced members of the Transocean rig crew. Transocean and Sperry Drilling rig personnel then missed a number of further signals that hydrocarbons had entered the well and were rising to the surface during the final hour before the blowout actually occurred. By the time they recognized a blowout was occurring and activated the rig's blowout preventer, it was too late for that device to prevent an explosion. By that time, hydrocarbons had already flowed past the blowout preventer and were rushing upward through the riser pipe to the rig floor.

Management Findings

The Chief Counsel's team concluded that all of the technical failures at Macondo can be traced back to management errors by the companies involved in the incident. BP did not fully appreciate all of the risks that Macondo presented. It did not adequately supervise the work of its contractors, who in turn did not deliver to BP all of the benefits of their expertise. BP personnel on the rig were not properly trained and supported, and all three companies failed to communicate key information to people who could have made a difference.

Among other things:

- BP did not adequately identify or address risks created by last-minute changes to well design and procedures. BP changed its plans repeatedly and up to the very last minute, sometimes causing confusion and frustration among BP employees and rig personnel.

- When BP did send instructions and procedures to rig personnel, it often provided inadequate detail and guidance.

- It is common in the offshore oil industry to focus on increasing efficiency to save rig time and associated costs. But management processes must ensure that measures taken to save time and reduce costs do not adversely affect overall risk. BP's management processes did not do so.

- Halliburton appears to have done little to supervise the work of its key cementing personnel and does not appear to have meaningfully reviewed data that should have prompted it to redesign the Macondo cement slurry.

- Transocean did not adequately train its employees in emergency procedures and kick detection, and did not inform them of crucial lessons learned from a similar and recent near-miss drilling incident.

What the men and women who worked on Macondo lacked—and what every drilling operation requires—was a culture of leadership responsibility. In remote offshore environments, individuals must take personal ownership of safety issues with a single-minded determination to ask questions and pursue advice until they are certain they get it right.

Regulatory Findings

The Commission's full report examines in depth the history of Minerals Management Service (MMS) regulatory programs and makes specific recommendations for regulatory reform of what is now the Bureau of Offshore Energy Management, Regulation, and Enforcement (BOEMRE). The Chief Counsel's team found that the MMS regulatory structure in place in April 2010 was inadequate to address the risks of deepwater drilling projects like Macondo. Then-existing regulations had little relevance to the technical and management problems that contributed to the blowout. Regulatory personnel did not have the training or experience to adequately evaluate the overall safety or risk of the project.

Chapter 1 | Scope of Investigation and Methodology

Nature of Report

O n May 21, 2010, President Barack Obama signed an Executive Order establishing the National Commission on the BP Deepwater Horizon Oil Spill and Offshore Drilling ("the Commission").[1] The Order directed the Commission to "examine the relevant facts and circumstances concerning the root causes of the Deepwater Horizon oil disaster."[2] The Order instructed the Commission to present a final public report of its findings to the President within six months of the date of its first meeting. The Commission first met on July 12, 2010.

The Commission appointed Fred H. Bartlit, Jr. as Chief Counsel to investigate and present to the Commission findings regarding the root causes of the Macondo explosion and blowout.[3] At public hearings on November 8 and 9, 2010, Bartlit and his team ("the Chief Counsel's team") presented preliminary findings to the Commission regarding the technical, managerial, and regulatory causes of the rig explosion and well blowout. Bartlit and the Chief Counsel's team emphasized at the time that their investigation was ongoing and that these findings were preliminary.

The Commission set forth its findings regarding the root causes of the blowout in a report that it released on January 11, 2011. Several of the Commission's findings were based on the work of the Chief Counsel. Given the factual and technical complexity of some of the underlying causes of the blowout, the Commission asked the Chief Counsel's team to issue a separate Chief Counsel's Report setting forth in greater detail their findings and conclusions regarding the technical, managerial, and regulatory causes of the blowout. This document is that report.

Certain sections of this Report are accompanied by video clips accessible on the Commission's website at www.oilspillcommission.gov. The video clips contain graphics and narration to better explain many of the concepts and findings contained in this Report. They are meant to supplement, not to replace, the Report itself.

Scope of Investigation and Report

The Chief Counsel focused his investigation on the technical, managerial, and regulatory causes of the blowout. This Report sets forth the conclusions of that investigation.

The Report does not discuss containment or response issues, except insofar as certain technical issues related to the blowout also bear on those issues. A separate Commission investigative team addressed these issues. The results of that team's work are reflected in Chapter 5 of the Commission's report and in several Commission staff papers, available on the Commission website.

Similarly, this Report does not extensively review or analyze the regulatory structure in place prior to the spill, but rather addresses the extent to which the absence of particular regulations or enforcement may have contributed to the blowout. This Report also does not broadly review or critique oil and gas industry practices prior to the spill, except to the extent such practices related directly to the Macondo blowout. Other members of the Commission's staff focused on these broader issues, all of which are addressed in the Commission's report and staff papers.

The Chief Counsel's team has worked to provide as exhaustive an analysis of the blowout's causes as is possible given the evidence currently available. Limitations in that evidence prevent the team from issuing findings on some issues.

The Chief Counsel's team cannot offer any final conclusions regarding whether and to what extent the Macondo well's blowout preventer (BOP) failed. The government has engaged the Norwegian engineering firm DNV (Det Norske Veritas) to perform a forensic analysis on the BOP and determine whether it worked as expected and, if not, how and why it failed. The Chief Counsel's team believes that its technical and management findings will stand following the outcome of the pending BOP tests. Information available from those on the rig on April 20 indicates that the BOP was activated, at best, only moments prior to mud overflowing onto the rig floor. And all available evidence shows that by the time it was activated, hydrocarbons were in the riser and expanding rapidly to the surface and the rig. It thus appears that the initial explosions and fire would have occurred, and 11 men would have died, regardless of the function and state of the BOP at the time of the blowout.

As of the date of this Report, DNV is still conducting its analysis. It would be premature for the Chief Counsel's team to issue conclusive findings concerning potential BOP failures before that analysis is complete.

The Chief Counsel's team based its findings on evidence and information available to date. In some instances, that evidence or information was not available when the Commission issued its findings on January 11. And to the extent that new or additional information is revealed regarding issues in this Report, the team may alter or supplement its findings.

Investigation Methodology

The Chief Counsel's team's investigation was extensive. The team collected documents from the companies involved, interviewed percipient and other witnesses, met repeatedly with the companies principally involved in the blowout, reviewed materials gathered by other investigative bodies, engaged and consulted experts, met with industry representatives, and met with representatives from non-industry interest groups. The Chief Counsel's team conducted hearings on November 8 and 9, 2010, during which it questioned, on the record, representatives from BP, Halliburton, and Transocean as well as industry experts, regulators, and the executive officers of Shell and Exxon.

Congress did not provide the Commission with subpoena power. The Chief Counsel's team therefore obtained information about the blowout through the cooperation of many of the companies and individuals involved, and through the cooperation of technical experts in the oil and gas industry. The team encouraged cooperation by conducting its investigation in a completely transparent manner, allowing all of the companies and individuals involved to review and refute the team's preliminary findings. With a few exceptions, these companies and individuals cooperated in an unprecedented fashion. The Commission also obtained the cooperation of other investigative bodies, including most notably the joint Coast Guard-BOEMRE investigation. The lack of subpoena power therefore did not meaningfully hinder the team's investigation.

Every offshore drilling incident deserves serious scrutiny, if only to identify lessons for the oil and gas industry as a whole. But the *Deepwater Horizon* incident involved special circumstances; not every offshore accident will merit the creation of a Presidential Commission. Government and industry should therefore create a standing organization or pre-existing structure that will facilitate future investigations similar to the one the Chief Counsel's team conducted. That entity should have all of the following: (1) pre-existing subject-matter expertise; (2) investigative experience and a clear investigative mandate; (3) a focus on finding facts rather than determining or evaluating legal liability; and (4) the power to compel testimony and the production of evidence. The oil and gas industry and Congress should consider creating such an organization or clarifying that an existing organization fulfills that role. The structure and organization of the National Transportation Safety Board may provide a useful model.

Structure of the Report

Chapter 2 provides introductory and background material, including a detailed explanation of how to drill a deepwater well. Chapter 3 presents a timeline of relevant events leading up to the blowout and a description of the parties and witnesses involved.

Chapter 4 is the lengthiest chapter in the Report. It sets forth the Chief Counsel's team's technical findings and conclusions regarding the technical causes of the blowout. Chapter 4 lists actual causes, potential causes that the Chief Counsel's team cannot rule out, and potential causes that the team *can* rule out even though they appeared credible in the months soon after the blowout. Chapter 4 also briefly describes management failures that may have contributed to each technical cause.

Chapter 5 identifies broader management failures that contributed to the blowout.

Finally, Chapter 6 addresses regulatory failures that may have contributed to the blowout. ◖

Chapter 2 | Drilling for Oil in Deepwater

Oil and Gas in Deepwater

How Oil and Gas Form

Offshore oil and gas reservoirs are formed from sediments deposited by rivers flowing from land into the ocean. Those sediments originate hundreds and sometimes thousands of miles away, in the mountains and broad uplands of an adjoining continent. If the sediments contain organic materials, and if geological processes later subject these sediments to intense pressure and heat, the organic materials can be transformed into liquid and gaseous hydrocarbons over the course of millions of years. The hydrocarbons may remain in the **source rock** where they originally formed, or they can be expelled from the source rock and into other, more porous rock layers. The hydrocarbons tend to migrate upward because they are lighter than other fluids in the pore spaces.

Figure 2.1. Schematic geological cross section.

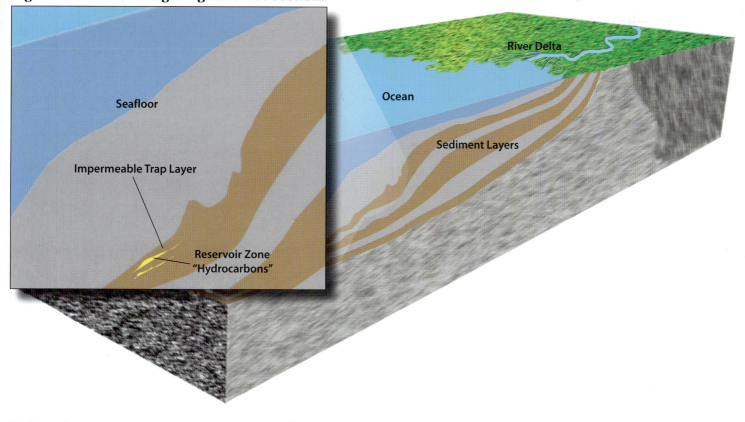

Hydrocarbons trapped beneath an impermeable layer.

TrialGraphix

If a path of porous rock layers leads to the surface, the hydrocarbons will emerge above ground in a seep or tar pit. If an impermeable **trap** layer instead blocks the way, the hydrocarbons can collect in porous rock beneath this seal (see Figure 2.1). Drilling for oil consists first of finding "reservoir zones" (or "pay zones") of trapped hydrocarbons and then drilling through the trap layers into the oil.

Porosity and Permeability. Rock layers vary according to two important measures: porosity and permeability. Porosity is the percentage of the rock that is made up of microscopic pore spaces— places where fluid can reside. Permeability is the ease with which fluid can flow through a rock and is a measure of how well the rock's pore spaces are connected.

The weight of sediments and fluids above the oil-bearing zone exert tremendous pressure on the trapped hydrocarbons. The pressure can be sufficient to force the hydrocarbons all the way to the surface once a well is drilled. Over time, the reservoir pressures can drop as hydrocarbons are extracted, and the hydrocarbons then may need to be pumped out of the well.

Deepwater Oil and Gas Reservoirs in the Gulf of Mexico

The Gulf of Mexico is a rich hydrocarbon province. The Mississippi and other coastal rivers have been eroding the North American continent and have deposited tens of thousands of feet of organic sediments in the Gulf over the past few tens of millions of years. The sedimentary section beneath the sea bed of the central Gulf of Mexico is made up primarily of sand and finer grained shale deposited slowly over time in successive layers.

The thick sedimentary section of the Gulf of Mexico has been the focus of offshore oil and gas exploration efforts since the 1930s. Over many decades, the waters of the Gulf have developed into one of the world's most prolific hydrocarbon provinces, yielding tens of billions of barrels equivalent of oil and natural gas. Today much of the submerged land in the Gulf is administered by the federal government and therefore considered part of the "outer continental shelf." (The National Commission's final report discusses the history of oil exploration in the Gulf of Mexico in Chapter 2.)

Early exploration in the Gulf of Mexico focused on what is known as "shallow water" oil deposits as opposed to "deepwater" deposits. Shallow water wells are typically drilled into sediments that lie atop the continental shelf where it extends into the ocean.

In the central Gulf of Mexico (south of Louisiana), the continental shelf slopes down gently to water depths of roughly 600 feet. At a water depth of about 600 feet, there is a marked transition from the continental shelf to continental slope. On the slope, water depth increases more quickly with distance from shore, and the sea bottom and underlying geology becomes more complex. Although there is no precise definition for the term, "deepwater" wells are typically drilled into sediments that lie on the continental slope and beyond.[*]

Deepwater wells involve markedly different conditions from shallow water wells. The water depth obviously increases, but the sea bottom and geological conditions become more complex too. The sedimentary layers in deeper water are also different. While there is generally continuous deposition of sand and shale on the shelf, deepwater sediments can be dominated by the deposition of **turbidites**. Turbidites are sediments that are deposited episodically during underwater avalanches or other discrete events. Such events can create thick layers of sand that can be very well sorted and display certain attractive reservoir properties: **high porosity**, the

---

[*] Different companies and governmental organizations have adopted definitions for deepwater ranging from 600 to 1,500 feet. The Commission's report used 1,000 feet as the starting point of deepwater, a definition that is used by many in the oil and gas industry.

fraction of rock volume available for holding oil and gas, and **high permeability**, the ability of fluids to move through the rock. These properties make turbidites some of the best oil and gas reservoirs. Individual turbidite layers in the Gulf of Mexico can be many tens of feet to more than several hundred feet thick.

In addition to changes in the underlying geology, the greatly increased water depth requires different drilling approaches. In water depths greater than a few hundred feet, wells are drilled using floating rather than bottom-based rigs.

In water depths greater than about 1,000 feet, it is increasingly impractical to conduct production operations from structures that are supported by the ocean floor, and floating facilities and subsea production systems dominate.

These fundamental changes in drilling methods and reservoir geology combine to define the transition to deepwater.

The Deepwater Opportunity, Attraction, and Challenge

Because of the complexities of deepwater operations, developing a major deepwater oil field can cost enormous sums of money—far more than shallow water development. To make such developments economically viable, oil companies must identify highly productive reservoirs and then install high-productivity wells and production systems. Deepwater turbidite reservoirs are ideal targets because of their high porosity and permeability. Good shallow water wells produce at rates of a few thousand barrels of oil a day. By contrast, deepwater wells commonly produce more than 10,000 barrels per day.

In the early stages of deepwater exploration, operators were surprised by the productivity of deepwater reservoirs. For instance, when Shell developed the Auger Field in the early 1990s, the platform for collecting oil from the wells in the field was originally designed to handle about 40,000 barrels per day of production. Shell was able to increase the platform's capacity to greater than 100,000 barrels per day, despite the fact that it had drilled less than half the wells it originally planned to develop in the Auger Field.

These kinds of reservoirs, which deliver high rates of flow for long periods of time, became the standard for deepwater developments. Drilling wells into these reservoirs became a critical factor for deepwater project success because these production rates justified the high cost of deepwater development.

Favorable geology, while critical, does not alone guarantee deepwater success. It was important in early deepwater developments to establish a "learning curve" where successful engineering practices could be developed and replicated in successive field developments.

In its early Gulf of Mexico deepwater developments in the 1990s, for instance, Shell was able to cut the per barrel development cost by nearly two-thirds over time. Further optimization by Shell and other operators continued to improve the economics of deepwater operations.

As a result, with oil prices and price outlook low in the late 1990s, major oil companies moved aggressively into the deepwater Gulf of Mexico. While these operations were expensive in absolute terms, the development cost per barrel of a carefully executed deepwater project became comparable to shallow water and even small onshore developments.

Deepwater Reservoir Pressures

Another feature of deepwater Gulf of Mexico reservoirs that contributed to overall well productivity also made drilling in deepwater significantly more dangerous: The oil and gas in deepwater reservoirs was often under very high pressure.

Pressure Gradient. In oil and gas drilling, pressure is usually described in terms of a pressure gradient measured from the surface and expressed as an equivalent density of a column of fluid. A "normal" gradient is similar to that produced by a column of seawater, or 8.6 pounds per gallon (ppg). Many deepwater reservoirs, however, are at pressures exceeding 12 ppg. These pressures are not uncommon in oil and gas exploration, but they represent a challenge in that they must be managed carefully.

Managing high pressures in deepwater presents unique challenges. The oil found in deepwater Gulf of Mexico reservoirs typically contains a significant amount of dissolved natural gas. As the oil comes to the surface, the decrease in pressure allows much of this gas to come out of solution. Deepwater Gulf deposits also commonly contain "free gas," which is natural gas that exists separate from the oil, either as a gas cap in the same reservoir sand or a separate gas-bearing zone.

The pressure in a deepwater reservoir can often exceed 10,000 to 15,000 pounds per square inch (psi), several hundred to more than 1,000 times the pressure at the surface. As oil and gas come to the surface, they expand. While the expansion of the oil is moderate, the gas expands in proportion to the drop in pressure. As a result, 10 barrels (bbl) at 5,000 feet could be greater than 1,000 bbl at the surface.

For these reasons, the subsurface pressure and the expansion of fluids flowing to the surface must be carefully managed. The oil and gas industry has developed tools and techniques for doing so. The next section discusses the specific technical tools and methods used to contain and control these pressures.

Rig personnel must be especially vigilant at a deepwater well; because of the pressures involved, it is critical that they detect and address hydrocarbon influxes into the well as early as possible. If they do not stop such influxes early, the rapid expansion of hydrocarbons as they near the surface can become difficult, if not impossible, to control.

How to Drill a Deepwater Well

There are three phases to safely extracting hydrocarbons from an offshore deepwater reservoir. The first phase is drilling. During this phase, rig crews drill and reinforce a hole from the seafloor down through the trap layers and into the reservoir zone. During drilling, it is important for rig crews to prevent hydrocarbons in the reservoir from entering the hole they are drilling, which is called the **wellbore**.

The second phase is completion. During completion, rig crews open the wellbore to allow hydrocarbons to flow into it and install equipment at the wellhead that allows them to control the flow and collect the hydrocarbons.

The third phase is production. In the production phase the operator actually extracts hydrocarbons from the well.

Figure 2.2. Drilling.

TrialGraphix

Typical onshore and offshore drilling consists of drilling mud, a drill bit, and circulating pressures to clear cuttings through the annulus to the surface.

This introduction to drilling a deepwater well focuses on the first phase—the actual drilling—and the concept of **well control**, which refers to the methods for controlling hydrocarbon flow and pressure in a well.

Drilling Overview

Offshore drilling is similar in many ways to drilling on land. Like their onshore counterparts, offshore rig crews use **drilling mud** and rotary drill bits to bore a hole into the earth (see Figure 2.2). Drillers pump the mud down through a **drill pipe** that connects with and turns the bit. The mud flows out of holes in the bit and then circulates back to the rig through the space between the drill pipe and the sides of the well (the **annulus** or **annular space**). As it flows, the mud cools the bit and carries pulverized rock (called **cuttings**) away from the bottom of the well. When the mud returns to the surface, rig equipment sieves the cuttings out and pumps the mud back down the drill string. The mud thus travels in a closed loop.

Pore Pressure and Fracture Pressure

In addition to carrying away cuttings, drilling mud also controls pressures inside the well as it is being drilled. The mud column inside a well exerts downward hydrostatic pressure that rig crews can control by varying the mud weight.

The crew monitors and adjusts the mud weight to keep the pressure exerted by the mud inside the wellbore between two important points: the pore pressure and the fracture pressure. The **pore pressure** is the pressure exerted by fluids (such as hydrocarbons) in the pore space of rock. If the pore pressure exceeds the downward hydrostatic pressure exerted by mud inside the well, the fluids in the pore spaces can flow into the well, and unprotected sections of the well can collapse.

An unwanted influx of fluid or gas into the well is called a **kick**. The **fracture pressure** is the pressure at which the geologic formation will break down or "fracture." When fracture occurs, drilling mud can flow out of the well into the formation such that mud returns are **lost** instead of circulating back to the surface.

Both pore pressure and fracture pressure vary by depth. The **pore pressure gradient** is a curve that shows how the pore pressure in the well changes by depth. The **fracture gradient** is a curve that shows how the fracture pressure in a well changes by depth (see Figure 2.3). Both gradients are typically expressed in terms of an equivalent mud weight.

MAINTAINING BALANCE

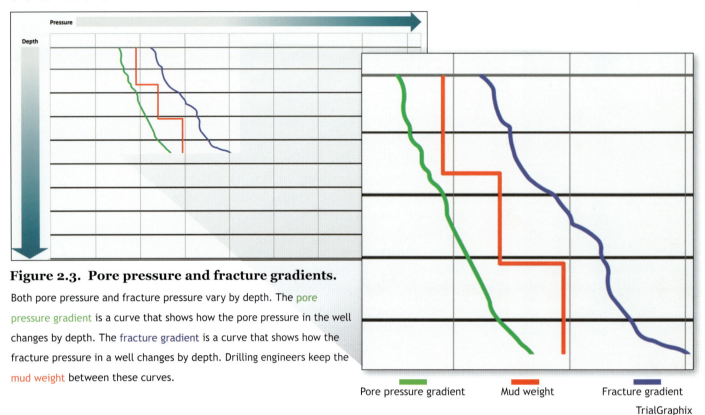

Figure 2.3. Pore pressure and fracture gradients.

Both pore pressure and fracture pressure vary by depth. The pore pressure gradient is a curve that shows how the pore pressure in the well changes by depth. The fracture gradient is a curve that shows how the fracture pressure in a well changes by depth. Drilling engineers keep the mud weight between these curves.

Pore pressure gradient   Mud weight   Fracture gradient

TrialGraphix

The pore pressure and fracture gradients define the boundaries of the drilling process. Drillers strive to keep the mud weight between these two curves.

LOT and FIT. There are two ways to determine a formation's fracture gradient: a **leak off test (LOT)** and a **formation integrity test (FIT)**. In a leak off test, the driller gradually increases the pressure on the formation and stops when the formation begins to give way. The driller can see this occurring by monitoring the pressure at the surface. In a formation integrity test, the driller gradually increases the pressure on the formation to a predetermined value less than the fracture pressure. This test stops before the formation actually begins to give way. In each case, the stopping point is recorded as the fracture gradient of the formation.

Achieving this goal would be simple enough if the pore pressure remained constant from the seafloor all the way down to the hydrocarbon zone. But pore pressure and fracture pressure vary. They typically increase with increasing depth but can sometimes decrease depending on the nature of the formation. As the well goes deeper, drillers typically must increase the weight of drilling fluid to balance increasing pore pressure.

Casing and Cement

At some point as the crew drills deeper, the pore pressure in the bottom of an open hole section will exceed the fracture pressure of the formation higher up in this open hole section. When this happens, the crew can no longer rely on mud to control pore pressure. If the crew increases the mud weight, it will fracture the formation higher up. If the crew keeps drilling but does not increase the mud weight, hydrocarbons or other fluids in the deeper formation will flow into the well.

Figure 2.4. Casting strings (greatly simplified).

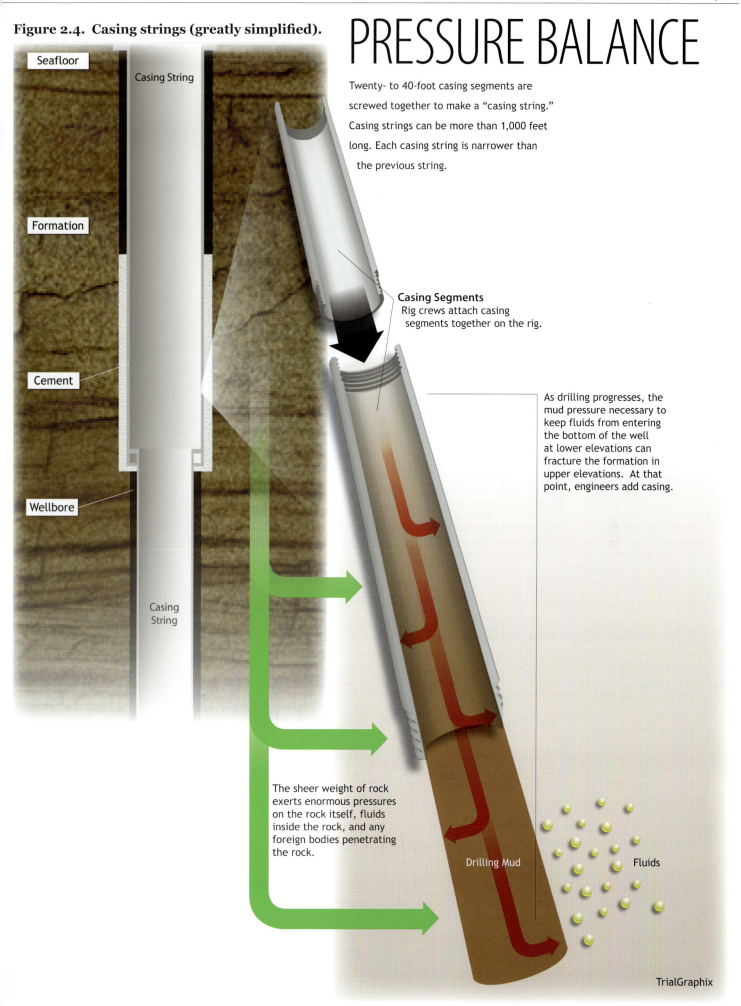

PRESSURE BALANCE

Twenty- to 40-foot casing segments are screwed together to make a "casing string." Casing strings can be more than 1,000 feet long. Each casing string is narrower than the previous string.

Seafloor

Casing String

Formation

Cement

Wellbore

Casing String

Casing Segments
Rig crews attach casing segments together on the rig.

As drilling progresses, the mud pressure necessary to keep fluids from entering the bottom of the well at lower elevations can fracture the formation in upper elevations. At that point, engineers add casing.

The sheer weight of rock exerts enormous pressures on the rock itself, fluids inside the rock, and any foreign bodies penetrating the rock.

Drilling Mud

Fluids

TrialGraphix

At this point, the crew stops and sets **casing**. Casing is high-strength steel pipe that comes in 20- to 40-foot sections that rig crews screw together (or "make up") on the rig to make a **casing string** (see Figure 2.4). Once placed in a well, the casing string serves at least two purposes. First, it protects more fragile sections of the hole outside the casing from the pressure of the drilling mud inside. Second, it prevents high-pressure fluids (like hydrocarbons) outside the casing from entering the well.

After a rig crew runs a casing string down a well, it must cement the casing string into place. Once it sets, the cement does two things. First, it seals the interior of the well (inside the casing) off from the formation outside the casing. Second, it anchors the casing to the rock around it, structurally reinforcing the wellbore to give it mechanical strength.

Drilling in More Detail

The Drilling Rig

Figures 2.5 and 2.6. Rig structures.

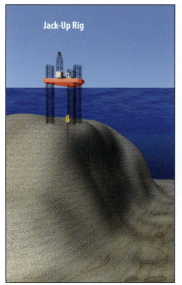
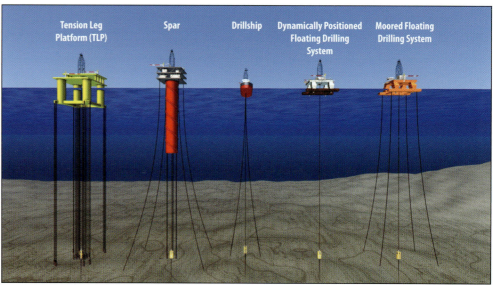

TrialGraphix

Various offshore drilling rigs.

Both offshore and onshore drill crews use drilling rigs to raise and lower drilling tools and casing down the well, pump fluids down the wellbore, and turn the drill bit. An offshore drilling rig must also provide the crew with a stable platform from which to work. There are several types of offshore drilling rigs.

A mobile offshore drilling unit (MODU) is able to move from location to location. In shallow water, rig crews can work from "jack-up" platforms that are towed onto location and then supported by mechanical legs lowered to the seafloor (see Figure 2.5). Deepwater operations require structures that float on the water's surface. Some are floating structures that are moored in place with cables attached to giant anchors. Others are drillships—vessels that carry drilling rigs and support drilling operations (see Figure 2.6).

A dynamically positioned (DP) semi-submersible like the *Deepwater Horizon* is yet another kind of rig that combines features of each of these other rig types. Once moved onto location, a DP rig holds itself in place above a drilling location using satellite positioning technology and directional thrusters.

Spudding the Well

Drilling in deepwater starts when the rig crew "spuds" the well by lowering a first string of casing down to the seafloor. This "conductor casing" is typically 36 inches in diameter or more, and serves as part of the structural foundation for the rest of the well. Welded to the top of the conductor casing is a wellhead assembly. The wellhead assembly remains above the seafloor and serves as an anchoring point for future casing strings. The rig crew lowers the conductor casing into place using the drill string and a "running tool" that attaches the drill string to the wellhead.

The Drill String. The **drill string** is made up of "joints" of **drill pipe** that are 20 to 40 feet in length. Each joint has a threaded male end and a threaded female end that is larger than the overall pipe diameter; these **tool joints** connect one joint of drill pipe to the next. Drill pipe itself comes in different diameters (typically ranging from 5½ to 6⅝ inches in deepwater). In addition to standard drill pipe, rig crews can use drill collars or heavyweight drill pipe as well. **Drill collars** have much thicker walls than standard drill pipe. Rig crews often use drill collars to add weight to the bottom part of a drill string. Doing this helps keep the drill string from kinking or breaking and puts more weight on the bit. **Heavyweight drill pipe** is of an intermediate weight; it minimizes the stress between the drill collars at the bottom of the drill string and standard drill pipe at the top.

The sediments in the first several hundred feet below the seafloor in the Gulf of Mexico are typically unconsolidated materials that lack cohesive strength. They are little more than watery mud. Accordingly, the rig crew does not need to use drilling mud or even a rotary drill bit to create a hole for the conductor casing. The weight of the drill string and conductor casing alone can be more than enough to drive the casing down into the mud at the seafloor. The rig crew helps this process along by pumping seawater down the drill string at high pressure to "jet" away the sediments at the bottom of the conductor casing. The jetted water then carries the sediments up the inside of the conductor casing and out through ports in the wellhead into the surrounding seawater.

Figure 2.7. Early drilling phases.

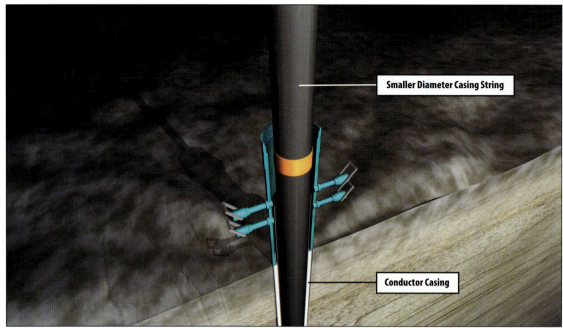

Smaller Diameter Casing String

Conductor Casing

TrialGraphix

Conductor in seafloor with early casing string attached.

Setting the Conductor Casing and Cementing Additional Early Casing Strings

Once the rig crew has jetted the conductor casing to its design depth, a second, smaller diameter casing string is installed, extending deeper into the seabed (see Figure 2.7). In a deepwater well, this second string is sometimes jetted into place, or large diameter drill bits might be used to drill the hole for it to be lowered into. If it is drilled, the hole diameter is slightly larger than the casing to leave room for the cement that secures it into place.

To cement the casing, a cementing crew pumps cement down the drill string. The cement flows down the drill string, out the bottom of the casing and back up against gravity into the annular space around the casing (between the casing and open hole). When cementing is complete, the cement fills the annular space around the casing, reinforcing the casing and creating the mechanical foundation for further drilling. This process continues as the hole is drilled using progressively smaller diameter casing and cementing each in place.

Cement Slurry. The **cement slurry** that the rig crew pumps down a well is a high-tech blend of dry Portland cement, water, and numerous dry and liquid chemical additives. Operators typically employ specialized cementing contractors to design the slurry, provide the raw materials for the slurry, and pump it into place. Cementing specialists can adjust the cement slurry composition to reflect the needs of each well. For instance, they can add "accelerators" to increase the rate at which the cement sets, or "retarders" to decrease it.

Figure 2.8. Wellhead assemblies.

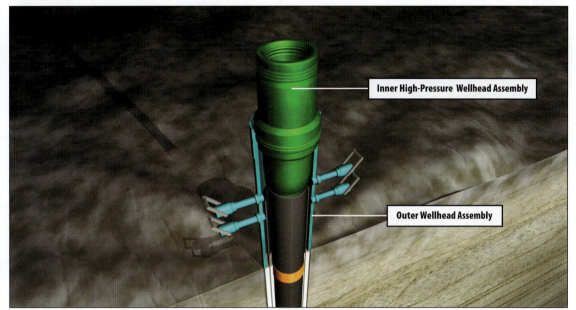

TrialGraphix

The Wellhead. A deepwater **wellhead** consists of a series of sophisticated interlocking components that are assembled together as the well is constructed. The outer portion of the wellhead is welded to the conductor casing and lowered to the bottom along with that casing. The outer wellhead accommodates multiple casing hangers that support the weight of early casing strings and seal the annular space at the top of those casing strings. Prior to lowering the BOP, drillers install an inner high-pressure wellhead assembly that is welded to a smaller diameter casing string. The high-pressure wellhead assembly interlocks with the outer wellhead assembly and includes fittings that allow the BOP to latch on to the integrated wellhead assembly. Like the outer wellhead assembly, the inner high-pressure wellhead assembly accommodates casing hangers inside it.

Lowering the Riser and BOP

When the sediments at the bottom of the well are strong enough that they can no longer be removed by jetting, the drilling crew must begin to use rotary drilling bits and may begin using drilling mud.

The term "mud" was once descriptive—early drilling fluids were simple mixtures of water and clay. Nowadays mud is a complex blend of oil- or water-based fluids and additives that serves many functions in a well. Unlike the seawater used during the jetting process, which is discharged into the surrounding sea after use, drilling mud must be recovered after it is pumped down a drill string—it is expensive ($100 per barrel or more) and can damage the surrounding ocean environment if released. Federal law generally prohibits the discharge of oil-based drilling mud into the ocean.

Figure 2.9. BOP components.

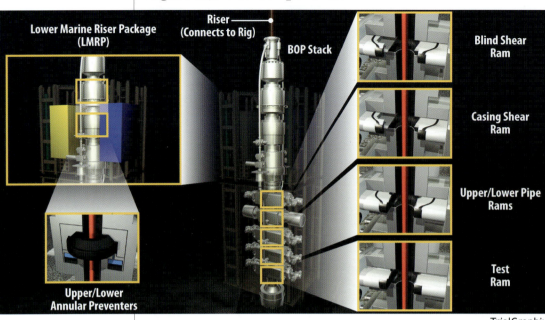

TrialGraphix

edges. The semicircular area is lined with a rubber seal. The pipe ram is designed to close around the drill pipe and seal off the annulus in the well below. Variable bore rams are a type of pipe ram with several concentric semicircular pieces; the concentric pieces allow the variable bore rams to seal around several different sizes of pipe.

A blind shear ram consists of two metal blocks with blades on the inner edges. It is designed to cut the drill string and seal off the annulus and the drill string in the well below. It can withstand and seal a substantial amount of pressure from below. Blind shear rams are designed to cut through drill pipe but will not cut through a tool joint (the place where two pieces of pipe are threaded together), casing hangers, or multiple pieces of pipe.

The casing shear ram is designed to cut through casing as it is being lowered into the wellbore and when there is no drill string in place. It does not seal the wellbore completely.

A test ram, if installed, sits at the bottom of the BOP stack. It is typically a pipe ram that is inverted—whereas the pipe ram is normally designed to hold pressure coming up from beneath it, the test ram is inverted and so holds pressure from above it. This allows the driller to test elements above the test ram.

The Blowout Preventer. The BOP as a whole is called the "BOP stack"; it consists of a series of annular preventers and rams stacked in vertical sequence on top of one another.[†]

The term lower marine riser package (LMRP) refers to the top part of the BOP stack that contains the annular preventers and the control pods (described further in Chapter 3).

An annular preventer is a large rubber element designed to close around the drill pipe and seal off the annulus. It is like a hard rubber donut. Upon activation, the annular preventer expands and fills the space within that part of the BOP; if there is something in the annular preventer (such as pipe), the annular preventer seals around it. If no drill pipe is in the hole, the annular preventer can close off and seal the entire opening.

A pipe ram consists of two mirror-image metal blocks with semicircles cut out of the inner

In order to switch from using seawater as a drilling fluid to using drilling mud, the rig crew must add several elements to the emerging well system. The first is a **blowout preventer**, or **BOP**. The BOP is a giant assembly of valves that latches on to the wellhead. The BOP stack serves as both a drilling tool and a device for controlling wellbore pressures. The BOP stack is connected back to the rig by the **lower marine riser package (LMRP)** and the **riser**. The riser is a

[†] Although not separately depicted in Figure 2.9, there are hydraulic, power, and communications lines (cables), as well as the choke, kill, and boost lines (pipes) running from the rig to the blowout preventer.

sequence of large diameter high-strength steel pipes that serves as the umbilical cord between the rig and the BOP during all remaining drilling operations. Once rig crews lower the BOP and riser system into place atop the wellhead, they perform the rest of their drilling operations through this system. The drill string, drilling tools, and all the remaining casing strings for the well go down into the well through the riser and the BOP.

Figure 2.10. Flow in a typical mud system.

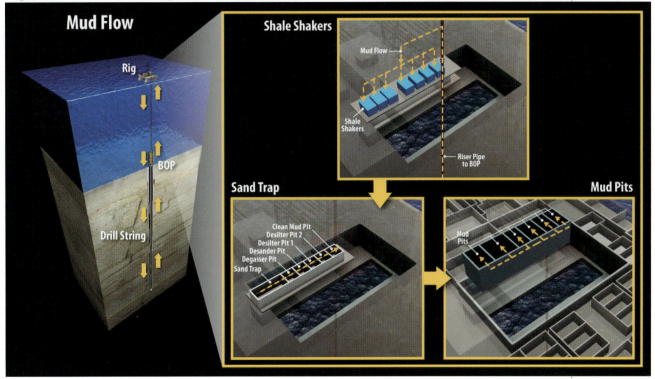

TrialGraphix

With the riser and BOP stack in place, the drilling crew can begin using the rig's drilling mud system (see Figure 2.10). The crew circulates mud down through the drill string, into the wellbore, back up the annular space around the drill string, up through the riser, and back to the rig. Once the mud reaches the rig, it goes through a series of devices including shale shakers and sand traps to remove cuttings and suspended debris from the mud. The mud then travels to mud tanks or **pits** where the mud is stored until being pumped back down into the well again.

An operator typically contracts with **mud engineers** to prepare drilling mud and operate the mud systems, and **mudloggers** to monitor the drilling mud and other drilling parameters. Mud engineers can add additional fluids or solid materials to the circulating mud in order to change its characteristics. Most importantly, they can add weighting agents such as barite to the circulating mud to increase the pressure on the wellbore below.

Engineers typically represent the density of drilling fluids in terms of pounds per gallon (ppg). Mudloggers regularly examine the mud and cuttings for clues to the nature of the geologic formation at the well bottom, and they check the mud to see if it contains hydrocarbons.

Well Logging. Well logging refers generally to the use of instruments to learn about the characteristics of a well during or after drilling operations. The oil and gas industry has developed many different types of instruments, or "logs." For example, **pressure while drilling logs** measure the pressure inside and outside the drill bit in real time. **Electric logs** measure the electric potential and resistivity of the formation, and can identify the boundaries between formations and the fluids within them. **Gamma ray logs** measure the natural radioactivity of rock. **Sonic and ultrasonic logs** can be used to measure the porosity and lithology of a formation. **Caliper logs** measure the size of the hole that has been drilled. **Temperature logs** measure temperature gradients in a well. **Cement evaluation logs** can help identify the amount and quality of cement in the annular spacer.

Setting Subsequent Casing Strings

Figure 2.11. Casing hanger.

TrialGraphix

Using the drilling mud system and rotary drill bits, the drilling crew drills ahead through the previously set casing strings. The rig crew extends the open hole below the existing casing strings as far as the pore pressure and fracture gradient allow and then sets subsequent smaller diameter casing strings inside the existing ones. Each new string of casing has a smaller diameter than the previous string because it must be run through the previous string. Some of these subsequent casing strings extend all the way back up to the wellhead. Others, called **liners**, attach to the bottom segment of previous casing strings. A **casing hanger** or **liner hanger** mechanically holds the casing in place (see Figure 2.11).

The basic method for installing a new casing string is the same whether that string will be hung from a hanger installed in the wellhead or hanger installed deeper in the well.

Figure 2.12. Wiper plugs.

TrialGraphix

Once the crew drills to a depth where a new casing string is needed, the rig crew removes the drill string from the well in a process called **tripping out**. Tripping out (or in) with the drill string is time-consuming; it typically takes a drilling crew an hour to trip in or out 1,000 feet, and tripping out of a deepwater well can be a day-long process. After tripping out, the drill crew attaches a **running tool** to the end of the drill string. The crew attaches the running tool to the casing hanger, which is in turn welded to the top of the casing. The drill crew then lowers the drill string, running tool, and casing string down the riser, through the BOP, and down into the well until the casing hanger is in position (either in the wellhead or the proper depth in the well).

Cementing Casing Strings

The process for cementing casing strings into place after installing the BOP is slightly different than cementing the early casing strings. Just as in earlier cementing steps, the rig crew pumps cement down the drill string and into place at the bottom of the well. However, because cement is typically incompatible with drilling mud, cementing crews employ two methods to keep the mud and cement separated as they flow down the well. The first involves separating the mud and cement with a water-based liquid **spacer** that is designed to be compatible with both oil-based drilling mud and water-based cement but that will prevent them from mixing. The second method involves further separating the spacer and cement with a plastic **wiper plug** that travels down the well between the spacer and the cement (see Figure 2.12).

Figure 2.13. Cementing.

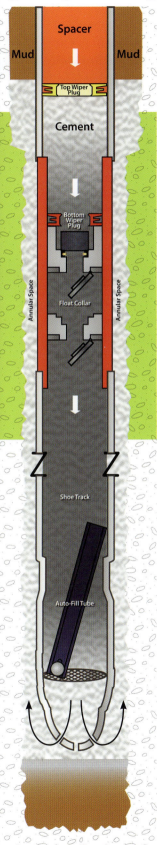

TrialGraphix

Figure 2.13 demonstrates cementing a casing string while using mud-based drilling techniques, the cementing crew starts by pumping spacer, followed by a "bottom" wiper plug, followed by a slug of cement, a "top" wiper plug, more spacer, and then drilling mud. The spacers, wiper plugs, and cement slug travel down in sequence. When the bottom plug reaches the **float valve assembly** near the bottom of the casing string, it ruptures, allowing the cement behind it to pass through. The cement flows through the float valves and out the bottom of the casing string. It then "turns the corner" and flows up into the annular space around the casing. When all of the cement has made it through the float valves, the top plug lands on top of the bottom plug. Unlike the bottom plug, the top plug is not designed to rupture. When it lands, it blocks the flow of mud, and the resulting pressure increase signals the end of the cementing process, at which time the crew turns off the pumps. Cement should fill the annular space around the bottom of the casing string and the portion of the casing between the bottom and the float valves (called the **shoe track**). Some companies even pump cement behind the top plug to improve the effectiveness of the cement job.

Float Collar. A "float collar" is a component installed at the bottom of a casing string. It typically consists of a short length of casing fitted with one or more check valves (called float valves). The float collar both (1) stops wiper plugs from traveling farther down the casing string, and (2) prevents cement slurry from flowing back up the casing after it is pumped into the annular space around the casing. During casing installation, the float valves are typically propped open by a short "auto-fill tube." The auto-fill tube allows mud to flow upward through the float collar as the casing string is lowered. Once the casing is in place, rig personnel "convert" the float collar. By circulating mud through holes in the auto-fill tube, the rig crew creates pressure that pushes the auto-fill tube down so that it no longer props the float valves open. Once the auto-fill tube is removed, the float valves "convert" to one-way valves that allow fluid flow down the casing but prevent fluid flow upward. Though a converted float collar should prevent cement slurry from flowing upward, it is typically not considered to be a barrier to hydrocarbon flow.

After the cement slurry has set (which takes many hours), the rig crew pressure tests it to ensure that it has sealed the casing in place. They then continue the drilling process by removing the running tool, installing a smaller diameter drill bit on the end of the drill string, and lowering it back down to the bottom of the well. The crew then uses the smaller diameter drill bit to drill through the float valves and the cement in the shoe track, creating a path for drilling to continue.

Figure 2.14. Perforating the production casing.

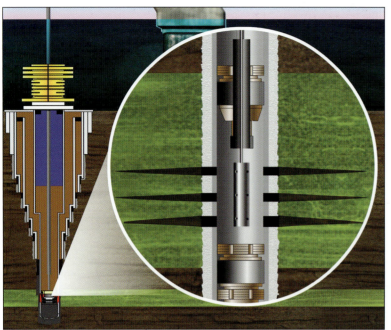

TrialGraphix

The Production Casing

If an operator drills a well purely to learn about the geology of an area and assess if oil or gas are present, the well is called an **exploration well**. If the operator uses the well to recover oil, it is called a **production well**. The bottomhole sections of exploration wells and production wells are different. Once an operator is finished drilling an exploration well, they typically fill the open bottomhole section with cement in a process called **plugging and abandoning**. By contrast, after drilling the final section of a production well, the operator typically installs a final string of **production casing** in the open hole section. The production casing extends past any hydrocarbon-bearing zones and down to the bottom of the well. After cementing the production casing into place, the operator can **perforate** the casing by shooting holes through it and the annular cement. This allows oil to flow into the well as shown in Figure 2.14.

Well Control

During drilling, casing, and completion operations, rig personnel must ensure that hydrocarbons do not migrate from the reservoir into the well. Well control is the process of monitoring the well and addressing any hydrocarbon influxes that are detected.

Primary Barriers—Barriers Inside the Well

To maintain well control, rig personnel must create and maintain **barriers** inside the well that will control subsurface pressure and prevent hydrocarbon flow. Some barriers are part of the well design itself while others are operational barriers that a drilling crew employs during the drilling process.

Drilling mud is a key operational barrier. As long as the column of drilling mud inside the well exerts pressure on the formation that exceeds the pore pressure, hydrocarbons should not flow out of the formation and into the well.

It is important to understand the following: If mud pressure exceeds pore pressure, the well is said to be **overbalanced**. If pore pressure exceeds mud pressure, the well is **underbalanced**, meaning that the mud pressure is no longer sufficient on its own to prevent hydrocarbon flow.

Physical components of the well also create barriers to flow. One is the casing installed in the well, along with the cement system in the bottom of the well. In a production casing string, the cement in the annular space and in the shoe track should prevent hydrocarbons in the formation from flowing up the annular space outside the production casing or up the inside of the well itself.

Rig personnel can use additional barriers inside the well to increase the redundancy of the barrier system. For instance, rig personnel can pump cement inside the final casing string of a well to create cement plugs at various depths inside the well. Rig personnel can also install metal or plastic mechanical plugs inside the well. Some mechanical plugs are designed to be removed and retrieved later in the drilling process while others are designed to be drilled out as necessary.

Secondary Barrier—The Blowout Preventer

A BOP stack is also a potential barrier. By closing various individual **rams** in a BOP stack, rig personnel can close off the well, thereby preventing hydrocarbon flow up the well and into the riser. When a BOP ram is closed, it becomes a barrier to flow. However, the rams do not close instantaneously—they take anywhere from 40 seconds to a minute to close once activated. **Accumulators** are tanks that contain pressurized hydraulic fluid used to close the BOP. Subsea accumulators on the BOP stack are constantly charged through a conduit line from the rig.

BOP rams can be activated in several ways: manually from the rig, robotically by remotely operated vehicles (ROVs), and automatically (when certain conditions are met). Each ram is activated separately.

Manual activation is generally done by the driller but also can be done by other rig personnel including the subsea engineer. BOP control panels are located on the bridge and in the drill shack. To manually activate a given BOP ram, a rig worker presses a button on the control panel corresponding to that ram. Electrical signals are sent from the control panels to subsea **control pods** on the BOP stack. The signals electrically open or close a solenoid valve, which in turn sends a pilot signal to activate the hydraulic pressure needed to operate the individual elements of a BOP stack. The control panel has a flow meter display that indicates how many gallons of hydraulic fluid are flowing into the ram, which helps the driller and subsea engineer to determine whether the ram is responding properly.

The BOP can also be activated by a mobile underwater robot (ROV) that can carry and use tools. An ROV can activate the blind shear ram through the control system or by pumping hydraulic fluid through "hot stab" ports located on the outside of the BOP stack.

Last, the BOP can be activated automatically. One automated system is the automatic mode function or **deadman** trigger. If the power, communications, and hydraulic lines running from the rig to the BOP are severed or otherwise lose functionality, circuits on the BOP stack will activate the blind shear rams to close off the well. An ROV can also create the conditions to activate the deadman by cutting power, communications, and hydraulic lines at the LMRP. Another automated system that activates the blind shear ram is the **autoshear**. A BOP system can be configured so that the autoshear activates where the rig is drifting off or driving off of its location. If the rig moves a sufficient distance, a rod between the LMRP and BOP stack is severed, and the autoshear activates. An ROV can also cut the rod between the LMRP and BOP stack to activate the autoshear system.

Figure 2.15. Barriers in a well.

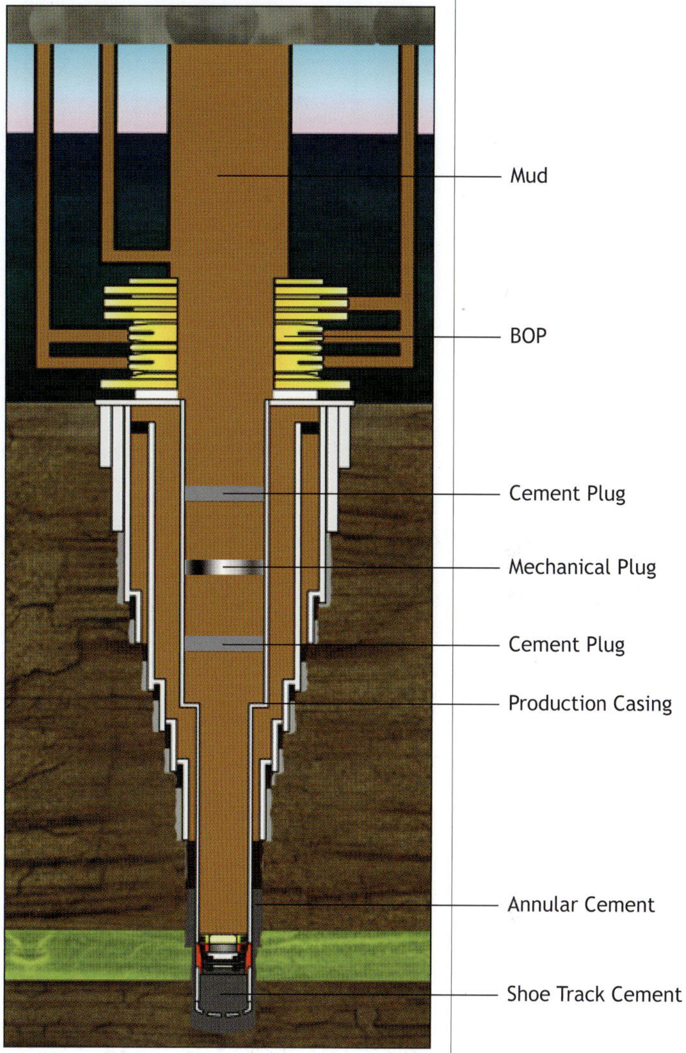

TrialGraphix

- Mud
- BOP
- Cement Plug
- Mechanical Plug
- Cement Plug
- Production Casing
- Annular Cement
- Shoe Track Cement

Figure 2.16. Simplified AMF control system schematic.

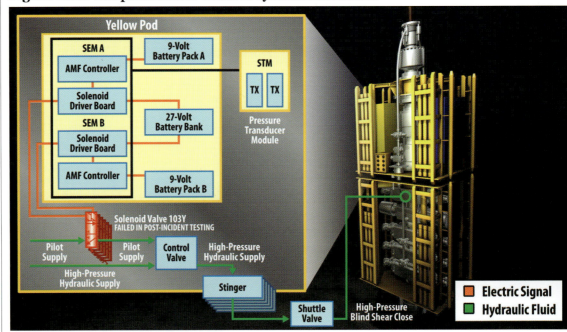

TrialGraphix

Activating the ram systems.

Well Monitoring

If the primary hydrocarbon barriers in a well (such as the weight of drilling mud) are inadequate to contain the reservoir pressures, a kick of hydrocarbons can flow into the well. During well operations, rig personnel must always monitor the well for such kicks and respond to them quickly. Their options for responding to a kick diminish rapidly as the kick progresses.

Rig personnel (primarily the driller) watch several different indicators to identify kicks. One is the amount of fluid coming out of the well. If flow out of the well exceeds flow in or the volume of mud in the mud pits increases anomalously, that may indicate that hydrocarbons are flowing into the well. Data from sensors that measure the gas content of returning drilling mud can also warn of hydrocarbon flow. Other indicators include unexplained changes in drill pipe or other pressures, and changes in the weight, temperatures, or electrical resistivity of the drilling mud.

Once rig personnel detect a kick, they must take action to control it. The driller has a number of options for dealing with a kick depending on its size and severity. In a routine kick response scenario, the driller activates an **annular preventer** or a **pipe ram** to seal off the annular space in the well around the drill pipe. The driller then pumps heavier mud into the well. He can do this either through the drill pipe or through the **kill line**—one of three separate pipes that run from the rig to the BOP. The heavier mud is called "kill mud," designed to counteract the pore pressure of the rock formation. Because the BOP has sealed off the annular space around the drill pipe, the driller opens the **choke line** (another one of the three separate pipes running from the rig to the BOP) to allow circulating mud to return to the rig. Once the weight of the heavier drilling mud overbalances the hydrocarbon pressure and any hydrocarbons that flowed into the well have been circulated out, the driller can reopen the BOP and safely resume operations. On modern rigs, kill lines can function as choke lines, and vice versa. A third pipe, a **boost line**, connects at the bottom of the riser and can help speed circulation of fluids.

If a kick progresses beyond the point where the driller can safely shut it in with an annular preventer or pipe ram, the driller can activate the **blind shear ram**. When the two elements of the blind shear ram close against each other, they simultaneously shut in the well and sever the drill string. ◗

The *Deepwater Horiz*

Chapter 3 | Background on the Macondo Well, the *Deepwater Horizon,* and the Companies Involved

The Macondo Well

Well Location

In February 2009, BP filed an exploration plan with the United States Minerals Management Service (MMS) indicating its intention to drill two exploration wells in Mississippi Canyon Block 252 (MC 252). Both wells were located 48 miles from shore in 4,992 feet of water, and both would be drilled to a total depth of 20,600 feet below sea level.[1] BP planned to drill both using a semi-submersible drilling rig. BP stated that it would take 100 days to drill each well and that it would begin the first well on April 15, 2009, and the second well one year later on April 15, 2010.

MMS approved BP's Exploration Plan in early April 2009. BP later revised the plan slightly in mid-April 2009 to include a larger anchor pattern for its rig, and MMS approved the revised plan on April 21, 2009. MMS approved BP's application for permit to drill (APD) the first of the two wells, the "A" location, on May 22, 2009.[2]

BP would drill the well in order to reach the Macondo prospect. The name "'Macondo" was the result of a charitable donation. BP had donated naming rights to the United Way, which in turn auctioned the rights to a Colombian-American group. That group chose the name of the fictional Colombian village in Gabriel García Márquez's novel *One Hundred Years of Solitude.*[3]

The Geology and Exploration Objectives

BP had decided to drill at Macondo after examining 3-D seismic data, offset well data, and other information about the area. The 3-D data had included a prominent "amplitude anomaly" that suggested the presence of hydrocarbon-bearing sands. This information, combined with offset well data and knowledge of the overall geological structure of the area, strongly suggested to BP that it might find hydrocarbon-bearing sands.[4]

BP defined its primary geologic objectives as mid-Miocene age turbidite sands buried 13,000 to 15,000 feet beneath the seafloor—18,000 to 20,000 feet below sea level. These sands were deposited on the ancient seabed some 12 million to 15 million years ago. BP's plan called for drilling the well to a total depth of 20,600 feet to penetrate this primary objective interval. From the beginning, BP planned to use the well as a long-term production well if it penetrated the objective sands.

Operators are not required to include pre-drilling estimates of potential oil and gas reservoirs. However, during the containment efforts following the blowout of the Macondo well, BP estimated the volume of oil at Macondo to be 110 million barrels.[5]

The *Deepwater Horizon*

BP had intended to use Transocean's *Marianas* to drill the entire Macondo well. The *Marianas* spudded the Macondo well on October 6, 2009. The crew of the *Marianas* drilled and set casing for the first 9,090 feet of the Macondo well but were forced to leave after the rig sustained damage from Hurricane Ida on November 9, 2009. The *Deepwater Horizon* took over and resumed drilling operations at Macondo in February 2010.

The *Deepwater Horizon* was a semi-submersible **mobile offshore drilling unit (MODU)**. Unlike fixed drilling platforms used in shallower water, MODUs can move from one location to another under their own power. Dynamically positioned MODUs utilize **dynamic satellite positioning technology** connected to powerful directional **thrusters** to maintain themselves in place over a subsea wellhead.

The *Deepwater Horizon* entered service in 2001. It was built by Hyundai Heavy Industries and owned by Transocean. It initially sailed under the flag of Panama and later the Marshall Islands. In 1998, BP signed a contract with Transocean securing the services of *Deepwater Horizon* from the time it first left the shipyard for a period of three years.[6] After the initial three years, BP extended the contract in annual increments.[7] At the time of the blowout, BP's contract required it to pay $533,495 per day, but under the contract BP was not obligated to pay for time in excess of 24 hours each month spent on certain equipment repairs.[8] With additional costs (fuel, expendables, and services), BP was paying approximately $1 million per day to operate the *Horizon*.[9]

Figure 3.1. *Deepwater Horizon.*

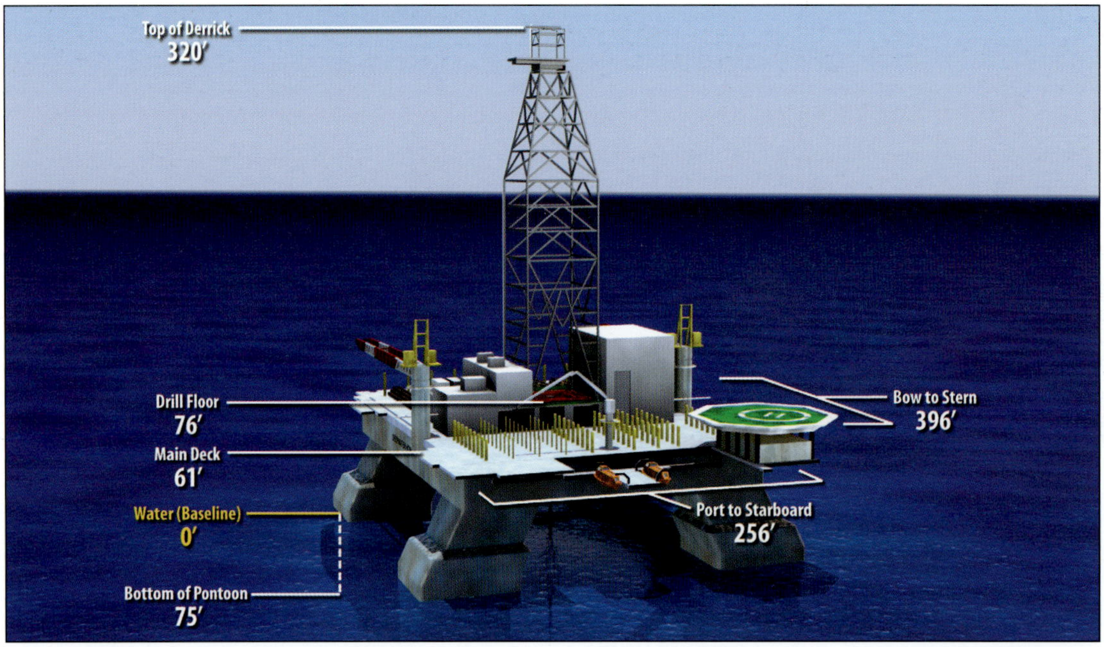

Top of Derrick
320'

Drill Floor
76'

Bow to Stern
396'

Main Deck
61'

Water (Baseline)
0'

Port to Starboard
256'

Bottom of Pontoon
75'

TrialGraphix

Figure 3.1 shows the basic dimensions of the *Deepwater Horizon* while drilling. The rig was 256 feet wide (from port to starboard) and 396 feet long (from bow to stern). The **main deck** sat 61 feet above the water's surface while drilling, with the **drill floor** another 15 feet above that. The **derrick** was 244 feet tall, towering a total of 320 feet above the ocean while drilling.

Figure 3.2 is a close-up of the main deck with pertinent parts labeled in yellow. The **drill floor** was located in the center of the main deck. The crew ran casing, drill pipe, and drilling tools through the **rotary table**, down the riser, and into the well. The **drill shack** was on the drill floor. It was a small windowed room that housed the drillers' and assistant drillers' chairs, along with well monitoring equipment and controls for the blowout preventer. The drillers and assistant drillers drilled and monitored the well from the drill shack.

The **mudloggers shack** was a structure installed on the starboard side of the drill floor. It was owned by Sperry Drilling and housed the mudloggers and their monitoring equipment. The **bridge** was located beneath the helipad on the front port (left) side of the rig. The bridge contained the dynamic satellite positioning system and was the helm of the rig when in transit. It also housed monitors and controls for the alarm systems, a second set of controls for the blowout preventer, and other rig and well monitoring equipment.

Figure 3.2. Main deck.

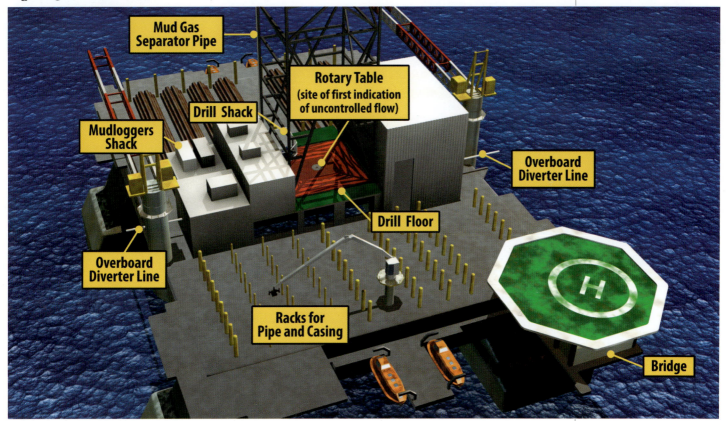

TrialGraphix

Two sets of pipes are also shown in Figure 3.2. The first comprises the port and starboard **overboard lines**, through which the rig crew could send fluids flowing up from the well overboard into the ocean. The second is the **mud gas separator pipe**, through which the rig could route mud returning from the well to remove small amounts of hydrocarbon gas before sending the mud on to the **mud pits**. The overboard lines and mud gas separator were part of the rig's **diverter system**, which provided the crew two alternative routes for diverting fluids coming up from the well.

Figure 3.3 shows the *Deepwater Horizon*'s three decks. Decks 2 and 3 housed the rig's living quarters, engine rooms, and other work areas, including the mud pits and moon pool shown in Figure 3.4. The *Deepwater Horizon* had 20 **mud pits**, which were tanks for holding drilling fluids such as mud. The **moon pool** was located directly beneath the drill floor. It was a wide opening in the bottom of the rig through which the crew could lower and raise large pieces of equipment to and from the ocean, such as the blowout preventer.

Figure 3.3. Decks.

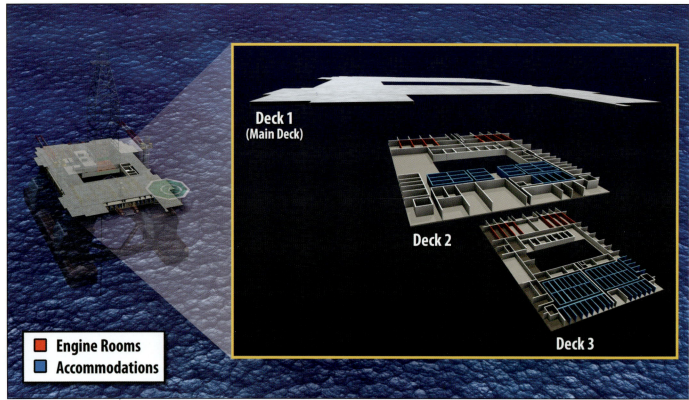

TrialGraphix

Figure 3.4. Mud pits and moon pool.

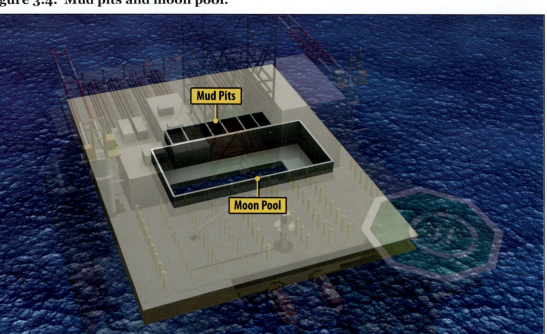

TrialGraphix

Figure 3.5. Subsurface portion of the *Deepwater Horizon*.

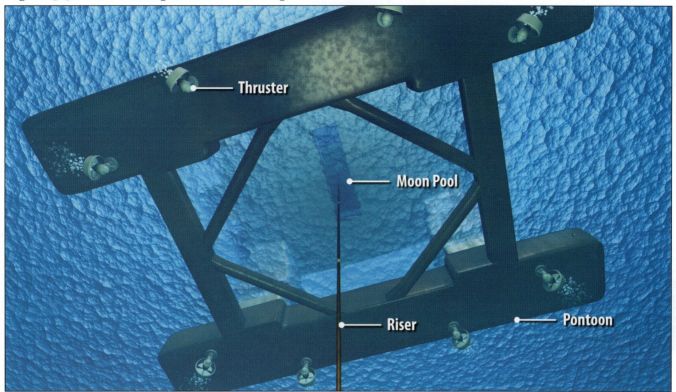

Thruster

Moon Pool

Riser

Pontoon

TrialGraphix

Figure 3.6. Blowout preventer.

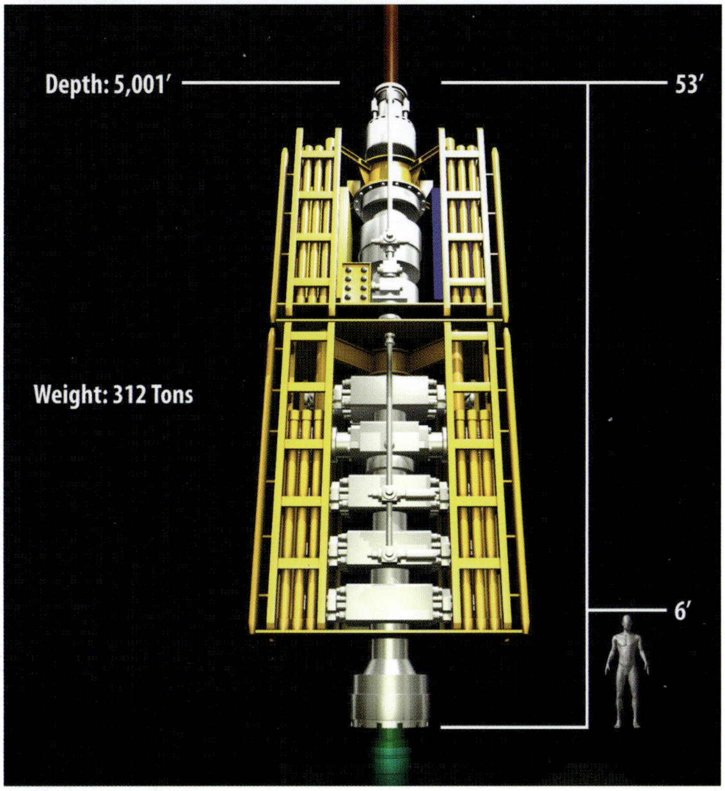

Depth: 5,001'

53'

Weight: 312 Tons

6'

TrialGraphix

Figure 3.5 depicts the subsurface portion of the *Deepwater Horizon*. The rig sat atop two enormous **pontoons** extending 30 feet below the ocean's surface that stabilized the rig and kept it afloat. The rig had eight directional **thrusters** for propulsion and to keep the rig in place over the wellhead. Figure 3.5 also depicts the **riser**, which would have extended from the rotary table on the drill floor through the moon pool and down to the blowout preventer on the ocean floor below.*

Figures 3.6 and 3.7 depict the *Deepwater Horizon*'s **blowout preventer (BOP)** sitting atop the wellhead on the ocean floor. As discussed in Chapter 2, the

---

* Although not separately depicted in Figures 3.5 and 3.6, there are hydraulic, power, and communications lines (cables), as well as the choke, kill, and boost lines (pipes) running from the rig to the blowout preventer.

Figure 3.7. Blowout preventer.

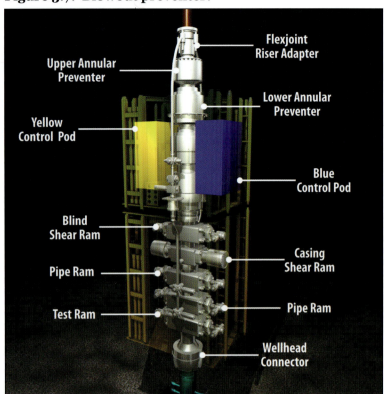

TrialGraphix

blowout preventer comprised a set of five vertically stacked **rams** and two vertically stacked **annular preventers** for closing in the well during routine well activities and emergency situations. As shown in Figure 3.6, the blowout preventer was more than five stories tall and weighed more than 300 tons. Figure 3.7 displays the various parts of the blowout preventer, including the blue and yellow **control pods**.

Companies and Individuals Involved in the Macondo Blowout

By purchasing the rights to drill in Block 252, BP became the legal **operator** for any activities on that block. Like most operators, however, BP neither owned the rigs that drilled Macondo nor "operated" them in the normal sense of the word.

Instead, the company's shore-based engineering team designed the well and specified in detail how it was to be drilled. BP employed a number of contractors to perform the physical work of actually drilling and constructing the well. As a result, on the day of the Macondo blowout, only seven of the 126 individuals on the rig were BP employees.[10] The following sections list the core members of BP's team as well as the principal contracting companies and their key employees.

BP is a large oil and gas company headquartered in the United Kingdom. With annual revenues of approximately $246 billion, BP is the world's fourth-largest company of any kind.[11] It is the world's third-largest energy company[12] and the largest producer of oil and gas in the Gulf of Mexico.[13] BP held more than 500 lease blocks in the Gulf of Mexico, and more than 1,600 employees worked for BP in the region.[14]

A number of different individuals and groups at BP had a hand in designing and supervising the construction of the Macondo well. The initial design of the Macondo well involved more than 25 professionals, ranging from drilling engineers to regulatory experts.[15] During the drilling of the well, reports about operations on the rig went out to about 80 BP employees and contractors.[16]

Daily activity at Macondo centered on a handful of BP employees (see Figure 3.8). BP had two well site leaders on the *Deepwater Horizon* at any given time. In Houston, BP had a wells team leader, an engineering team leader, an operations engineer, and two drilling engineers.

BP was in the process of reorganizing its management structure at the time of the blowout to clarify reporting relationships for engineers. The reorganization complicates the task of identifying the precise lines of authority and areas of responsibility, both at the time of and in the months leading up to the blowout. In addition, because of the reorganization, many of the

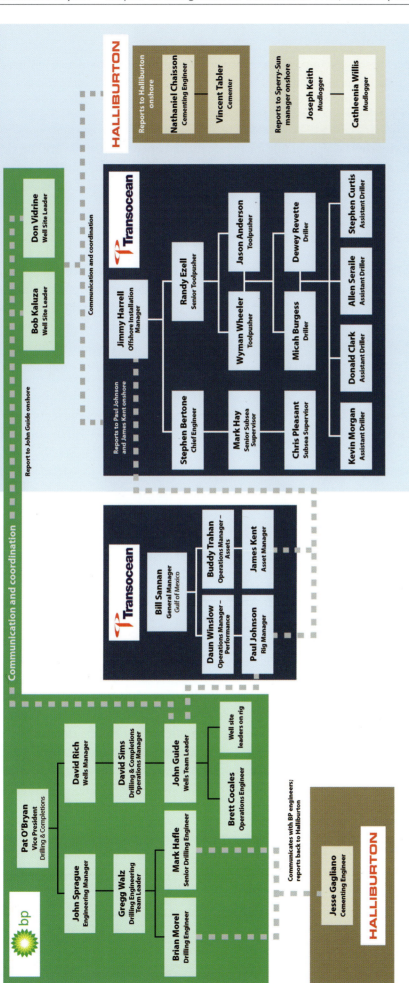

Figure 3.8.
Deepwater Horizon's
organizational structure.
TrialGraphix

managers overseeing the Macondo team had only a few months of experience in their respective positions at the time of the blowout.[17]

Wells Team Leader

The **wells team leader** is accountable for the safety and operations of the drilling rig.[18] The wells team leader for Macondo was John Guide. He supervised the well site leaders on the rig and an operations engineer in Houston. Ian Little managed John Guide until David Sims took over from Little in March 2010.[19]

Well Site Leaders on the Drilling Rig

The top BP employees stationed on the rig were the **well site leaders**. The well site leaders served as the company's eyes and ears, and made important decisions regarding the course of drilling operations. At any given time, two well site leaders served on the rig, splitting responsibility according to 12-hour shifts. The two-man team worked on the rig for several weeks at a time and then returned to shore for a similar period.

At the time of the blowout, Bob Kaluza and Donald Vidrine were BP's two well site leaders on the *Deepwater Horizon*. Lee Lambert, who was in training to become a well site leader, was also present. Kaluza was onboard as a temporary replacement for Ronnie Sepulvado, an experienced well site leader who had worked on the *Deepwater Horizon* since it set sail in 2001 but who had left the rig early to attend a training program. Murry Sepulvado (Ronnie Sepulvado's brother) and Earl Lee were the prior regular BP well site leaders for the *Deepwater Horizon*. They were not on the rig at the time of the blowout.

Engineering Team Leader

The **engineering team leader** is accountable for well design.[20] The engineering team leader supervises drilling engineers. David Sims was the engineering team leader until about a month before the blowout.[21] In March, Sims moved from engineering to an operations role and handed his responsibilities over to Gregg Walz.

Other BP Engineers

Other BP engineers closely involved with the Macondo project included Brian Morel, Mark Hafle, and Brett Cocales. Morel and Hafle were **drilling engineers** who designed wells and shepherded designs through BP's processes, ensuring that they complied with internal guidelines.[22] They planned the Macondo casing program and set out the steps to drill the well.[23] Morel and Hafle reported first to Sims, and then to Walz when he took over from Sims in March. Cocales was an operations engineer responsible for planning and preparing *Deepwater Horizon*'s future activity. Cocales focused on end-of-well operations and preparations for future work.[24] He reported to Guide.

Anadarko and MOEX

Anadarko and MOEX were BP's partners at Macondo. Anadarko Petroleum is an independent oil exploration company and owned a 25% share of the Macondo well.[25] MOEX Offshore 2007, an American subsidiary of the Japanese oil company Mitsui Oil Exploration, owned a 10% share of the well.[26] The partners shared the costs to drill the well and expected to share profits from production.

Transocean is the world's largest contractor of offshore drilling rigs.[27] BP has entered into several contracts with Transocean to secure the long-term services of certain Transocean rigs, including the *Deepwater Horizon*. Transocean crews performed most of the basic work of drilling the Macondo well.

OIM and Master

Two different Transocean employees were in charge of the rig at different times. Captain Curt Kuchta, Transocean's **master**, was in charge when the rig was moving from location to location. Once the rig arrived at a given site and began drilling or drilling-related operations, Jimmy Harrell, Transocean's **offshore installation manager (OIM)**, took over.

Senior Toolpusher and Toolpushers

The **toolpushers** on a Transocean rig are drilling managers who direct and supervise day-to-day drilling operations. The toolpushers stationed on the rig when the well blew out were Jason Anderson and Wyman Wheeler. Anderson was on duty and lost his life. Wheeler was severely injured.

The **senior toolpusher** on a Transocean rig is a senior drilling operations supervisor, second only to the OIM in the chain of command. The senior toolpusher on the rig was Miles "Randy" Ezell.

Rig Floor Personnel

The Transocean employees who served primarily on the rig floor included drillers, assistant drillers, floorhands, and roustabouts. **Drillers** and **assistant drillers** worked in the drill shack and were responsible for operating drilling machinery and monitoring and controlling the well. The drillers stationed on the rig when the well blew out were Dewey Revette and Micah Burgess. Revette was on duty and lost his life. The assistant drillers stationed on the rig when the well blew out were Donald Clark, Stephen Curtis, Patrick Kevin Morgan, and Allen Seraile. Clark and Curtis were on duty. Both lost their lives. **Floorhands** and **roustabouts** are the rig's labor force for drilling operations. Three floorhands lost their lives in the blowout. They were Shane Roshto, Karl Kleppinger, and Adam Weise.

Other Transocean Employees

In all, Transocean had 79 employees onboard the *Deepwater Horizon* when it blew out, including welders, technicians, radio operators, and other specialized personnel. Aaron Burkeen, a crane operator, and Roy Wyatt Kemp, a derrickhand,[†] both lost their lives.

[†] A derrickhand works from the rig's derrick to assist with drilling operations.

HALLIBURTON

Halliburton is one of the world's largest oil field services providers and owns several other oil field services companies, including Baroid and Sperry Drilling. Halliburton designed and pumped the cement for all of the casing strings in the Macondo well.

Jesse Gagliano was Halliburton's lead cementing specialist for the project. Gagliano worked closely with BP and had an office in BP's building near the offices of BP's engineers. Halliburton also sent several individuals to the rig to actually perform cementing work. Those individuals included Nathaniel Chaisson, Vincent Tabler, Christopher Haire, and several foamed cement technicians. Tabler and Haire were on the rig at the time of the blowout.[28]

Other Important Contractors and Suppliers

Cameron is a Houston-based company that manufactures well drilling equipment and well construction components. Cameron manufactured the *Deepwater Horizon*'s blowout preventer.

Dril-Quip is a Houston-based manufacturer of components used in the construction of oil wells. Dril-Quip manufactured the wellhead assembly used at Macondo, including the casing hanger, seal assembly, and lockdown sleeve components. A Dril-Quip technician named Charles Credeur was on the *Deepwater Horizon* when Macondo blew out.

M-I SWACO, a Schlumberger subsidiary, is a Houston-based company that provides drilling fluids and drilling fluid services. M-I SWACO provided drilling mud and spacer used at Macondo, and its personnel operated the *Deepwater Horizon*'s mud system. M-I SWACO had five mud engineers on the rig the day it blew out: Gordon Jones, Leo Lindner, Blair Manuel, Greg Meche, and John Quebodeaux. Jones and Manuel lost their lives.

Schlumberger is a multinational company that delivers a variety of oil field services through its own employees and through subsidiaries including M-I SWACO. BP hired Schlumberger to run cement evaluation logs for the primary cement job on the final Macondo production casing. Schlumberger also provided well logging services used in the evaluation of the Macondo well.

Sperry Drilling, a Halliburton subsidiary, delivers oil field services. At Macondo, BP employed Sperry Drilling to collect data from sensors mounted on the rig and to provide trained personnel to monitor and interpret the data, including monitoring the well for kicks. Sperry Drilling had two of these "mudloggers" on the rig the day it blew out: Joseph Keith and Cathleenia Willis.

Weatherford is a Houston-based manufacturer of well construction components. It manufactured float valves and centralizers used at Macondo. Four Weatherford technicians were on the rig the day it blew out. ◆

Chapter 4 | Technical Findings

The Chief Counsel's team's overall technical findings are straightforward. The Macondo well blew out because the cement that BP and Halliburton pumped down to the bottom of the production casing on April 19 failed to seal off, or "isolate," hydrocarbons in the formation. As rig personnel replaced heavy drilling mud in the well and riser with seawater on April 20, they steadily reduced the pressure inside the well. At approximately 8:50 p.m., the drilling fluid pressure no longer balanced the pressure of hydrocarbons in the pay zone at the bottom of the well. At this point, the well became "underbalanced."

Once the well was underbalanced, hydrocarbons began to flow into the annular space around the production casing. In oil field terms, the Macondo well was "taking a kick." Those hydrocarbons flowed down through the annular space to the bottom of the well, into the production casing through the "shoe track," then up the well and into the riser. As they traveled up the well, the hydrocarbons expanded at an ever-increasing rate and the kick escalated into a full-scale blowout. Transocean's rig crew did not respond to the kick before hydrocarbons had entered the riser, and perhaps not until mud began flowing out of the riser onto the rig floor. Within 10 minutes of the rig crew's first response, hydrocarbon gas from the well ignited, triggering the first explosion.

Underlying Technical Causes

Behind this simple story is a complex web of human errors, engineering misjudgments, missed opportunities, and outright mistakes. Chapter 4 of the Chief Counsel's Report divides technical analysis of the blowout into 10 subchapters. Each subchapter presents the Chief Counsel's team's findings on specific technical issues.

- Chapter 4.1 presents the basis for the Chief Counsel's team's conclusions regarding the precise flow path of hydrocarbons during the blowout.

- Chapter 4.2 explains a number of the well design decisions that BP's engineering team made at Macondo and presents several findings regarding the impact of those decisions. The Chief Counsel's team finds that BP's decision to use a long string production casing increased the difficulty of achieving zonal isolation during the cement job. While the decision did not directly cause the blowout, it increased the risk of cementing failure. The Chief Counsel's team also finds that BP's decisions to include rupture disks and omit a protective casing from its well design complicated post-blowout containment efforts.

- Chapter 4.3 presents findings regarding the final cement job at Macondo. The cement job failed to isolate hydrocarbons. While it may never be possible to determine precisely why, the Chief Counsel's team identified a number of risk factors and other issues that could have contributed to cement failure. The rig crew, cement contractors, and engineering team do not appear to have fully appreciated these risk factors.

- Chapter 4.4 presents findings regarding pre- and post-blowout testing of the foamed cement slurry design used at Macondo. The Chief Counsel's team finds that the foamed cement used at the well was very likely unstable and that this could have been a major contributing factor to overall cement failure.

- Chapter 4.5 presents findings regarding the temporary abandonment procedures that BP developed and employed at the Macondo well. The Chief Counsel's team finds that those procedures reduced the number of barriers that would be present in the well when it became underbalanced, and significantly and unnecessarily increased the risk of a blowout.

- Chapter 4.6 presents findings regarding the negative pressure test conducted on April 20. The Chief Counsel's team finds that the test clearly showed that the cement had failed to isolate hydrocarbons. BP and Transocean rig personnel both failed to interpret the test properly and instead reached a consensus that the test had demonstrated well integrity.

- Chapter 4.7 explains that the Transocean crew and Sperry-Sun mudloggers missed warning signs of a kick on the evening of April 20. The Chief Counsel's team finds that data from the rig show signs of an anomaly as early as 9:01 p.m. Some of the signs went unnoticed; others the crew detected. But even after rig personnel detected the anomaly, they did not identify it as a kick until after hydrocarbons had entered the riser. If rig personnel had identified the kick earlier, they could have prevented the Macondo blowout.

- Chapter 4.8 presents findings regarding the crew's response to the blowout after it occurred. The Chief Counsel's team finds that the crew might have mitigated the size and impact of the fires and explosions on April 20 if they had immediately diverted flow during the blowout overboard rather than to a mud gas separator system that was incapable of handling that extreme flow volume.

- Chapter 4.9 presents findings regarding the rig's blowout preventer, or BOP. Hydrocarbons had entered the riser well before the crew attempted to activate the BOP, and even a perfectly functioning BOP could not have prevented the explosions that killed 11 men on April 20. Nevertheless, BOP failures may have contributed to the magnitude of the oil spill. While BOP forensic testing is ongoing, the Chief Counsel's team presents findings regarding maintenance history and certain BOP failure theories.

- Chapter 4.10 presents findings regarding the role of rig maintenance in the blowout. The Chief Counsel's team finds that Transocean did not maintain its BOP according to manufacturer recommendations. And the Chief Counsel's team cannot rule out that this may have contributed to BOP failures. While the Chief Counsel's team found some indications of other maintenance problems on the *Deepwater Horizon*, it does not find that any of these contributed to the blowout.

Underlying Management Causes

Each of these chapters also presents management findings that relate specifically to the technical findings in the chapter. The Chief Counsel's team finds that management failures lay at the root of all of the technical failures discussed in this Report. Chapter 5 discusses management failures in detail. ♦

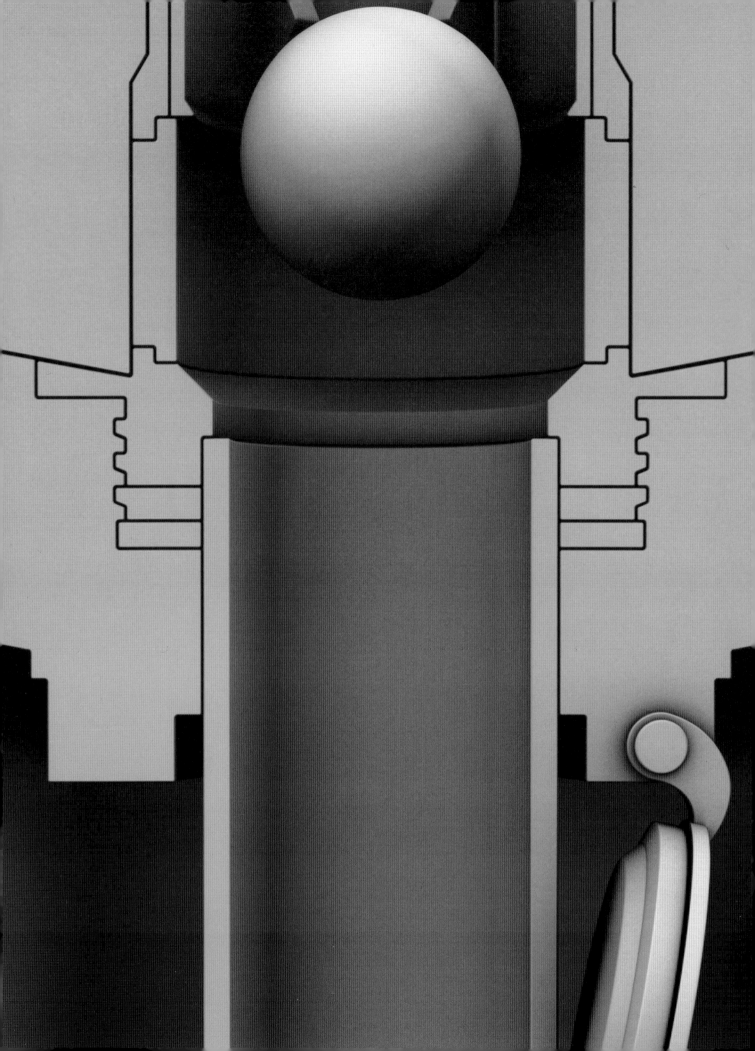

Chapter 4.1 | Flow Path

Before addressing potential technical causes of the blowout, the Chief Counsel's team presents its findings regarding the flow path of hydrocarbons from the well. These findings form an important background to the subsequent technical analyses. Because different kinds of well failures cause hydrocarbons to flow through different paths, these findings can help to refine theories about what caused the blowout.

Figure 4.1.1. Possible flow paths for hydrocarbons.

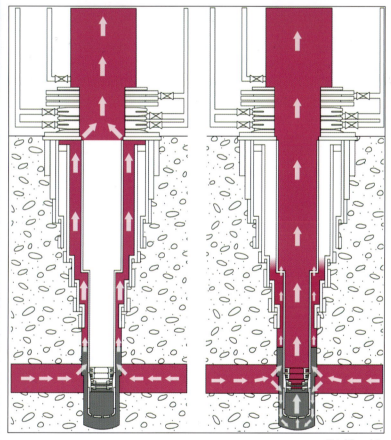

TrialGraphix

Hydrocarbons can reach the surface by traveling up the annulus and through the seal assembly (left). Hydrocarbons can also enter and migrate up the inside of the production casing, through a number of possible flow paths (right).

The Chief Counsel's team finds that hydrocarbons came to the surface by traveling through the inside of the production casing, as seen on the right side of Figure 4.1.1. It is almost certain that hydrocarbons entered the production casing because of a failure of the shoe track cement. However, the Chief Counsel's team cannot entirely rule out the possibility that hydrocarbons may have entered the production casing from the annulus through a breach in the production casing somewhere near the bottom of the casing.

The analysis in this section reflects information currently available to the Chief Counsel's team. The team recognizes that various parties continue to gather additional information that may be relevant to flow path analysis.[1]

Potential Flow Paths

For the Macondo blowout to have occurred, hydrocarbons must have traveled from the formation into the wellbore and then up to the surface through the blowout preventer (BOP) and the riser. The fact that hydrocarbons entered the wellbore at all means, at the very least, that the annular cement did not isolate the pay zones.[2] For hydrocarbons to have traveled up to the surface, they must either have gone up the annulus and through the seal assembly at the wellhead or into and up through the production casing.

Flow up the Annulus and Through the Seal Assembly

The **seal assembly** is in the wellhead. It seals the interface between the **casing hanger** for the production casing and the inside of the high-pressure wellhead housing. A **lockdown sleeve** locks the casing hanger and seal assembly in place so that hydrocarbons traveling up the wellbore during production do not lift them up.

Figure 4.1.2. Flow through the seal assembly.

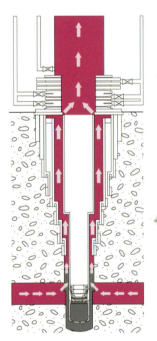

As Figure 4.1.2 illustrates, there are small **flow passages** through the casing hanger connecting the annulus to the inside of the wellhead.[3] The flow passages permit mud in the annulus to flow into the wellhead and up into the riser, thereby allowing the crew to circulate drilling fluids through the annulus even after the crew has set the production casing in place. The flow passages remain open prior to and during the final cement job.

The crew sets the seal assembly atop these flow passages to seal them off once there is no longer a need to circulate fluids in the annulus. At Macondo, the crew set the seal assembly shortly after pumping the bottomhole cement job.

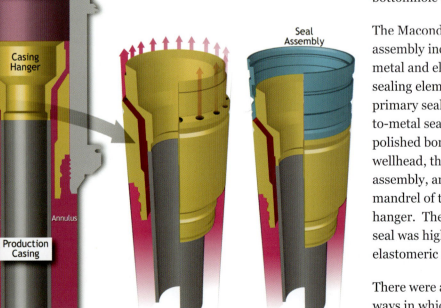

The Macondo seal assembly included both metal and elastomeric sealing elements. The primary seal was a metal-to-metal seal between the polished bore of the wellhead, the seal assembly, and the polished mandrel of the casing hanger. The secondary seal was highly resilient elastomeric material.

There were at least two ways in which hydrocarbons could have flowed up the annulus and through the seal assembly.

First, there could have been a leak through the flow passages. This might have occurred because debris obstructed the seal area during the setting process, the seal failed to expand and set properly, or the seal dislodged after it was set.[4]

Second, because the lockdown sleeve had not yet been set at the time of the blowout, pressure and forces from the well below could have lifted the casing hanger up and out of place in the wellhead. Several forces could have generated such uplift, alone or in combination:

- upward pressure in the annulus that exceeded the weight of the production casing;[5]
- sustained flow of high-temperature hydrocarbons that caused the metal production casing to expand and lengthen;[6]
- sufficiently forceful hydrocarbon flow; and
- nitrogen gas that escaped from unstable foamed cement (explained in Chapter 4.4).[7]

If the casing hanger lifted up as a result of net upward pressure in the annulus, the casing would have dropped back down once pressurized fluids escaped and the pressure equalized. That lifting and dropping motion would have occurred repeatedly, resulting in intermittent flow through the seal assembly. Repeated up-and-down movement could also dislodge the shoe track cement, creating an easier path for continuous flow.

Flow up the Inside of the Production Casing

Hydrocarbons could have traveled into and up through the production casing in two different ways.

First, the cement in the shoe track could have failed, creating a path for hydrocarbons to flow into the open bottom end of the production casing. Those hydrocarbons would also have had to bypass two mechanical float valves (explained in Chapter 4.3).

Second, hydrocarbons in the annulus could have flowed into the production casing through an opening in the casing. That opening could have been a breach in the 9⅞-inch × 7-inch tapered crossover joint,[8] a leak in the threads of a casing joint,[9] or a hole in the casing wall, as illustrated in Figure 4.1.3.

Figure 4.1.3. Flow up the production casing.

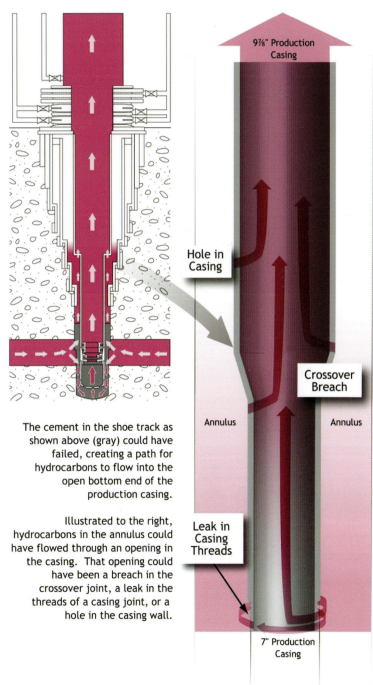

The cement in the shoe track as shown above (gray) could have failed, creating a path for hydrocarbons to flow into the open bottom end of the production casing.

Illustrated to the right, hydrocarbons in the annulus could have flowed through an opening in the casing. That opening could have been a breach in the crossover joint, a leak in the threads of a casing joint, or a hole in the casing wall.

TrialGraphix

Expert and Investigator Opinions on Flow Path Scenarios

Each of the four general flow path scenarios described above are plausible during a blowout. Hydrocarbon flow up through the annulus is a more common problem[10] that has "long plagued the petroleum industry."[11] But hydrocarbons have also been known to flow through shoe track cement and breaches of casing.[12]

Experts involved in the Macondo containment operations initially speculated that flow had come up through the annulus and the seal assembly.[13] But based on the evidence now available, expert opinion has shifted to favor the scenario in which flow came up through the inside of the production casing.[14]

BP internal investigators have concluded that hydrocarbons came up through the shoe track, based in large part on post-blowout well flow modeling.[15] Transocean internal investigators have expressed agreement with this finding.[16] Halliburton representatives, by contrast, continue to posit a theory in which seal assembly liftoff contributed to or caused annular flow.[17] Halliburton has also speculated that there may have been a breach in the production casing.[18]

The Chief Counsel's team finds that hydrocarbon flow came up through the production casing, most likely due to a failure of the shoe track cement.[19]

Forensic Evidence Suggests That Hydrocarbons Did Not Flow up the Annulus and Through the Seal Assembly

On September 5, 2010, BP removed the *Deepwater Horizon*'s blowout preventer from the Macondo wellhead and replaced it with the blowout preventer from the *Development Driller II*, one of the rigs drilling the two relief wells. With a new blowout preventer and riser in place, the crew of the *Development Driller II* performed a series of forensic operations in and through the upper portions of the Macondo production casing.[20]

If hydrocarbons had flowed up the annulus and through the seal assembly, one would have expected to see at least the following two things:

- hydrocarbons should have been present throughout the annular mud; and
- the outside surfaces of the casing hanger and seal assembly should have been eroded by sustained high-volume flow through the flow passages.[21]

If the casing hanger had lifted up, one would further expect the casing hanger not to have been seated properly in the wellhead housing after the blowout. The evidence does not bear out these expectations.

No Significant Presence of Hydrocarbons in the Annulus

Post-blowout operations analyzing the density of the fluid in the upper annular space suggest that the annular space contained insufficient hydrocarbons to support an annular flow path theory.[22]

Perforation of the Production Casing

On October 7, BP perforated the 9⅞-inch production casing midway down the well (from 9,176 to 9,186 feet), creating a path from the inside of the production casing into the annulus.[23] BP did this in order to determine the density of the fluids in the annular space.

If the annulus had been filled with gaseous hydrocarbons (which are low in density, generally 7 ppg or less[24]), high-density drilling mud (14.3 ppg[25]) inside the production casing would have flowed into the annulus until the densities in the annulus and production casing had equalized.[26] This would have led the crew of the *Development Driller II* to observe two signs: lost mud returns and a significant decrease in drill pipe pressure caused by the decrease in density of the fluid column in the production casing.

Rig personnel did not observe either of those signs. Following perforation, they observed only a slight decrease in drill pipe pressure (from 250 to 143 psi[27]), indicating that the fluids in the annulus were similar in density to the mud in the production casing.[28] (The bottomhole cementing procedure before the blowout left 14.17 ppg drilling mud in the annulus.[29]) After perforation, rig personnel monitored the well for 10 minutes and recorded no change in returns; the well was static.[30]

Both of these observations suggested that the fluids present in the annulus after the blowout were the drilling fluids that BP and Halliburton had left in the annulus before the blowout.[31] If hydrocarbons had flowed through the annulus, they would have flushed those drilling fluids out of the annulus during the course of the blowout.

Sampling of the Annular Fluid

Subsequently, in mid-October, the *Development Driller II's* crew cut the production casing midway down the well (at 9,150 feet),[32] detached the production casing hanger from the wellhead,[33] and lifted the cut portion of the casing up 15 feet.[34] The crew then circulated the annular fluid up to the rig by pumping mud down into the production casing, around the corner of the cut portion, and up through the annulus into the riser, taking mud samples intermittently during the circulation.[35] Those samples ranged from 13.0 to 14.3 ppg in density.[36] Once again, those density measurements were consistent with the density of the drilling fluids that BP and Halliburton had left in the annulus at the end of the bottomhole cement job before the blowout.[37] This indicated again that hydrocarbons likely had not flowed through the annulus.[38]

No Erosion on the Outside of the Casing Hanger and Seal Assembly

A tremendous volume of oil and gas flowed out of the well at a tremendous rate during the course of the blowout.[39] If that flow had traveled through the annulus, past the casing hanger, and through the seal assembly, it would have severely eroded the casing hanger and seal assembly.

On October 13, BP recovered the production casing hanger and seal assembly from the Macondo wellhead.[40] Neither piece of equipment showed any signs of damage in locations where annular flow would have caused serious erosion. Instead, the relevant areas were totally undamaged.

Figure 4.1.4. Exterior of the Macondo production casing hanger and seal assembly.

- Figure 4.1.4. Exterior of the Macondo production casing hanger and seal assembly. The outside surfaces of the Macondo casing hanger and seal assembly show no damage (left). They have no erosion-induced channels. Instead, they resemble the condition of brand-new equipment (right).

- The white square placed on the casing hanger before it was set remains. If hydrocarbons had flowed past that area, they almost certainly would have removed this mark.[41]

- The 18 flow passages in the casing hanger show no signs of erosion.[42] If hydrocarbons had flowed through those passages at the velocities estimated for this blowout, they likely would have eroded and enlarged the holes.[43]

- The rubber elastomeric element of the seal assembly (removed post-incident and circulated out into the shaker[44]) still retains its original shape, including a protrusion that one would expect to have been eroded away by annular hydrocarbon flow.[45]

Dril-Quip

By contrast, the interior of the BOP[46] (through which hydrocarbons definitely flowed) showed serious erosion, as did the interior of the casing hanger, seen in Figure 4.1.5.[47]

Figure 4.1.5. Interior of the Macondo production casing hanger compared to new equipment.

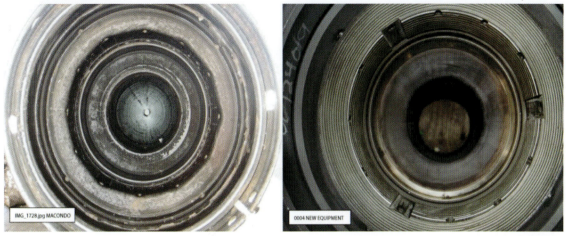

Dril-Quip

Macondo Equipment   **New Equipment**

This is strong evidence that hydrocarbons progressed up the inside of the production casing, not up the annulus past the casing hanger and through the seal assembly.[48]

No Detachment of the Casing Hanger

Post-blowout operations on the production casing hanger and seal assembly also suggest that the casing hanger and seal assembly remained in precisely the same place they had been set before the blowout. That observation is inconsistent with the theory that upward forces in the well lifted the casing hanger out of the wellhead. If the casing hanger had been lifted out of place, vented pressure, and then dropped back down, one would almost certainly expect the metal edges of the casing hanger and seal assembly to show damage and expect the casing hanger to have landed in a different position than the one in which it had originally been set.

No Apparent Damage to Metal Edges

The casing hanger and seal assembly contain a series of circular metal lips (as shown in Figure 4.1.6) that protrude and fit inside a corresponding profile on the inside of the wellhead housing. The parts fit together very precisely to create a metal-to-metal seal. If the casing hanger had lifted out of place, it would have caused significant damage to these metal lips. Post-blowout photographs of the casing hanger and seal assembly show no such wear.[49]

Figure 4.1.6. Undamaged metal edges of the casing hanger and seal assembly.

Dril-Quip

Macondo Equipment   **New Equipment**

Casing Hanger Properly Seated

In order to set a casing hanger, rig personnel normally lower the casing hanger into the wellhead. When in the correct position, a load transfer ring pops into place to support the load of the casing.[50] The crew must lower the casing hanger *slowly* to avoid missing the correct landing spot.

If the casing hanger had lifted up and dropped down during the blowout, it is highly likely that such movement would have been neither gentle nor slow. As a result, the load ring probably would have passed by its intended seat, and the casing hanger would not have reseated properly in its original position.[51]

On September 9, the crew of the *Development Driller II*, along with representatives from Dril-Quip (the manufacturer of the casing hanger), ran a **lead impression tool**.[52] The tool indicated that the 9⅞-inch casing hanger was "seated properly" in the 18¾-inch high-pressure wellhead housing, where it had been placed prior to the blowout.[53] Because none of the post-blowout operations would have reconnected the casing hanger, this is strong evidence that it never disconnected, and the casing hanger did not lift up during the blowout.[54]

Lead Impression Tool. A **lead impression tool** is a small block with soft metal (usually lead). Rig personnel lower it into the wellhead and take an impression to identify the internal profile of the wellhead, including the elevation of the casing hanger.[55]

Passing Post-Blowout Positive Pressure Test

On September 10, the crew of the *Development Driller II* conducted a positive pressure test on the production casing and saw no significant change in pressure or flow.[56] (Chapter 4.6 describes a positive pressure test in detail.) This is inconsistent with the casing hanger liftoff theory. A positive pressure test examines the pressure integrity of the casing hanger and seal assembly for a sustained period of time. If the casing hanger had lifted up or the seal assembly had leaked, the crew of the *Development Driller II* likely would have observed a significant decrease in pressure or return flow from the well, or both.[57]

Successful Installation of the Lockdown Sleeve

Finally, on September 11, the crew of the *Development Driller II* successfully installed and pressure tested a lockdown sleeve in the Macondo wellhead.[58] The fact that BP was able to install a lockdown sleeve after the blowout suggests that the casing hanger was properly seated in the wellhead.[59] In order for the lockdown sleeve to properly set onto the casing hanger, the casing hanger itself must be properly seated in its high-pressure housing.[60]

Circulation of Fluids During the Pre-Blowout Cement Job

Despite the evidence described above, Halliburton argues that "hydrocarbons may have already been present in or even flowing into the annulus before the production casing cement job was complete."[61] The company bases its hypothesis on the "discernable drop in surface pressure at the conclusion of the cement job" that occurred on April 20 (illustrated in Figure 4.1.7).[62]

Halliburton's argument is unpersuasive for several reasons.

First, the observed fluctuation in surface pressure can be explained by the wellbore geometry at Macondo.[63] Macondo had a **tapered** production casing string—9⅞ inches from wellhead to 12,488 feet below sea level, tapering to 7 inches from 12,488 feet below sea level to the bottom of the casing. In wells with a tapered production casing (and hence a tapered annulus), "each discrete volume of fluid will grow in column height as it travels down the well [past the crossover joint] and shrink as it comes up the well [past the crossover joint]."[64] As a result, the hydrostatic pressure differential between the casing and the annulus will change over the course of the cement job (as it did at Macondo).

Second, the drop in surface pressure did not appear particularly anomalous at the time. In fact, Halliburton's own pre-job cementing model predicted that pressure would decrease by some amount.[65] The Chief Counsel's team has not identified any evidence to suggest that rig personnel monitoring the Macondo cement job thought that the pressures they were seeing were abnormal.[66]

Finally, the cement job pressure readings cannot alone support a theory of annular flow (a point that Halliburton concedes[67]), and the other evidence discussed above is inconsistent with annular flow.

Figure 4.1.7. Halliburton post-cement-job report.

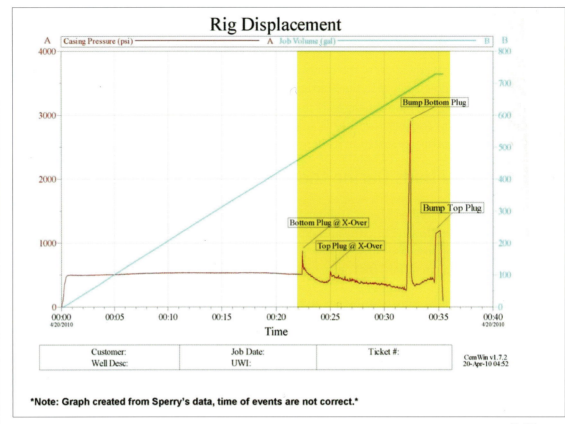

Halliburton

Hydrocarbons Appear to Have Flowed Into and up the Production Casing

Post-blowout inspection of the production casing hanger and seal assembly retrieved from the Macondo well shows severe erosion on the inside of the casing hanger (shown in the left-side photo in Figure 4.1.8). Serrations near the top of the casing hanger—normally ⅛-inch deep—are almost completely abraded away.[68] Threads that normally run around the inside of the casing hanger are flattened.[69] The slot that normally interrupts the threads—¼-inch deep when new—appears as an almost nonexistent indentation.[70] These observations all suggest that hydrocarbons came up through the production casing.

Figure 4.1.8. Erosion of the inside of the casing hanger.

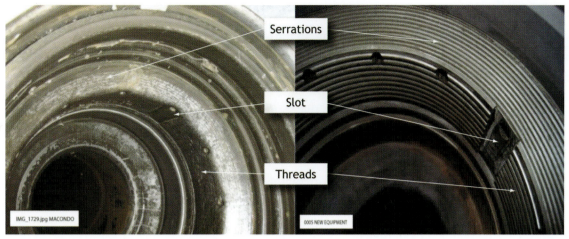

Dril-Quip

Macondo Equipment   **New Equipment**

The remaining question is precisely how hydrocarbons entered the inside of the production casing. Currently available evidence leads the Chief Counsel's team to conclude that hydrocarbons almost certainly entered the production casing through the shoe track. At the same time, the Chief Counsel's team cannot rule out the possibility that hydrocarbons entered the production casing from the annulus through a breach in the side of the casing string.

Hydrocarbons Likely Entered the Production Casing Through the Shoe Track

Problems With the Primary Cement Job Could Have Compromised the Shoe Track Cement

The bottomhole cement job at Macondo involved an unusual number of risk factors. Some were inherent in the conditions at the well; others developed during the course of the design and execution of the bottomhole cement job. This includes a cement slurry that may have been unstable, uncertainties with regard to cement placement (because of doubts about float conversion and centralization), and concerns over cement contamination (as a result of limited pre-cementing circulation and low cement volume and flow rate). Chapter 4.3 discusses these risks in more detail.

The Float Valves Would Not Have Provided an Independent Barrier to Flow Through the Shoe Track

It is not clear whether the float valves in the Macondo well converted prior to the pumping of the bottomhole cement job. A failure to convert these two-way valves into one-way valves would have allowed the cement to flow back in the wrong direction and therefore could have compromised the bottomhole cement job. Even if they had converted, the float valves may not have closed fully due to malfunction or debris. In any case, float valves are not typically considered independent barriers to hydrocarbon flow. Chapter 4.3 discusses these issues in more detail.

Evidence From the Static Kill Operation Suggests Flow Through the Shoe Track

Data from the August 4 static kill operation on the Macondo well suggest that flow came up through the shoe track. In the static kill operation, BP planned to pump 13.2 ppg mud into the well, from the top of the wellbore to the bottom, monitoring pressures along the way.[71] Before doing so, the company modeled expected pressures and volumes for several flow path scenarios, including flow up the annulus and flow up the production casing (with the drill pipe in different positions).[72] Pressures observed during the operation more closely matched flow up the production casing.[73]

The static kill data analysis has several shortcomings. First, BP performed its analysis with imperfect knowledge of the wellbore geometry and without knowing whether there was debris or other obstructions in the well.[74] Second, the observed pressures matched the modeled pressures only up to a certain point and then diverged.[75] Third, it is unlikely that the pressure observations were sensitive enough to distinguish a casing breach near the bottom of the production casing (such as near the float collar).[76]

Analysis of the static kill data is still ongoing and subject to future revision.

The Chief Counsel's Team Cannot Rule Out the Possibility of Flow Through a Breach in the Production Casing

Figure 4.1.9. 16-inch casing and rupture disks.

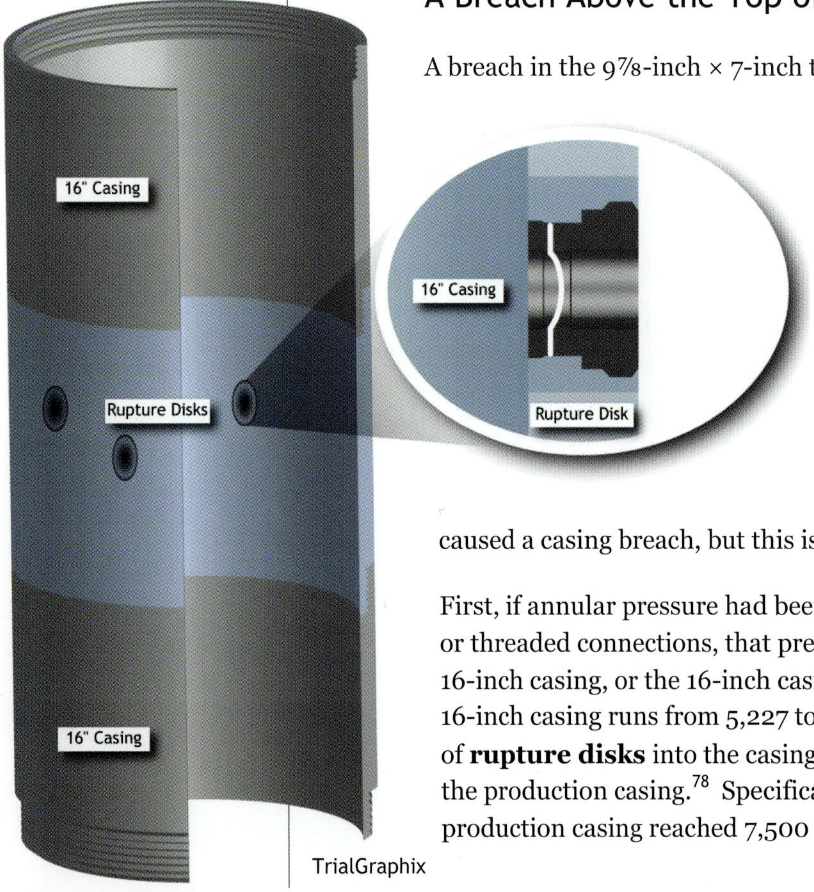

Hydrocarbons may have entered through a breach in the production casing, although the Chief Counsel's team considers this scenario unlikely.

A Breach Above the Top of Cement Is Unlikely

A breach in the 9⅞-inch × 7-inch tapered crossover joint or anywhere above the top of the annular cement is unlikely. If hydrocarbons went from the formation into the annulus and then through such a breach, one would expect to observe hydrocarbons in the annular space. As explained above, there is no evidence of a significant hydrocarbon presence in the annulus.

A Breach as a Result of External Pressure Is Unlikely

External pressure in the annulus (caused by hydrocarbon flow or nitrogen gas) could have caused a casing breach, but this is unlikely for at least two reasons.

First, if annular pressure had been sufficient to cause a breach in the production casing or threaded connections, that pressure should first have caused rupture disks in the 16-inch casing, or the 16-inch casing itself, to burst (shown in Figure 4.1.9). The 16-inch casing runs from 5,227 to 11,585 feet below sea level.[77] BP installed three sets of **rupture disks** into the casing wall. The rupture disks were designed to fail before the production casing.[78] Specifically, if pressure between the 16-inch casing and the production casing reached 7,500 psi, the rupture disks should have burst outward.[79]

16" Casing

16" Casing

Rupture Disks

16" Casing

Rupture Disk

TrialGraphix

This pressure is, by design, less than the 11,140 psi that the production casing and its threaded connections are designed to withstand.[80] Even if the rupture disks did not function as designed, the 16-inch casing probably would have failed in some manner once pressures significantly exceeded 6,920 psi.[81] But it appears that neither the rupture disks nor the 16-inch casing failed. Chapter 4.2 discusses this issue in more detail.

Second, there is no evidence to date that the production casing was designed improperly, or that crew members improperly made up one or more casing joints before sending them downhole. A Weatherford representative was on the rig, monitoring the makeup of the casing, tracking torques and turns through a computer program, and verifying that all of the connections were up to standard.[82] Furthermore, the Weatherford daily log and data from the computer program do not show any mishaps in casing makeup for most of the production casing.[83] (The integrity of connections made up onshore—including the reamer shoe, centralizer subs, float collar, and crossover joint—remains unconfirmed.[84]) While members of the rig crew inadvertently dropped and damaged some pipe when making up the 7-inch portion of the casing,[85] the evidence shows that they subsequently replaced the damaged joints before sending them downhole.[86]

A Breach Below the Top Wiper Plug as a Result of Internal Pressure Cannot Be Ruled Out

The Chief Counsel's team cannot completely rule out a casing breach below the top plug, though it is unlikely.[87] If such a breach occurred prior to the cement job, it could have jeopardized the placement of the bottomhole cement.

Testimonial evidence shows that in the day before the blowout BP personnel were concerned about a possible casing breach. (Chapter 4.3 discusses these facts in more detail.) On April 19, after attempting to convert the float equipment and establishing circulation, one witness recalls well site leader Bob Kaluza saying, "I'm afraid that we've blown something higher up in the casing joint."[88] Kaluza was presumably referring to the possibility that the unusually high 3,142 psi pressure that BP directed the rig crew to apply to convert the float valves created a breach in the production casing.[89] BP and rig personnel subsequently observed lower-than-expected circulating pressures, which could be consistent with mud being circulated through a breach in the casing and back up to the rig through the upper part of the annulus, rather than out the bottom of the casing and up the entire annulus. Kaluza expressed his concern to BP drilling engineer Brian Morel, who was also on the rig.[90] Morel relayed the concern to BP wells team leader John Guide, who was onshore.[91] Meanwhile, Morel also emailed Weatherford sales representative Bryan Clawson, "Yah we blew it at 3140, still not sure what we blew yet."[92]

After discussing the issue, the BP Macondo team determined that if there were a casing breach, they could not fix it at that point in the operations.[93] They also concluded that they would detect any such breach in later well integrity pressure tests and could take remedial measures at that time.[94] There is no evidence that anyone actually revisited the issue prior to the blowout.

BP personnel may not have detected a casing breach near the float collar. After the cement job, rig personnel performed a positive pressure test on the well to test the integrity of the production casing. But a positive pressure test does not test the casing below the top wiper plug.[95]

(Chapter 4.6 discusses positive pressure tests in more detail.)[*] After the blowout, BP conducted a static kill operation on the well and observed pressure data consistent with shoe track flow. But the modeled and observed pressure and volume data were not sensitive enough to distinguish a casing breach near the bottom of the production casing (such as near the float collar) from flow through the shoe track cement.[96] And although a Weatherford log tracking the makeup of the production casing showed no mishaps, the log did not contain data on the integrity of connections made up onshore—including the float collar.[97]

Technical Findings

The Annular Cement Did Not Isolate the Hydrocarbon Zones

The Chief Counsel's team finds that the cement in the annular space did not isolate the hydrocarbon zones. This finding calls into question the quality of the bottomhole cement job. Chapters 4.3 and 4.4 identify possible shortcomings in that cement job including mud contamination, improper cement placement, and cement slurry instability.

Hydrocarbons Came to the Surface by Traveling Through the Production Casing

The Chief Counsel's team finds that hydrocarbons came to the surface through the inside of the production casing. This finding calls into question BP's temporary abandonment procedure and design. Chapter 4.5 discusses the risks attendant to the temporary abandonment.

The Shoe Track Cement Probably Failed

The Chief Counsel's team finds that flow almost certainly came up through the shoe track of the production casing. Cement in the shoe track should have blocked this flow. This finding again calls into question the quality of the bottomhole cement job. Chapter 4.3 discusses possible reasons for shoe track cement failure. ◖

---

[*] Rig personnel also performed a negative pressure test on the well. A negative pressure test does test the integrity of the casing down through the shoe track as well as the shoe track cement. But rig personnel misinterpreted the negative pressure test. Chapter 4.6 discusses this in more detail.

Chapter 4.2 | Well Design

BP's engineering team made a number of important well design decisions that influenced events at Macondo. Among other things, the engineers (1) decided to use a long string production casing, (2) installed rupture disks in the well, and (3) decided to avoid creating trapped annular spaces by omitting a protective casing and leaving annular spaces open to the surrounding formation. The Chief Counsel's team finds that these decisions complicated pre-blowout cementing operations and post-blowout containment efforts.

Deepwater Well Design

Wells are drilled for a reason: either to explore for oil and gas, appraise an earlier discovery, or create a development well in an existing oil field. By the time the well is designed, subsurface geologists and geophysicists will have identified subsurface objectives, usually using seismic reflection data. They will also have prepared—in as much detail as possible—a geologic prognosis describing lithology, pressure, and fluid content as a function of depth. If there are other wells nearby, the geologists and geophysicists will have used data from those wells to inform their prognosis.

The design team that plans the well must determine how best to achieve the well's objectives while managing potential drilling hazards. The hazards can include a variety of geologic features. For instance, porous gas-bearing intervals ("shallow gas")—sand layers containing pressurized gas or water, or unstable formations— may occur in the first few thousand feet below the seabed. Geologic faults and low-pressure hydrocarbon-bearing sands (depleted by nearby oil production) can also present hazards. Sudden variations in subsurface pore pressure can pose hazards as well. Operators must also consider man-made hazards such as nearby oil and gas development infrastructures (wells, platforms, pipelines) and ship traffic.

In many cases the design team can identify drilling hazards in advance and avoid them. But some geologic hazards, such as high pore pressures and hydrocarbon deposits, are impossible to avoid. Indeed, they are closely associated with the drilling objectives—oil companies often target high-pressure hydrocarbon reservoirs. High pore pressures are a common feature of the deepwater Gulf of Mexico environment, and often signal the presence of oil and gas.

Drilling engineers must therefore keep several key issues in mind as they design a deepwater well.

TrialGraphix

Figure 4.2.1

Artist's rendering of the Macondo well from rig to rathole.

Pore Pressure and Fracture Gradients

Drilling engineers must design wells to manage intrinsic risks. Specifically, they must develop drilling programs that will manage and reflect the pore pressure and fracture gradients at a given drilling location as shown in Figure 4.2.2. (Chapter 2 describes these concepts in more detail.) The design team must specify the kinds of drilling fluids that will be used and the number and type of casing strings that will extend from the seafloor to the total depth of the well. The drilling fluids and casing strings must work together to balance and contain pore pressures in the rock formation without fracturing the rock.

Figure 4.2.2. Narrow drilling margins.

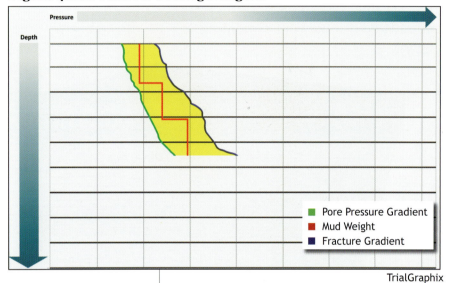

Pore Pressure Gradient
Mud Weight
Fracture Gradient

TrialGraphix

Creating this plan can be difficult if engineers have limited information about subsurface geology and if actual pore pressures vary significantly from predictions.[1] This is often the case in exploration wells or in the first well in a new field. The problem frequently crops up in the Gulf of Mexico, which is prone to having a narrow window between the pore pressure and fracture gradients as well as zones of pore pressure repression (where the pore pressure gradient suddenly reverses and decreases with depth).[2]

Because drilling conditions often differ significantly from predictions, engineers often design and redesign a deepwater well as the well progresses. They work constantly to keep two factors within tolerable limits: **equivalent static density** (ESD) and **equivalent circulating density** (ECD). ESD refers to the pressure that a column of fluid in the wellbore exerts when it is static (that is, not circulating). ECD refers to the *total* pressure that the same fluid column exerts when it *is* circulating. When drillers circulate fluids through a well, ECD exceeds ESD because the force required to circulate the fluids exerts additional pressure on the wellbore.

In planning the well, engineers will design a mud program to keep both ESD and ECD below the rock's fracture gradient. Drillers monitor these parameters carefully as they work.

Barriers to Flow

As discussed in Chapter 2, operators typically employ redundant **barriers** to prevent hydrocarbons from flowing out of the well before production operations. One important barrier in any well is the mud and drilling fluid system in the wellbore. When properly designed and operated, the drilling fluid system should balance the pressure of any hydrocarbons in the well formation. Engineers can also use other kinds of barriers during drilling and completion. Those barriers include cemented casing, mechanical and cement plugs, and the blowout preventer (BOP). Sound industry practice—and BP's own policy—generally requires an operator to maintain two verified barriers along any potential flow path.[3]

Annular Pressure Buildup

If an operator plans to use a given well to produce oil in the future (rather than merely to learn about subsurface geology), its design team must consider the environmental and mechanical stresses that the well will experience over its lifetime. The casing and completion program must ensure that these stresses do not compromise well integrity over the life of the well, which could be as long as several decades.

In deepwater production wells, engineers pay special attention to a phenomenon called **annular pressure buildup** (APB). Figure 4.2.3 illustrates that during production activities, high-temperature hydrocarbons travel up from the pay sands through production tubing installed inside the production casing. The flow of hydrocarbons heats up the well. As a result, fluids and gases in the annular spaces of the well expand. If the well design creates annular spaces that are enclosed, the fluids and gases trapped within those spaces will exert increasing pressure on the well components as they heat up. In some cases, the pressure can become high enough to collapse casing strings in the well and to force the operator to abandon the well.

Managing annular pressure buildup in a deepwater well requires careful planning and design. Engineers can use a number of design features to manage annular pressures or mitigate the risks of casing collapse. These include rupture disks, compressible fluids in the annular space, and insulated production tubing. Finally, they can design wells in ways that avoid creating trapped annular spaces at all.

Figure 4.2.3. Annular pressure buildup (APB).

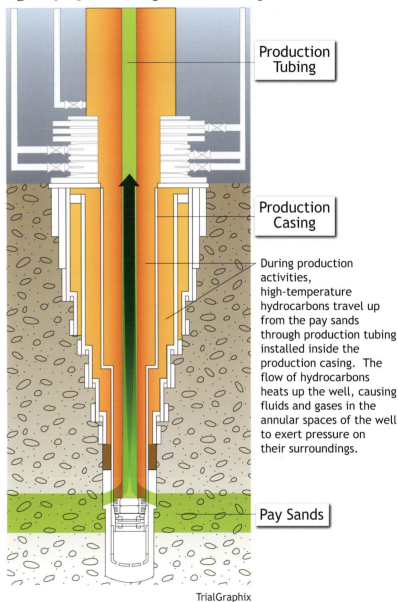

Production Tubing

Production Casing

During production activities, high-temperature hydrocarbons travel up from the pay sands through production tubing installed inside the production casing. The flow of hydrocarbons heats up the well, causing fluids and gases in the annular spaces of the well to exert pressure on their surroundings.

Pay Sands

TrialGraphix

The Macondo Well Design

Even before it began drilling Macondo, BP believed that the well might encounter a substantial hydrocarbon reservoir.[4] But BP also recognized that it might also encounter a number of hazards, including shallow gas sands, overpressures, and depleted reservoir zones, as well as the expected oil and gas in the mid-Miocene objective reservoir. BP chose the particular drilling location for Macondo to penetrate the objective section while avoiding shallow gas sands that it had identified. BP identified potential minor drilling hazards beneath 8,000 feet below sea level: thin gas-charged sands and depleted (low-pressure) zones.[5]

Using seismic imagery, BP had a high degree of confidence that the formation below contained a significant accumulation of oil and gas.[6] BP therefore planned the Macondo well as an exploration well that it could later complete and turn into a production well.[7]

Figure 4.2.4. Offset wells and seismic data.

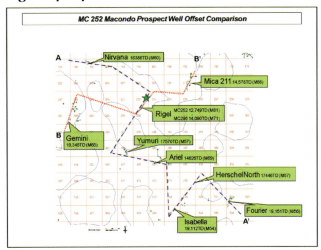

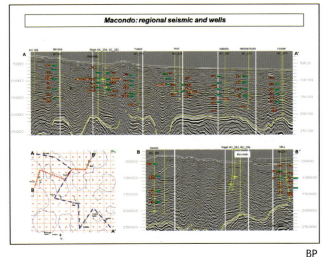

BP

The green star indicates Macondo's location.

BP drilling engineer Brian Morel and senior engineer Mark Hafle had the primary responsibility for the Macondo well design work.[8] They worked with a number of BP engineers and geoscientists to develop their plans.[9] Geologists and petrophysicists from BP's Totally Integrated Geological and Engineering Resource (TIGER) team helped develop a pore pressure profile for the well based on other wells in the vicinity ("offset wells") as shown in Figure 4.2.4.[10] A BP casing and tubular design team independently reviewed the well design.[11] Fluid experts and rock strength experts checked the geomechanical aspects of the well.[12] And because the well was being designed as a producer, BP completion engineers also provided input during the design process.[13] The completion engineers recommended, among other things, an analysis of the well's potential for annular pressure buildup and possible mitigation measures.[14]

In June 2009, the initial Macondo well design underwent peer review.[15] The reviewers concluded that the Macondo design team "did a lot of good work," that the initial design was "[r]obust" and "supported by good data and analysis," and that "all major risk[s] [were] addressed and mitigations developed."[16] Over the course of the next year, the Macondo engineering team would update its drilling program several times. But three key design features never changed.

Rupture Disks

All of BP's Macondo well designs included three sets of rupture disks in the 16-inch casing.[17] The 16-inch casing was the longest piece of pipe outside of the production casing. The **rupture disks** (or burst disks) would relieve annular pressure before that pressure could build up high enough to cause a collapse of the production casing or the 16-inch casing.

The disks worked in two ways as shown in Figure 4.2.5. First, if pressure between the 16-inch casing and the production casing reached 7,500 pounds per square inch (psi), the rupture disks would *burst outward* and release that pressure.[18] Because the production casing was rated to withstand 11,140 psi of pressure, this would prevent annular pressure from rising to the point at which it could collapse the production casing.[19] Second, if pressure *outside* of the 16-inch casing (that is, between the 16-inch casing and the other larger casing strings outside it) exceeded 1,600 psi, the rupture disks would *collapse inward* to release that pressure.[20] Because the 16-inch casing was rated to withstand 2,340 psi of pressure, this would prevent pressure outside the 16-inch casing from rising to the point at which it could collapse the 16-inch casing.[21]

Once ruptured, the disks would leave small holes in the 16-inch casing through which pressure could bleed into the surrounding rock formation.[22]

Figure 4.2.5. Rupture disks.

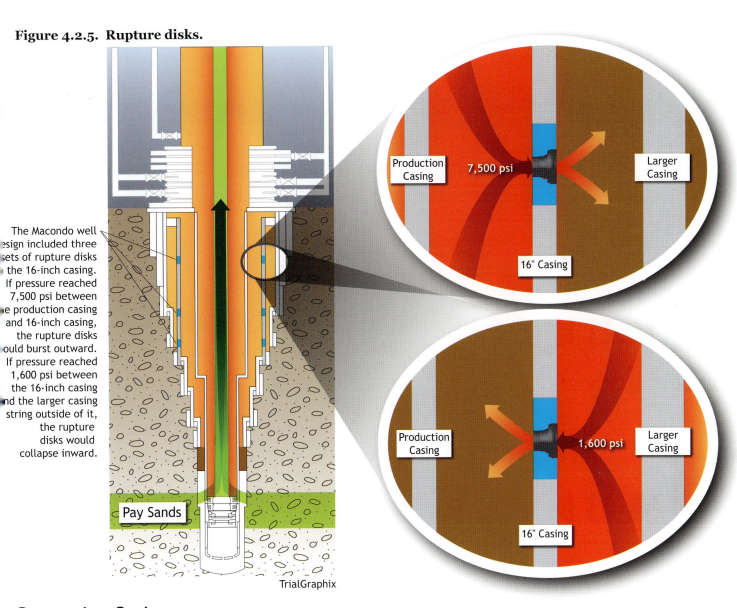

The Macondo well design included three sets of rupture disks in the 16-inch casing. If pressure reached 7,500 psi between the production casing and 16-inch casing, the rupture disks would burst outward. If pressure reached 1,600 psi between the 16-inch casing and the larger casing string outside of it, the rupture disks would collapse inward.

Production Casing   7,500 psi   Larger Casing

16" Casing

Production Casing   1,600 psi   Larger Casing

16" Casing

TrialGraphix

Protective Casing

BP's well design consistently and deliberately omitted a **protective casing**. A protective casing is an intermediate casing string outside the production casing that runs from deep in the well all the way back to the wellhead.[23] A protective casing supplies a "continuous pressure rating" for the interval that it covers (as shown in Figure 4.2.6) and seals off potential leak paths at the tops of previous liner hangers.[24]

It is common industry practice to use a protective casing whenever running a long string production casing.[25] But the Macondo team never planned for a protective casing[26] because installing such a casing would also have negated their efforts to mitigate annular pressure buildup.[27] Specifically, it would have sealed off the rupture disks and the previously open annuli in the casing design.

Figure 4.2.6. Protective casing.

Protective casing (yellow) provides a "continuous pressure rating" for the casing interval that it covers (gray).

TrialGraphix

Figure 4.2.7. Casing options in deepwater drilling.

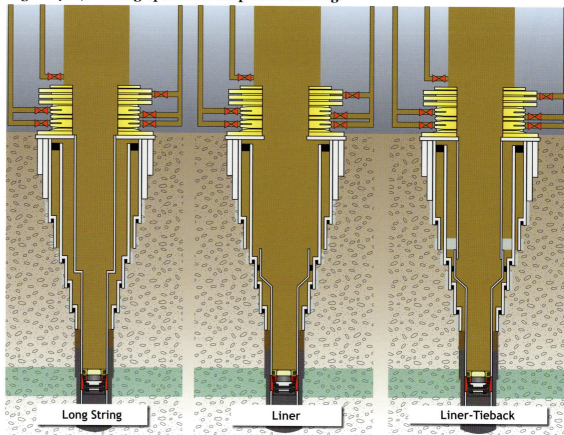

Long String   Liner   Liner-Tieback

TrialGraphix

Long String Production Casing

Third, BP's Macondo well design called for a long string production casing, or **long string**, stretching from the bottom of the well all the way to the wellhead. This was true of the initial well design as well as the final well design.[28]

As shown in Figure 4.2.7, the alternative to a long string production casing would have been a **liner**. A liner is a shorter string of casing hung from a casing hanger lower in the well. In order to connect the liner back to the wellhead, BP would eventually have had to install a **tieback**—a string of casing pipe stretching between the top of the liner on one end to the wellhead on the other end. Setting the tieback adds two annular flow barriers to the well design.

In the weeks just prior to the blowout, BP briefly considered using a liner instead of a long string at Macondo. There is no evidence that the Macondo team ever considered having the *Deepwater Horizon* crew install the tieback before temporarily abandoning the well.[29] They presumably would have left that job for a completion rig.

Drilling the Macondo Well

BP encountered a series of complications while drilling the Macondo well. This included two previous kicks, a ballooning event, lost circulation events, and trouble determining pore pressures (as shown in Figure 4.2.8). Together, these issues made Macondo "a difficult well."[30]

Kicks and Ballooning

Twice prior to April 20, the Macondo well experienced an unwanted influx into the wellbore, or a "kick." On October 26, 2009, the well kicked at 8,970 feet. The rig crew detected the kick and shut in the well. They were able to resolve the situation by raising the mud weight and circulating the kick out of the wellbore.[31] On March 8, 2010, the well kicked again, at 13,305 feet.[32] The crew once again detected the kick and shut in the well.[33] But this time, the pipe was stuck in the wellbore.[34] BP severed the pipe and sidetracked the well.[35]

On March 25 the Macondo well also had a ballooning, or "loss/gain," event. The rig lost fluids into the formation. When the crew decreased the pressure of the mud in the wellbore, the rig then received an influx of fluids from the formation.

Lost Circulation During Drilling

A major risk at Macondo was the loss of drilling fluid into the formation, called **lost circulation** or **lost returns**.[36] At various points in February, March, and April, the pressure of drilling fluid exceeded the strength of the formation, and drilling fluid began flowing into the rock instead of returning to the rig.[37] Lost circulation events are common in offshore drilling. The *Horizon* rig crew generally responded with a standard industry tactic: It pumped thick, viscous fluid known as **lost circulation material** into the well and thereby plugged the fractures in the formation.

The *Horizon* crew successfully addressed repeated lost circulation events while drilling the Macondo well.[38] The events occurred frequently and at various depths, and sometimes lasted several days: once in mid-February, four times in March, and three times in April.[39] In total, BP lost approximately 16,000 barrels of mud while drilling the well, which cost the company more than $13 million in rig time and materials.[40]

Uncertain Pore Pressures Affect the Well Design

The kicks, ballooning, and lost circulation events at Macondo occurred in part because Macondo was a "well with limited offset well information and preplanning pressure data [were] different than the expected case."[41] Given BP's initial uncertainty about the pore pressures

Figure 4.2.8. Timeline of drilling events.

October 6, 2009
Marianas spuds the well
5,067' below sea level

6,217'

7,937'

October 26, 2009
8,968' Kick at 8,970 feet

November 9, 2009
Hurricane Ida damages
Marianas at 9,090 feet
January 31, 2010
Deepwater Horizon arrives
and resumes drilling on
February 10, 2010

11,585'

March 8, 2010
13,145' Kick at 13,305 feet

March 25, 2010
15,103' Ballooning at 15,113 feet

17,168'

April 9, 2010
Total depth at 18,360 feet
18,360'

Lost Returns

February 17-21, 2010
12,350 feet

March 2, 2010
11,587 feet

March 3-5, 2010
11,575 feet

March 21, 2010
13,150 feet

March 31, 2010
17,163 feet

April 3, 2010
17,761 feet

April 4-7, 2010
18,260 feet

April 9, 2010
18,193 feet

TrialGraphix

of the rock, the company had to adjust its well design as it drilled the well and gained better pore pressure information.

This was particularly true after the March 8 kick. According to contemporaneous communications among BP engineers, the "kick and change in pore pressure...completely changed" the forward design[42] and did so "rapidly."[43] "Due to well pressure uncertainty, it [was] unknown how many more liners [BP would] need to set before getting to TD."[44] Accordingly, the Macondo team decided to proceed more conservatively and set casing strings shallower in the well.[45] They installed an intermediate 11⅞-inch liner (at 15,103 feet) that had been set aside as a contingency in the original plan.[46] They then set an additional liner, 9⅞ inches in diameter, above the reservoir (at 17,168 feet).[47] And they planned for yet another smaller casing size in the final hole section.[48]

Rig Crew Calls Total Depth Early Due to Narrow Drilling Margin

The last of the lost circulation events occurred on April 9, after the rig had begun to penetrate the pay zone.[49] At 18,193 feet below sea level, the drilling mud pressure exceeded the strength of the formation, and the rig crew observed lost returns. The point at which the formation gave way— when ESD was approximately 14.5 pounds per gallon (ppg)—came as a surprise to the Macondo team.[50] The crew had to stop drilling operations until they could seal the fracture and restore mud circulation. They pumped 172 barrels of lost circulation material down the drill string, hoping to plug the fracture.[51] The approach worked, but BP's onshore engineering team realized the situation had become delicate.[52] In order to continue drilling, they had to maintain the weight of the mud at approximately 14.0 ppg in order to balance the pressure of hydrocarbons pushing out from the formation. But drilling deeper would exert even more pressure on the formation. Engineers calculated that drilling with 14.0 ppg mud would yield an ECD of nearly 14.5 ppg— presenting the risk of once again fracturing the rock and losing returns.[53] At that point, "it became a well integrity and safety issue."[54] The engineers had "run out of drilling margin."[55] The well would have to stop short of its original objective of 20,600 feet.

Rig personnel were able to carefully drill ahead an additional 167 feet and called total depth at 18,360 feet. In that sense, drilling was successful: BP reached the targeted reservoir zone and was able to run a comprehensive suite of evaluation tools.[56]

ECD Concerns Influence Final Production Casing Design

BP engineers then began preparing to install a production casing. BP had Halliburton run a series of computer models to help plan for cementing the production casing.

March 23 Meeting Considers Both Long String and Liner Production Casing

On March 23, Hafle, Morel, and in-house BP cementing expert Erick Cunningham met with Halliburton cementing engineer Jesse Gagliano to discuss ECD concerns in the modeling.[57] The team was trying to decide what size production casing to install and cement at the bottom of the well.[58] Earlier that month, the engineers had modeled both long string and liner production casing designs on two sizes of pipe—7⅝-inch and 7-inch.[59] They were concerned the 7⅝-inch pipe would create a narrow annulus and increase friction to the point that the formation would break.[60] According to Halliburton's models, a smaller 7-inch pipe reduced ECD significantly.[61] Though no decision was made as to casing design or diameter, the group decided to find out how much 7-inch pipe was available should they decide to use that size production casing at the bottom of the well.[62]

April Meetings Finalize Well Design

BP and Halliburton continued to meet and review Halliburton's computer models of the production casing. The team met on April 9 but decided Halliburton's model was inaccurate because it predicted an ESD of 13.9 ppg, which was erroneously low because the weight of the mud in the wellbore was itself heavier than 13.9 ppg.[63] Gagliano created a new model, but on April 12 BP drilling and completions operations manager David Sims determined the ESD in this model was now too *high*[64] and requested that Cunningham review and lend his expertise to the well plan.[65]

At that point, the team considered running a liner instead of a long string in the production interval. The Macondo team believed that ECD would be lower in running the liner.[66] But BP engineering manager John Sprague raised additional technical concerns and requested a review of annular pressure buildup issues related to running a liner.[67]

The potential for a last-minute switch had BP engineers scrambling. Morel asked casing design specialist Rich Miller for a "quick response" on the annular pressure buildup review.[68] "Sorry for the late notice," he added, "this has been a nightmare well which has everyone all over the place."[69] Miller replied, "We have flipped design parameters around to the point that I got nervous," but with respect to annular pressure buildup issues related to the liner he determined "[a]ll looks fine."[70]

Although the onshore engineers had not yet decided the final casing parameters, the rig crew was still supposed to set the casing in a few days, so BP wells team leader John Guide instructed the BP well site leaders on the rig to ready the equipment necessary to run either a liner or a long string.[71] BP had a number of boat and helicopter runs to the rig over the next several days, trying to coordinate the logistics of equipment and people necessary for the upcoming casing and cement jobs. Well site leader Don Vidrine complained to Guide about the last-minute changes. "[T]here [have] been so many last minute changes to the operation that the WSL's have finally come to their wits end," Guide recounted. "The quote is 'flying by the seat of our pants.'"[72]

Transocean also expressed concern to Guide about the long string/liner decision being made "very late in the day."[73] The contractor needed sufficient advance notice to verify logistics and, in particular, that the rig's equipment was fit to handle the final casing string's weight.[74]

Engineers Decide to Run Long String at April 14 Meeting

On April 14, Hafle, Morel, Cunningham, BP operations engineer Brett Cocales, and drilling engineering team leader Gregg Walz met to review Halliburton's ECD modeling.[75] The group identified another limitation of the model—they determined that its data inputs did not reflect the actual latest data acquired during the well logging process.[76] After reassessing well conditions with Cunningham,[77] the team decided they could successfully run and cement a long string.[78]

Several factors appear to have motivated the decision to install and cement a long string production casing:[79] a desire to stick with the original design basis of the well,[80] a desire to mitigate future annular pressure buildup by avoiding a trapped annulus,[81] a desire to eliminate an extra mechanical seal that could leak during production,[82] and a desire to save $7 million to $10 million in future completion costs.[83]

The team made the decision official in a **management of change** (MOC) document—part of BP's process for documenting changes in well design.[84] According to the MOC, the long string provided the best "well integrity case for future completion operations," "the best economic case" for the well, and could be cemented successfully with careful cement job design.[85] The document also discussed the risk that the primary bottomhole cement would not act as a barrier (as discussed in Chapter 4.3).[86] Senior BP managers—including Sims, Walz, Guide, Sprague, and others—reviewed the management of change document and approved.[87]

Technical Findings

Choosing a Long String Production Casing Made the Primary Cement Job at Macondo More Difficult

Operators in the Gulf of Mexico routinely use long string production casings in deepwater wells.[88] But BP's decision to use a long string at Macondo triggered a series of potential problems, particularly with the bottomhole cement job.

Figure 4.2.9. Cementing a long string vs. cementing a liner.

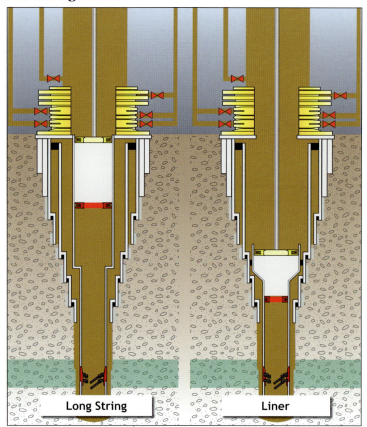

TrialGraphix

The lost circulation event at the pay zone in early April led the company's engineers to carefully analyze whether they could circulate cement successfully around the production casing (or liner) without fracturing the already delicate formation. Because cementing a liner is typically easier than cementing a long string, the decision by BP engineers to stay with the long string design further complicated an already complex cement procedure in several ways.[89]

First, the use of a long string increased the risk of cement contamination. Cementing a long string instead of a liner required cement to travel through a larger surface area of casing before reaching its final destination, as shown in Figure 4.2.9. That increased surface area translates into increased exposure of cement to the film of mud and cuttings that adheres to the casing.[90] That risk was exacerbated by the fact that the long string production casing was tapered, making it more difficult for wiper plugs to reliably wipe clean.[91]

Second, using a long string eliminated the possibility of rotating or otherwise moving the casing in place during the cement job. Rig personnel could have rotated a liner, which would have improved the likelihood of a quality cement job.[92] But it is more difficult to rotate a long string than it is to rotate a liner, so choosing that design eliminated one option for mitigating cementing risks.

Third, cementing a long string typically requires higher cement pumping pressure (and higher ECD) than cementing a liner.[93] To compensate for that pressure increase in a fragile wellbore like the one at Macondo, BP engineers made other adjustments to the cement job. As Chapter 4.3 explains, some of the adjustments the engineers made to reduce ECD increased the risk of cementing failure. If BP engineers had chosen to use a liner, they not only could have obtained lower ECDs, but also may have been able to ignore ECD entirely. This is because the liner hanger includes a mechanical seal that serves as a barrier to annular flow.[94] By relying on that seal, engineers can design a more robust primary cement job—they can, for instance, deliberately exceed ECD limits, risk lost returns, and then plan to remediate cement problems later without having to rely on the cement as a barrier to flow.[95]

Fourth, it is harder to remediate a cement job at the bottom of a long string than it is to remediate one at the bottom of a liner. With a liner, rig personnel can remediate the cement job, before completing the setting of the liner, by lifting the stinger above the liner hanger and pumping additional cement over the top of the liner hanger.[96] That method is more effective and less complex than remediating a long string.[97] With a long string, rig personnel must perform a squeeze job (as defined in Chapter 4.3). A squeeze job is complicated and time-consuming—it can take several days.[98] And BP classifies the time spent squeezing as nonproductive time,[99] an undesired disruption that the company expects its employees to minimize.[100]

BP's Design Efforts to Mitigate the Risk of Annular Pressure Buildup Compromised Containment Operations

BP's decision to install rupture disks at Macondo and not to use a protective casing complicated its containment efforts and may have delayed the ultimate capping of the well. (Commission Staff Working Paper #6, titled "Stopping the Spill: the Five-Month Effort to Kill the Macondo Well," discusses these issues in more detail.) Had BP's design omitted the disks and included the casing, the company would have had increased confidence about the Macondo well's integrity. This, in turn, may very well have allowed the company to shut in the well earlier.

In BP's early analyses of its failed late-May top kill attempt, the company concluded that the rupture disks in the 16-inch casing may have collapsed inward during the initial blowout.[101] The disks could have collapsed if hydrocarbons had entered the annular space between the 16-inch casing and the production casing. Those hydrocarbons would have been much lighter than the heavy drilling mud that would have been in the annular space outside the 16-inch casing. That weight difference would have generated a pressure differential significant enough to collapse the rupture disks.[102]

Based on this theory, as well as pressure readings and visual observations from the field,[103] BP concluded that its top kill operation may have failed because the mud it pumped down the well had flowed out through the collapsed rupture disks rather than remaining within the well as intended.[104] Although BP vice president of engineering Paul Tooms emphasized several months later that rupture disk collapse was just one of several theories that could have explained the top kill results,[105] BP presented the theory to the government as the most likely scenario and changed its subsequent containment strategy to reflect it.[106] Although the government remained skeptical of certain elements of BP's analysis,[107] it too believed the rupture disks may have collapsed and that emergency workers needed to consider that possibility when moving forward.[108]

Before the top kill operations, BP had told Interior Secretary Ken Salazar and Energy Secretary Steven Chu that if the top kill failed, the company might try next to cut the riser, remove the lower marine riser package, and install a second blowout preventer on top of the existing one to shut in the well.[109] But BP and others deemed this approach unwise after theorizing that the rupture disks had collapsed.[110] If hydrocarbons had entered the annular space between the production casing and 16-inch casing and the rupture disks had collapsed, capping the well might divert hydrocarbon flow out the rupture disks and sideways into the rock formation around the well. This would have caused a "subsea blowout" in which hydrocarbons would have flowed up to the surface through the rocks below the seafloor. It would have been nearly impossible to contain that flow. To avoid this situation, BP and the government temporarily stopped trying to shut in the well.

A few weeks after the top kill operation, in mid-June, BP and the government revisited the idea of shutting in the well, this time using a tight-fitted capping stack. Although BP was prepared to install the capping stack in early July,[111] it appears that the government delayed installation for a few days to further analyze the stack's impact on the risk of a subsea blowout.[112] The government's team insisted on monitoring for signs of a subsea blowout using several different methods. BP eventually used ships and remotely operated vehicles (ROVs) to gather visual, seismic, and sonar information about the area around the well. It also used wellhead sensors to monitor acoustic and pressure data. All of these efforts were aimed at determining whether the Macondo well lacked the integrity to prevent oil from flowing sideways into the rock.[113] The government and BP were also concerned that closing the capping stack could increase pressures inside the well sufficiently to create new problems or burst the rupture disks (if they had not already collapsed).[114]

Management Findings

BP Appears to Have Sought the Long-Term Benefits of a Long String Without Adequately Examining the Short-Term Risks

BP engineers displayed a strong and perhaps unwarranted bias in favor of using a long string production casing.

Industry experts have stated that successfully cementing a long string casing is a more difficult enterprise than cementing a liner. BP's own engineers appear to have agreed—they considered using a liner as a means of mitigating the risks of losses during cementing. (Chapter 4.3 discusses this issue in more detail.) BP asked Halliburton to run numerous computer cementing models in an effort to find a way to make the long string casing a viable option. They appear to have approached the problem by trying to find a way to make a long string work instead of asking what design option would best address the cementing difficulties they faced.

BP has argued that its team preferred to use a long string casing because a long string offers better long-term well integrity than a liner-tieback. This may be so. But because the Macondo team did not adequately appreciate the risks of a poor cement job (as described in Chapter 4.3), they could not adequately have compared the risks and benefits of using a long string casing at Macondo. BP engineers appear to have been reluctant to switch to a liner for other reasons as well. They had already obtained peer review and approval of the long string design. And the long string approach costs substantially less than the liner.

BP's Special Emphasis on the Risk of Annular Pressure Buildup Overshadowed Its Identification and Mitigation of Other Risks

BP made several of the well design decisions discussed above in order to mitigate the risk of annular pressure buildup. Proper well design requires consideration of annular pressure buildup if the company plans to use the well for production.[115] But BP was particularly sensitive to the issue because of its experience at the Marlin platform at the Atlantis field.[116] BP attempted to mitigate the risk of annular pressure buildup in its Marlin wells by leaving the casing annuli open to the surrounding formation. But in late 1999, one of those wells nevertheless collapsed due to annular pressure buildup.[117] Debris or sediments had apparently plugged the opening in the relevant annulus. The event was a major loss for BP because casing collapse essentially destroys a well.[118]

In the aftermath of the Marlin incident, BP made it a top priority to minimize the risk of annular pressure buildup in its wells.[119] It created a dedicated group of design specialists who analyzed annular pressure buildup issues for every production well and recommended design features to mitigate those risks.[120] BP also developed standard guidance instructing its engineers to leave annuli open as part of a deepwater well's design.[121] And it encouraged the use of rupture disks as a primary annular pressure buildup mitigation measure.[122]

BP's focus on and approach to annular pressure buildup concerns effectively de-emphasized other risks and discouraged certain well design approaches. Because the Macondo team planned the well as a producer, they made several design decisions to mitigate the risk of annular pressure buildup.[123] These included adding rupture disks in the 16-inch casing, omitting a protective casing (which would have created a trapped annulus), leaving an open annulus below the 9⅞-inch liner, and using a long string production casing instead of a liner.[124] As described above, those design features complicated the cement job as well as post-blowout containment efforts.

While BP's methods of mitigating annular pressure buildup created risks, there were alternatives. For example, BP could have used insulated production tubing to protect the well from the heat generated during production. This might have allowed the company to omit burst disks and include a protective casing. BP could also have pumped compressible fluids (such as nitrogen foamed spacer or syntactic foam) into any trapped annular spaces to mitigate the risk of annular pressure buildup rather than designing its well to eliminate such spaces. This approach would have allowed BP to use a liner-tieback without worrying that the tieback would create a trapped annulus.[125] ◆

Chapter 4.3 | Cement

Well Cementing

Cement performs several important functions in an oil well. It fills the annular space between the outside of the casing and the formation. In doing so, it structurally reinforces the casing, protects the casing against corrosion, and seals off the annular space, preventing gases or liquids from flowing up or down through that space. A cement job that properly seals the annular space around the casing is said to have achieved **zonal isolation**.

The cementing process is procedurally and technically complex. This chapter first describes the steps in the cementing process, the ways in which cement can be evaluated and remediated, and methods for laboratory cement slurry testing. It then describes the Macondo cementing operation in detail. Finally, it sets out the Chief Counsel's team's technical and management findings regarding the Macondo cementing process. The Chief Counsel's team finds that the Macondo cement failed to achieve zonal isolation. While the Chief Counsel's team cannot be sure why the cement failed, the team has identified several risk and other factors that may have contributed to cement failure, either alone or together.

The Cementing Process

The cementing process involves pumping cement down the inside of a casing string until it flows out the bottom and back up into the annular space around the casing string. Achieving zonal isolation requires several things.

- First, the cement should fill the annular space in the zone to be isolated and also a specified space above and below that zone.

- Second, cement flowing into the annular space should displace all of the drilling mud from that space so that no gaps or uncleared **channels** of mud remain behind. If mud channels remain after the cement is pumped, they can become a flow path for gases or liquids from the formation. Good mud removal is critical for a successful cement job.[1]

- Third, the cement should be formulated so that it sets properly under wellbore conditions.

Although each cement job presents unique challenges, the principal steps involved in pumping cement at Macondo were the same as those for most deepwater wells. The following subsections describe the process in simplified form. These sections describe the process for running and cementing a production casing—the last casing string to be run in the well once a hydrocarbon-bearing zone has been penetrated. The process generally applies to running and cementing shallower casing strings and liners as well.

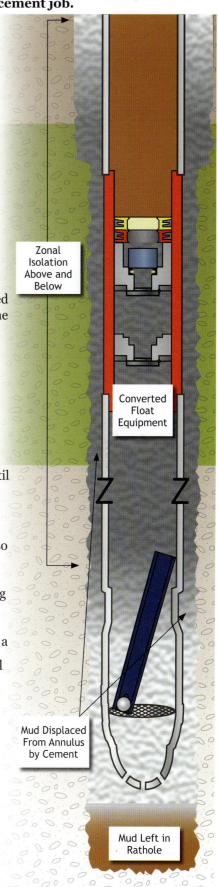

Figure 4.3.1. Typical completed cement job.

Zonal Isolation Above and Below

Converted Float Equipment

Mud Displaced From Annulus by Cement

Mud Left in Rathole

TrialGraphix

Figure 4.3.2. Sample caliper log data showing open hole diameter by depth.

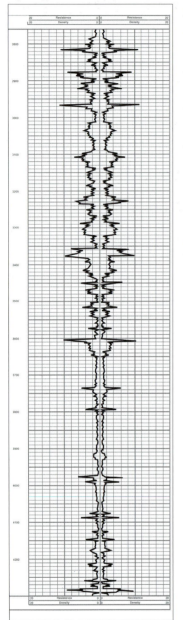

TrialGraphix

Primary Cementing. **Primary cementing** refers to an operator's initial attempt to seal a casing with cement. By contrast, **remedial cementing** refers to subsequent cementing efforts undertaken if the primary cement does not achieve zonal isolation.

Logging and Mud Conditioning

After rig personnel finish drilling a well that will be completed as a production well, they typically **condition** the mud in the wellbore and then log the wellbore itself before lowering the final production casing and performing the final cement job.

During drilling operations, mud engineers manipulate the characteristics of drilling mud in the wellbore to optimize the removal of cuttings and to maintain hydrostatic pressure in the well. At the end of drilling operations, the mud is normally circulated to homogenize its properties and modify those properties as necessary to facilitate wellbore logging and eventual mud removal. That circulation process is called mud conditioning. Drillers normally circulate the mud in order to remove cuttings from the mud and ensure that it displays uniform and appropriate density and viscosity characteristics.[2] American Petroleum Institute (API) recommendations state:

> Well preparation, particularly circulating and conditioning fluids in the wellbore, is essential for successful cementing. Many primary cementing failures are the result of fluids that are difficult to displace and/or of inadequate wellbore conditioning.[3]

Logging refers to the process of examining and recording the characteristics of the wellbore (first discussed in Chapter 2). Prior to running a production casing string, drillers typically examine the open section of the wellbore with an extensive suite of logging tools that use electric, sonic, and radiologic sensors to measure the physical characteristics of the formation and any fluids it might contain in order to learn as much as possible about the nature of the hydrocarbon-bearing formation.[4] One such tool, shown in Figure 4.3.2, is a **caliper log**, which measures the diameter of the wellbore. Because the wellbore diameter can vary significantly as a result of normal drilling variations, these data can be an important input in designing and modeling a primary cement job.

Lowering the Production Casing String in Place With Centralizers

After logging is complete, rig personnel lower the production casing into place. During this process, they may install **centralizers**, shown in Figure 4.3.3, which serve an important role in the cementing process.

When the cementing crew pumps cement (or any other fluid) down the production casing and back up the annular space around it, the cement tends to flow preferentially through paths of least resistance. When the casing is not centered in the wellbore, the wider annular space becomes the path of least resistance,[5] shown in Figure 4.3.4. Cement tends to flow up through those spaces. This can seriously compromise mud removal and leave channels of mud behind in the narrower annular spaces.[6] Because of this problem, cementing experts consistently emphasize the importance of keeping the casing centered in the wellbore.[7]

Figure 4.3.3. Centralizer.

TrialGraphix

Centralizers help keep the casing as close to the center as possible. They come in a variety of designs. Centralizer **subs**, shown in Figure 4.3.5, may be screwed between casing sections while bow spring centralizer **slip-ons** are attached to the outside of existing casing using collars. Sometimes **stop collars** (so named because they stop the centralizer from sliding up or down the casing) are separate pieces from the centralizer; sometimes they are integrated into the centralizer itself.[8]

Engineers measure the degree to which a pipe is centralized in a wellbore by calculating the "pipe standoff ratio."[9] A perfectly centered casing has a standoff ratio of 100% while a casing that touches the walls of the wellbore has a standoff ratio of 0%. Although the industry rule of thumb is to achieve a standoff of 75%,[10] cementing experts state that operators should achieve the highest possible standoff in order to facilitate mud displacement from the annular space.[11] Engineers must calculate the standoff not only at each centralizer location, but also between the centralizers. Casing can bend and sag between centralizers, dramatically lowering the standoff in the intervals between them.[12]

Figure 4.3.4. Top view of off-centered casing.

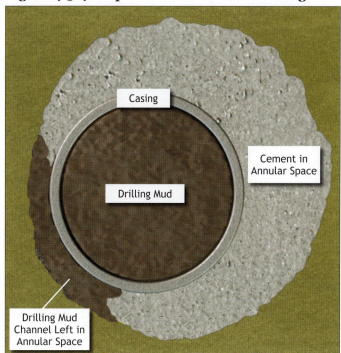

Casing

Cement in Annular Space

Drilling Mud

Drilling Mud Channel Left in Annular Space

TrialGraphix

Float Valves and Float Valve Conversion

Illustrated in Figure 4.3.6, float valves are one-way valves (also called check valves) installed at or near the interior bottom end of a casing string. Once operational, float valves permit fluid (such as mud or cement) to flow down through the inside of the casing while preventing fluids from flowing in the reverse direction back up the inside of the casing. By doing so, float valves prevent cement that is pumped down through the casing, into the shoe track, and up into the annular space from flowing back up through the valves once the cement is in place, an occurrence known as "reverse flow" or "u-tubing."[13]

Figure 4.3.5. Centralizer sub.

Weatherford

Shoe and Shoe Track. The **shoe** refers to the bottom of the casing. The **shoe track** is the section of the casing between the shoe and the float valves above it.

A **float check** examines whether the float valves are working properly—that is, preventing cement from flowing back up through the valves due to u-tube pressure. **U-tube pressure** is created by the differential hydrostatic pressure between the fluid column inside the casing and the fluid column in the annulus. In cases where the cement density is close to drilling mud density, the u-tube pressure may be very small—too small to induce backflow or to be detected at the rig. The smaller the density differential between the cement and mud, the smaller the u-tube pressure and its expected effects.[14]

Float valves are important during the cementing process but can interfere with the process of lowering a casing string. As the casing string is lowered, it is generally preferable that mud be allowed to flow up the inside of the casing string. Otherwise, the casing will, as it descends, force mud down the well and back up through the annular space, greatly increasing the pressure that the casing string exerts on the formation as it is lowered.[15]

Figure 4.3.6. Float valve conversion.

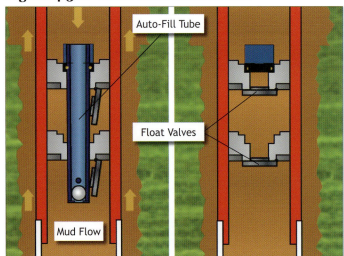

TrialGraphix

To allow mud to flow into the casing string while it is being lowered, operators typically use an **auto-fill tube**. An auto-fill tube is a hollow tube that extends through and props open the two float valves, allowing mud to flow up through the casing while the casing is being run into the well. Once rig personnel finish lowering the casing, they **convert** the float valve assembly by pushing the auto-fill tube down and out of the float valves. This allows the float valves to close, converting them into one-way valves before cementing begins.

Wellbore Conditioning

After converting the float valves, rig personnel normally circulate mud through the newly installed casing and wellbore again. Like the earlier mud circulation process, this has at least two benefits. First, it cleans the casing, drill pipe, and wellbore of cuttings, gelled mud, and other debris that can interfere with good cement placement and performance.[16] Second, the mud flow conditions the mud itself by breaking its gel strength, decreasing its viscosity, and increasing its mobility.[17]

Figure 4.3.7. Full bottoms up.

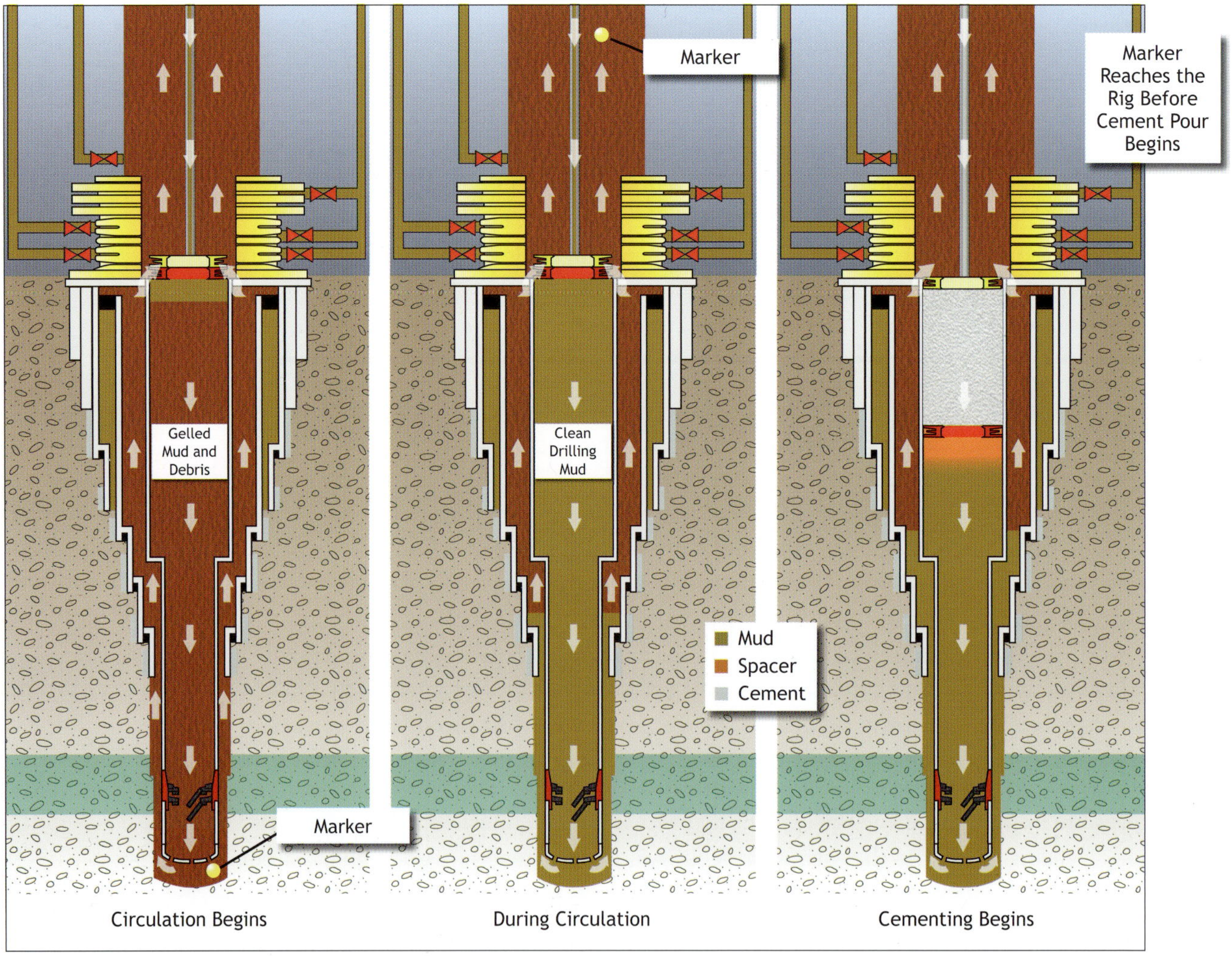

Marker

Marker Reaches the Rig Before Cement Pour Begins

Gelled Mud and Debris

Clean Drilling Mud

Marker

Mud
Spacer
Cement

Circulation Begins

During Circulation

Cementing Begins

TrialGraphix

Under optimum conditions, operators prefer to circulate enough drilling mud through the casing after landing it to achieve what is known as a full **bottoms up**.[18] Circulating bottoms up means that the rig crew pumps enough mud down the well so that mud originally at the well bottom returns back to surface[19] as shown in Figure 4.3.7. The extended circulation required to do this confers a third benefit in addition to the two described above: It allows rig crews to physically inspect mud from the bottom of the well for the presence of hydrocarbons before cementing.

Pumping Cement

After completing the pre-cementing mud circulation, rig personnel pump cement down the well, then pump additional drilling mud behind the cement to push (or **displace**) the cement into the desired location at the bottom of the well. As they pump the cement, rig personnel must ensure that the oil-based drilling mud does not contaminate the water-based cement. The oil and gas industry has developed a variety of techniques to ensure that this does not occur. Rig personnel at Macondo used a common approach called the "two-plug method."[20] The two-plug method uses rubber **darts** and **wiper plugs** to separate the cement from the drilling mud as the cement travels down the well.

Rig personnel begin the cement pumping process by pumping water-based **spacer fluid** down the drill pipe. They then drop a **bottom dart** into the drill pipe, followed by the cement, then a **top dart** and more spacer fluid. After pumping the final spacer fluid down the drill pipe, rig personnel resume pumping drilling mud to push the spacer-dart-cement-dart-spacer train down the drill pipe.

Figure 4.3.8. Wiper plugs cause cement contamination.

When the bottom dart reaches the end of the drill pipe, it fits into and launches a **bottom wiper plug** from the running tool that attaches the drill pipe to the production casing. The bottom plug then travels down inside of the production casing, separating the cement behind it from the spacer fluid and drilling mud ahead. Similarly, when the top dart reaches the end of the drill pipe, it launches a **top wiper plug** from the running tool. The top plug also travels down the inside of the production casing and separates the cement from spacer fluid and drilling mud behind.

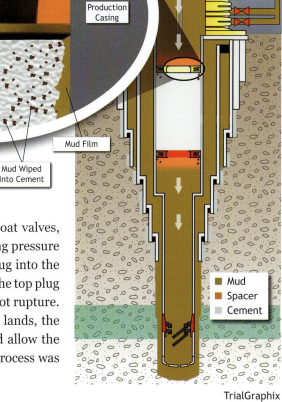

Top Wiper Plug

Production Casing

Cement

Mud Film

Mud Wiped Into Cement

Mud
Spacer
Cement

The rig crew continues to pump mud down the drill pipe to displace the cement into position. Eventually, spacer fluid reaches the float valves and flows through the valves. After the spacer flows through the float valves, the **bottom plug** lands on top of the float valves, where it stops. Circulating pressure causes the bottom plug to rupture, allowing cement to pass through the plug into the shoe track. After all of the cement flows through the ruptured bottom plug, the top plug lands on top of the float valves. Unlike the bottom plug, the top plug does not rupture. It instead blocks further flow of fluids down the well. When the top plug lands, the cement should be in place. Rig personnel stop pumping drilling mud and allow the cement to set in a process called **waiting on cement**. If the cementing process was

TrialGraphix

Figure 4.3.9. Lift pressure.

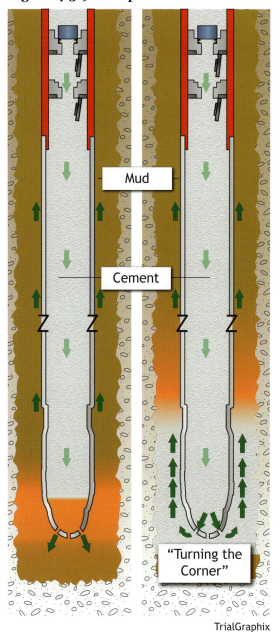

Mud

Cement

"Turning the Corner"

TrialGraphix

Pressure increases to lift cement in the annulus.

designed and executed properly, the cement should at this point fill the shoe track and should cover the hydrocarbon zone in the annular space outside the production casing.

Even if rig personnel execute a two-plug cementing process precisely according to plan, cement can still be contaminated by drilling mud. As the wiper plugs travel down the casing, they wipe a film of mud away from the casing walls. The bottom plug removes most of the mud film but not all of it. The remaining mud film can contaminate the cement between the plugs as shown in Figure 4.3.8. The top plug also wipes the casing, but instead of wiping mud out of the way of the cement, it wipes that mud *into* the back portions of the cement flow.

The casing shoe track is designed to provide room for contaminated cement at the tail end of the pumping process. Absent a shoe track, that contaminated cement would travel into the annular space, potentially compromising zonal isolation.

Cement Evaluation

It is not easy for rig personnel to be sure about the progress or final result of a cement job at the bottom of a deepwater well. Cement does its work literally miles away from the rig floor, and there is no way to observe directly if the cement slurry arrives at its intended location, let alone whether it is contaminated or otherwise compromised. As a result, rig personnel cannot know whether the cement will isolate the well from the hydrocarbons in the reservoir as they pump the cement.

Because cementing is difficult to observe directly, the oil and gas industry has developed a number of methods for evaluating cement jobs indirectly. And because proper cementing is critical to well integrity, the API calls proper cement evaluation "indispensable."[21] But each of the various methods of cement evaluation has limitations, and the API standard on cement evaluation therefore notes:

> Anyone who wants to competently evaluate the quality of a cement job must thoroughly understand all the variables, assemble and comprehend the relevant pieces of information, and reach the proper judgment.[22]

By understanding the full set of variables at play for a particular cement job, the right mix of tools can be employed to evaluate the cement.

Volume and Pressure Indicators

While pumping a cement job, a cementing crew knows only how much cement and mud they have sent down the well and how hard the pumps have been working to push it. Using these volume and pressure readings, the rig crew looks for three general indicators of success during the job: full returns, lift pressure, and on-time plug landing.

A cementing crew gets **full returns** when the volume of mud returning from the well during a cement job equals the volume of fluids (spacer, cement, and mud) pumped down into the well. To determine whether they are getting full returns, the cementing crew monitors mud tank volumes.

Figure 4.3.10. Bumping the plugs.

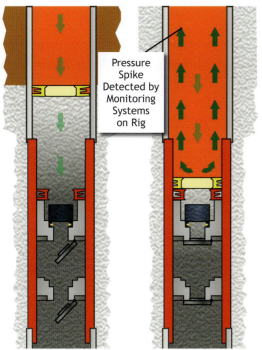

Pressure Spike Detected by Monitoring Systems on Rig

TrialGraphix

If the volume of fluid flow into the well equals the fluid flow out, the crew can infer that the well is behaving properly as a closed and leak-free container. If flow out is less than flow in, the crew has **lost returns** or **lost circulation**, and can infer that mud and/or cement has flowed into the formation.[23] The crew cannot tell *where* the rock fractured, however, and where the mud might have gone.[24]

Lift pressure, shown in Figure 4.3.9, is a steady increase in pump pressure that begins when the cement flows out the bottom of the well casing and "turns the corner" to flow upward against gravity. The pressure increases because cement is generally heavier than drilling mud (and has a different viscosity). If the cementing crew observes a steady pressure increase at the appropriate time after pumping cement down into a well, they can infer that the increase is lift pressure and that cement has arrived at the bottom of the well and has begun flowing upward into the annular space. Seeing the expected lift pressure also allows the crew to infer that cement is not being lost into the formation.

Finally, the rig crew can also watch pressure gauges to infer whether the wiper plugs used to separate the cement from surrounding drilling mud have **landed** or **bumped** on time at the bottom of the well as shown in Figure 4.3.10. By calculating the volume of the inside of the well and the rate at which they are pumping fluids into it, cementing crews can predict when the bottom plug and top plug should land. They then watch the rig's pressure gauges for telltale pressure spikes that indicate when the plugs actually land. If the pressure spikes show up when expected, the cementing crew can infer that the plugs landed properly, that cement arrived at the bottom of the well and flowed out of the shoe track into the annulus, and that substantial volumes of mud did not contaminate the cement as it moved down the well. If the pressure spikes do not appear on time, that suggests problems. For instance, large volumes of mud may have bypassed one or both of the wiper plugs. (Some volume of mud always bypasses the plugs; the plugs do not wipe the casing walls perfectly.)[25]

While pressure and volume indicators can suggest that a cement job has gone as planned, they do not give cementing crews any direct information about the location and quality of the cement at the bottom of the well. In particular, they do not indicate whether there has been channeling in the annulus or shoe track, or the location of the **top of cement** (TOC) in the annulus.[26] These indicators also are not sensitive to all of the issues that can cause cement to fail.

Cement Evaluation Logs

Because pressure and volume readings during the cement job are imperfect indicators of cementing success, the oil and gas industry has also developed tools for more directly examining a cement job after it is pumped. These cement evaluation tools generate data, or "logs," known as cement evaluation logs. Technicians commonly lower cement evaluation tools down inside the well on a **wire line.**[27] Once the tools reach an area that has been cemented, sensors in the tools probe the integrity of the new cement, measuring whether and to what extent the cement has

Figure 4.3.11. Cement bond log tool.

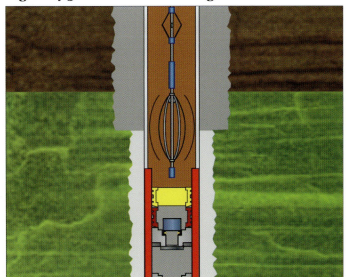

TrialGraphix

filled the annular space between the cement and the formation.[28]

The most basic element in a cement evaluation system is the **cement bond log** tool.[29] The cement bond log tool works by measuring the well casing's response to acoustic signals. The tool includes an acoustic transmitter and receiver that are separated from each other by several feet of distance. The transmitters emit bursts of acoustic waves, and the receivers record the reverberations from those waves[30] as illustrated in Figure 4.3.11. Because steel casing, set cement, and fluids all respond differently to the waves, a technician can use the recordings to evaluate the quality of the cement job, just as one can discern a muffled bell from a free-swinging bell by ringing it.[31]

Modern cement evaluation systems combine the fairly straightforward cement bond log with variable-density logs,[32] ultrasonic imaging tools, and flexural attenuation logs.[33] By interpreting the combined data from these tools, a technician can assess the amount and quality of the cement in the annular space,[34] including the TOC and the location and severity of channels in that cement.[35]

Although modern cement evaluation logs have become increasingly sophisticated and reliable, they still have limits.[36] First, they are not easy to read; it takes an experienced technician to properly interpret the data. Second, very low-density cement, such as cement produced with nitrogen foam technology, can be difficult to evaluate with these tools.[37] (The density of the foamed cement at Macondo was not low enough to cause evaluation difficulties, however.[38]) Third, cement evaluation tools must be adjacent to annular cement in order to examine it. That means that the tools cannot evaluate cement in the shoe track or in the annular space below the float equipment. Float equipment and the shoe track cement block the tools from physically accessing those areas. Fourth and finally, cement evaluation logs work best after cement has completely hardened—a process that can take more than 48 hours.[39] Consequently, operators typically do not run cement evaluation logs until completion operations.

Additional Methods

There are other methods to evaluate a cement job in addition to cement evaluation logs and pressure and volume indicators. In particular, a negative pressure test assesses whether a bottomhole cement job contains pressures outside the well and seals the well off from formation pressure. Chapter 4.6 of this report discusses negative pressure tests in detail.

Remedial Cementing

If cement evaluation reveals problems with the primary cement job, rig personnel can **remediate** the primary cement after pumping it. At a well like Macondo, the most common method for remediating the primary production casing cement is called **squeeze cementing**.

Figure 4.3.12 illustrates that squeeze cementing first involves perforating the production casing to provide access to the annular space around it. Rig personnel perforate the casing by lowering a tool that uses shaped explosive charges to punch holes through the casing and into the formation. Rig personnel then pump, or "squeeze," cement under pressure through the holes. In a properly

Figure 4.3.12. Remedial cementing—squeeze job.

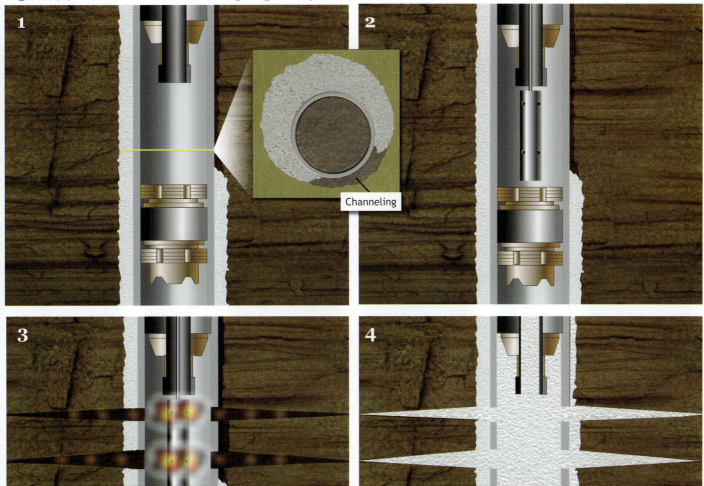

TrialGraphix

From upper left: 1) Poor centralization has led to channeling; 2) a bridge plug is placed below the remediation area, and a packer is positioned above; 3) a perforation gun is lowered and fires shells through the casing and into the formation; 4) cement is pumped into the area, forced through the perforations, and into the formation, creating a seal.

executed squeeze job, the remedial cement then flows into the annular spaces where the primary cement has failed, filling in any channels and isolating zones as necessary.

Cement Slurry Testing

Cement hardens as a result of chemical reactions that depend on pressure and temperature. In the field, cement slurries are normally mixed at ambient temperature and pressure, then exposed to increasing temperatures and pressures as they are pumped down the well. These increasing temperatures and pressures can not only alter the chemical and physical properties of the liquid slurry and cured cement, but also can affect the cement curing process itself. Because every well presents a different combination of cementing conditions, it is critical for a cementing company to test a cement slurry design against expected conditions in the particular well before pumping it into that well.

Cement slurries consist of a number of ingredients, including dry Portland cement (which itself is a combination of several chemical compounds), water, and various dry and liquid chemical additives. Cementing personnel adjust the concentrations of these ingredients to suit the particular needs of a given well. Cement slurry designs thus vary from well to well. To complicate matters further, many of the ingredients used in a cement slurry are made from naturally occurring materials, and their precise chemical composition depends on their source.[40] The liquid chemical additives may vary from batch to batch, and the mix water composition can vary depending on its source. This means that each batch of cement slurry is different. Finally, the constituents of a given cement slurry also may degrade in storage upon exposure to heat, humidity, and atmospheric gases such as carbon dioxide. To address this variability, cementing companies usually perform their pre-job testing with representative samples of the actual ingredients that will be pumped into the well.

Pilot and Pre-Job Testing

A cementing company typically conducts at least two rounds of cement testing prior to pumping a challenging or uncertain cement job. First, it performs "pilot tests" substantially in advance of pumping the job in order to develop an appropriate cement slurry design (the recipe). At the time of the pilot tests, the operator gives the cementing company the best available information about the downhole conditions. That information may be incomplete, especially in the case of an exploratory well (such as Macondo). Sometime prior to pumping the cement, when the operator has learned the actual downhole job conditions, the cementing company typically performs pre-job tests using the actual cement ingredients that have been stored on the rig and will be pumped downhole. These pre-job tests are meant to confirm that the cement design will perform successfully during the upcoming job.

Laboratory Tests

To isolate hydrocarbons at the bottom of a well, the cement must display several attributes. First, as the cement is pumped into place at the bottom of the well, it must remain in a pumpable fluid state and not thicken prematurely. Second, once in place, it must set and develop strength within a reasonable time period. And third, the set cement must be sufficiently strong to provide casing support and zonal isolation. To check these things, cementing companies typically run a number of tests to evaluate a cement design during pilot and pre-job testing. The API has published recommended procedures for running these tests.[41]

Cement Test. Cement tests examine various properties of the cement slurry and the set cement, and investigate the curing process. **Thickening time** tests determine how long the cement slurry will remain pumpable (before starting to set up) under the temperature and pressure conditions in the wellbore. **Compressive strength** tests determine the length of time required for the cement slurry to develop sufficient strength to provide casing support and zonal isolation. **Rheology** tests examine various cement slurry flow properties. The slurry viscosity and yield point affect the pumping pressure required for slurry placement and the displacement efficiency by which drilling fluid is removed from the annular space. The yield point also provides information concerning slurry stability—the ability of the slurry to keep solids in suspension and prevent fluid-phase separation. **Static gel strength** is a measure of the degree to which an unset cement slurry develops resistance to flow when at rest. **Free-fluid** tests directly examine slurry stability.

As cement slurry travels down a well, it encounters increasing heat and pressure. Laboratory technicians sometimes stir the slurry at elevated temperatures (and sometimes at elevated pressures) to simulate these conditions in order to better understand how the cement will behave when it reaches its intended location. This practice is known as **cement conditioning** (not to be confused with mud conditioning, described above).

Modeling the Cementing Process

Before pumping cement, engineers can also model the cementing process using computer simulation programs. Engineers run these simulations using data about wellbore and casing geometry, mud conditioning, the number and placement of centralizers, and the volume, pumping rate, and characteristics of the various fluids pumped down the well. The simulations, in turn, predict various things about the cementing process such as the pressure that will be required to pump cement.

Engineers routinely use cement simulations to model the complex process of mud displacement from the annular space. Predicting mud displacement is important for at least two reasons. First, if the cement flow does not displace mud and spacer from the annular space, those materials may create a flow path for hydrocarbons. Second, and relatedly, poor mud displacement increases the potential for gas to flow into the cement column as it sets.[42] This gas flow can itself cause channeling and further compromise zonal isolation.

As the oil and gas industry develops deeper wells and more complicated well designs, engineers rely increasingly on computer modeling to predict mud removal. Operators and cementers can use these models to predict the impact of changing parameters such as cement flow rate and centralizer placement. By doing so, they can optimize these interrelated parameters for individual well conditions rather than relying on rules of thumb to guide their decisions. At the same time, the fluid mechanisms of mud displacement, gas flow, and other cementing phenomena are exceedingly complex. Computer simulations cannot model these phenomena precisely. In addition, even the best computer models depend entirely on their input data; if the input data are inaccurate, the modeling results will be inaccurate as well.

Preparing for the Macondo Cement Job

Lost Returns at Macondo

BP and Halliburton designed crucial features of the Macondo cement job in response to the April 9 lost returns event (when drilling mud flowed out of the wellbore and into the formation) described in Chapter 4.2. Although BP engineers successfully restored mud circulation by pumping 172 barrels of heavy, viscous "lost circulation" fluids down the drill pipe,[43] they also realized the situation had become delicate. Based on data from the lost circulation event, the engineers calculated that they had to maintain the weight of the mud in the wellbore at approximately 14.0 pounds per gallon (ppg) in order to maintain well control.[44] Drilling ahead with that mud weight would exert even more pressure on the formation, raising the equivalent circulating density (ECD). BP engineers calculated that drilling with 14.0 ppg mud in the wellbore would yield an ECD of nearly 14.5 ppg—an increase that the engineers believed could induce lost returns again.

The engineers concluded they had "run out of drilling margin" and that they could no longer drill to their planned total depth of 20,600 feet below sea level.[45] Instead, they cautiously drilled ahead from 18,193 to 18,360 feet in order to extend the wellbore beyond the pay zone. Optimally, engineers prefer to drill far enough beyond the pay zone to ensure that the float collar and shoe track will both be entirely below the pay zone. Among other things, this allows the operator eventually to use logging tools to evaluate all of the cement in the annular space in the pay zone. In March, before the April 9 lost circulation event, a BP engineer stated that BP planned an extended shoe track at Macondo.[46]

Wellbore Logging and Conditioning

After drilling, BP directed Schlumberger to run a series of logs to collect data from the well. Between April 10 and 15, 2010, Schlumberger technicians evaluated the formation to determine its porosity and permeability, and gathered fluid and core samples from the well. The logging data led BP to conclude that it had drilled into a hydrocarbon reservoir of sufficient size (at least 50 million barrels[47]) and pressure that it was economically worthwhile to install a production casing. Schlumberger also ran a caliper log to determine the exact diameter of the wellbore.[48]

On April 16, before running the final 9⅞-inch × 7-inch long string production casing, the rig crew circulated the open wellbore bottoms up.[49] They did not record any mud losses during this process.[50] The crew inspected mud from the bottom of the well and found that it contained 1,120 gas units on a 3,000-unit scale.[51] This was not an unusual amount of gas because the mud at the bottom had been sitting in place in the well for about a week at that point.[52] After circulating on April 16, gas eventually decreased to 20 to 30 units.[53]

Designing the Macondo Cement Job

BP's cement planning focused heavily on reducing the risks of further lost returns. BP recognized that if the formation fractured again during cementing, it could compromise the cement job and force the rig crew to conduct remedial cementing operations. BP engineers focused particular attention on ensuring that the ECD during cementing would not exceed the threshold that they believed would induce further losses. In order to minimize the ECD during cementing, BP: (1) reduced the volume of cement that would be pumped, (2) reduced the rate at which the cement would be pumped, and (3) used nitrogen foamed cement for reduced density.[54]

Cement Volume

Wellbore conditions are rarely optimal, and it is difficult to be sure precisely where cement has flowed during a cement job. Engineers can therefore improve the odds of achieving zonal isolation by increasing the volume of cement in the well design. Pumping more cement is a standard industry safeguard against uncertain cementing conditions. It reduces the risk of contamination by diluting the amount of contaminants in the cement. It also decreases the impact of errors in cement placement.

MMS Cement Volume Requirements

At the time of the Macondo blowout, MMS regulations included very few requirements that related to the cement design process at Macondo. One of those requirements concerned the volume of cement for a primary production casing cement job. According to 30 C.F.R. § 250.421: "As a minimum, you must cement the annular space at least 500 feet above the casing shoe and 500 feet above the uppermost hydrocarbon-bearing zone."

In other words, MMS required that the TOC in the annular space of the production casing be at least 500 feet above the "uppermost hydrocarbon-bearing zone."

BP's Internal Guidelines

BP's Engineering Technical Practice 10-60 (ETP 10-60), titled "Zonal Isolation Requirements during Drilling Operations and Well Abandonment and Suspension," lists the company's internal engineering design rules for cementing. ETP 10-60 states:

> **1.3** Zonal Isolation design criteria for cementing of primary casing strings to meet well integrity and future abandonment requirements, shall meet one of the following:
>
> - 30 m TVD [total vertical depth] (100 ft TVD) above the top of the distinct permeable zone where the top of cement (TOC) is to be determined by a proven cement evaluation technique (Section 5.3).
> - 300 m MD [measured depth] (1000 ft MD) above the distinct permeable zone where the hydraulic isolation is not proven except by estimates of TOC (Section 5.3). For each well the actual TOC shall be recorded along with the method used for this determination. Where the actual TOC is below the plan, the TOC shall be reviewed with stakeholders for its impact on future well integrity, operability, suspension and abandonment operations.[55]

Section 5.3 of ETP 10-60 distinguishes a "proven cement evaluation technique" from an "estimate" of TOC by stating that "to accurately assess TOC and zonal isolation cement sonic and ultrasonic logs should be used." By contrast, the ETP states that temperature logs (which can detect the heat exuded by cement) and cement column backpressure measurements can be used to "estimate" TOC. This means that unless a BP engineering team plans to run sonic and ultrasonic logs, it should design the cement job so that there is 1,000 feet of cement above the highest distinct permeable zone in the well.

In addition to zonal isolation, BP also considers annular pressure buildup (APB) in planning TOC.[56] The high temperatures caused by bringing hydrocarbons to the surface during later production can cause pressure buildup in the annular space. If trapped, the annular pressure will build up and can potentially collapse the inner casing string on itself and ruin the well. One way drillers avoid this is by allowing annular pressure to escape into the formation. By not cementing all the way up to the next liner—which necessarily means a lower TOC and lower volume of cement—the drillers allow a route for escape.[57] It is likely that APB concerns were a factor in determining TOC and cement volume at Macondo.[58]

Macondo Cement Volume

After the early April lost returns events, the BP Macondo team decided to limit the height of the cement column in the annulus. They had little room to maneuver: A higher cement column in the annulus would have exerted more pressure on the fragile formation below, increasing the ECD of the cement job and risking further lost returns.

Driven by ECD concerns, BP's engineering team focused its attention on determining where TOC should be. While the main hydrocarbon reservoir zone at Macondo began at 18,100 feet,[59] BP estimated that the "top HC [hydrocarbon] zone" began at 17,803 feet.[60] BP engineers decided to pump only as much cement above that zone as MMS required.[61] On or about April 14,[62] they determined that TOC should be 17,300 feet below the ocean surface—503 feet above the top hydrocarbon zone and 830 feet above the main hydrocarbon zone.[63]

On April 14, BP senior drilling engineer Mark Hafle initiated a formal management of change review of the plan to set the production casing.[64] He marked the document as a high priority and asked that its approval be completed by the next day.[65] Hafle incorporated the design decision regarding TOC in the management of change document. The document discussed the risk that the primary bottomhole cement would not act as a barrier: "If losses occur during the cement job, possible cement evaluation, remedial cement operations, dispensations and/or MMS approvals will be required prior to performing TA operations due to a lower than required Top of Cement in the annulus. Possible hydrocarbon zones could be left exposed in the annulus with only the casing hanger seal as the single barrier for the TA."[66] In the event that occurred, the document went on to note, "A perf[oration] and squeeze operation could be performed to add a second barrier in the annulus."[67] BP drilling and completions operations manager David Sims reviewed the management of change document and commented that the "[c]ontent looks fine."[68] BP drilling engineer team leader Gregg Walz, BP wells team leader John Guide, BP engineering manager John Sprague, and others also reviewed the document—all approved.[69]

Keeping TOC to a minimum necessarily reduced the total volume of cement that Halliburton pumped down the well. Several other features of the Macondo well also limited the total amount of cement that could be pumped:

- the relatively short distance the well had been drilled below the main pay sands;
- the relatively narrow annular space between the production casing and the formation; and
- BP's decision not to pump any cement behind the top plug.[70]

Halliburton calculated that it should pump approximately 51 barrels of cement (about 60 barrels after foaming) down the well in order to fill the shoe track and the annular space up to BP's specified TOC.[71] BP engineers recognized that this was a relatively small volume of cement that would provide little margin for error.[72]

Cement Flow Rate

Just as increased mud flow rate improves wellbore conditioning, higher cement flow rates tend to increase the efficiency with which cement displaces mud from the annular space. Cement must be pumped fast enough so that it will scour mud from the side of the wellbore instead of merely flowing past. The API notes that "[h]igher pump rates introduce more energy into the system allowing more efficient removal of gelled drilling fluid."[73] However, increased pump pressure required to move the cement quickly would mean more pressure on the formation (ECD) and an increased risk of lost returns.[74]

One way in which BP reduced the risk of lost returns at Macondo was by lowering the rate of cement flow. BP pumped cement down the well at the relatively low rate of four barrels or less per minute.[75] This was a lower rate than called for in earlier drilling plans,[76] but BP did inform Halliburton of the change and Halliburton's computer models accounted for the reduced flow rate.

Use of Nitrogen Foamed Cement

One very direct way to reduce the amount of pressure that a column of cement exerts on the formation below is to use lightweight cement. While there are several ways to generate lightweight cement, BP and Halliburton chose to use nitrogen foamed cement. Cementing personnel create nitrogen foamed cement by injecting inert nitrogen gas into a base cement slurry. This produces a slurry that contains fine nitrogen bubbles. Because nitrogen gas weighs so little compared to cement, the nitrogen bubbles make the overall cement mixture less dense than the base cement slurry.

BP and Halliburton jointly decided to use foamed cement technology at Macondo. (Chapter 4.4 discusses the choice in more detail.) This would reduce the weight of the middle portion of the Macondo cement slurry from the base slurry density of 16.74 ppg down to a foamed slurry density of 14.50 ppg.[77]

While using foamed cement slurry brought certain benefits, it brought risks as well. Chapter 4.4 explains in more detail how an unstable foamed cement slurry can fail to provide zonal isolation. A BP cementing expert specifically advised one of the Macondo engineers in March that cementing the production casing using foamed cement would "present[] some significant stability challenges for foam, as the base oil in the mud destabilizes most foaming surfactants and will result in N_2 [nitrogen] breakout if contamination occurs."[78] To guard against this possibility, the expert advised the team to pump non-foamed cement ahead of the foamed cement. This would create a "cap slurry" on top of the foamed slurry in the annular space that would mitigate the risk of foam instability.[79]

Planning for and Installing Centralizers at Macondo

BP procured only six centralizers for its production casing ahead of time, even though its plans had originally called for a greater number. Shortly before running the casing, however, Halliburton's modeling revealed that BP would need more centralizers to prevent channeling. In response, BP decided at the last minute to purchase 15 more centralizers and send them out to the rig. But unlike the six centralizer subs that BP had purchased earlier, these additional centralizers were slip-on centralizers with separate stop collars. Once BP realized this, it reversed itself and decided not to use them, reasoning that the risks of using them outweighed the risks of channeling.

API's Centralization Guidance

While the API recognizes the importance of centralization, it has no recommended specific standoff ratio for casing. Rather, the API encourages drillers to determine the appropriate standoff ratio based on individual well conditions. Nor does the API have any recommendation or standard for how far above the pay zone casing should be centralized. [80]

BP's Centralization Guidance

BP's official technical guidance instructs engineers to design centralization programs to ensure there is at least 100 feet of "centrali[z]ed pipe" above the "permeable zone" in the event a cement bond log is not run.[81] The technical guidance does not provide any further detail on the number or type of centralizers that should be used or the overall standoff that should result. BP in-house cementing expert Erick Cunningham explained that the guidance does not provide specific instruction on the number of centralizers that must be used to create a "centralized pipe." A casing could have centralizers on every joint or every three joints; both could be considered "centralized pipe" depending on the particular well. Cunningham stated that the only way to predict the effect of centralizer placement on mud displacement is through computer modeling.[82]

Macondo Team's Early Centralizer Plans

The Macondo team's September 2009 well plan included enough centralizers to likely satisfy BP's internal technical guidance. That plan's formula would have required the team to install at least 16 production casing centralizers given the then-planned total depth of 20,200 feet.[83] BP then produced another well plan in January 2010. Its formula would have called for at least 11 centralizers on the production casing.[84] Given the ambiguity of BP's technical guidance, it is unclear whether the January 2010 plan would have satisfied BP's internal requirements.[85] Both of these plans were based on a deeper well depth and larger casing diameter than BP eventually used at Macondo.

The Macondo Team Procured Six Centralizers for the Production Casing

On March 31, BP drilling engineer Brian Morel emailed a Weatherford sales representative, Bryan Clawson, and asked for "7-10" centralizer subs.[86] Clawson emailed Morel to say that Weatherford could only supply six centralizers immediately, explaining that it would take up to 10 days to manufacture more. Though it is common for Weatherford to manufacture centralizers to order, Morel did not ask Clawson to do so, even though Weatherford could at that point have made additional subs in time.[87] Instead, the BP team decided that six centralizers would be sufficient.[88] These six centralizer subs that Morel ordered were ultimately the only centralizers that the Macondo team used.

The Macondo Team Decided to Increase the Number of Centralizers to Address Potential Channeling Problem

During the long string decision-making process, Halliburton cementing engineer Jesse Gagliano had run a cementing model that predicted that the long string could be cemented successfully. Though Gagliano was a Halliburton employee, he worked at BP's Houston campus, and his office was on the same floor as those of BP's Macondo team.[89] Gagliano's April 14 model assumed proper centralization (by assuming a 70% standoff ratio) instead of calculating standoff based on centralizer placement plans.[90] It also assumed optimal wellbore size and geometry because BP did not yet have caliper log data from the well.[91] The April 14 model report did not predict significant channeling.[92]

On April 15, BP provided additional data to Gagliano from the Schlumberger logs, including caliper data, that could improve the accuracy of his cementing predictions. Based on the new data, Gagliano modeled the cementing process again, this time without assuming optimal centralization.[93] His new model predicted that using only six centralizers would result in lower standoff ratios and that this would be inadequate to ensure good mud removal and avoid mud channeling.[94] It also predicted that the mud channeling would increase the height of the cement column in the annulus (measured as TOC). That, in turn, would increase the effective pressure that the cement column would exert on the well formation below (ECD).[95]

That afternoon, Gagliano alerted Walz and BP operations engineer Brett Cocales to his predictions. Although Guide was out of the office, BP's engineering team acted on the information. The team was already concerned that the ECD during cementing operations could lead to lost returns during cementing and viewed lost returns as the biggest risk they faced during the cement job.[96] Based on Gagliano's predictions of increased ECD, Walz sought and obtained agreement from Guide's superior, Sims, to procure more centralizers and fly them to the rig immediately.[97] It appears that Walz and the BP team were concerned at this point about

Figure 4.3.13. Centralizer sub (top) and slip-on centralizer with stop collars (bottom).

Weatherford

the impact that channeling might have on ECD and were not directly concerned about the impact channeling might have on zonal isolation. [98]

Gagliano ran and distributed two additional cementing models from the afternoon into the evening of April 15 to evaluate the impact of adding additional centralizers.[99] His first model predicted that there would be reduced channeling with 10 centralizers, but still a significant amount. He emailed the model to the team, writing what he had already warned them about in earlier conversation: "Updating [the model with caliper and other data] now shows the cement channeling and the ECD going up as a result of the channeling. I'm going to run a few scenarios to see if adding more centralizers will help us or not."[100] Morel, who was on the rig and unaware that the team had made the unusual decision to fly centralizers to the *Deepwater Horizon*, responded that it was "too late" to get any more centralizers to the rig.[101] Gagliano's second model showed even less channeling with 21 centralizers. Both models showed that increasing the number of centralizers at Macondo would reduce the potential for gas migration in the annular space, though the centralizers' effect on gas flow was apparently of minor concern to the team compared with its effect on ECD.[102]

Sitting in the Houston conference room with Gagliano, Cocales carried out Walz's instructions to secure additional centralizers. Cocales called Clawson and ordered 15 additional Weatherford centralizers, the most that could be sent on a single helicopter.[103] BP also arranged for a Weatherford technician to accompany the centralizers and oversee the installation.[104] These 15 centralizers were leftovers from another BP project called Thunder Horse. Unlike the six centralizer subs already on the *Deepwater Horizon*, however, the Thunder Horse centralizers were slip-on centralizers as shown in Figure 4.3.13. BP's engineering team assumed that the Thunder Horse centralizers had integrated stop collars.[105] But the centralizer schematics that Clawson sent to Cocales on April 15 (and that Cocales forwarded to the rest of the BP engineering team) showed that the stop collars would be separate from the centralizers.[106]

Figure 4.3.14. Gregg Walz April 16, 2010 email to John Guide about centralizers.

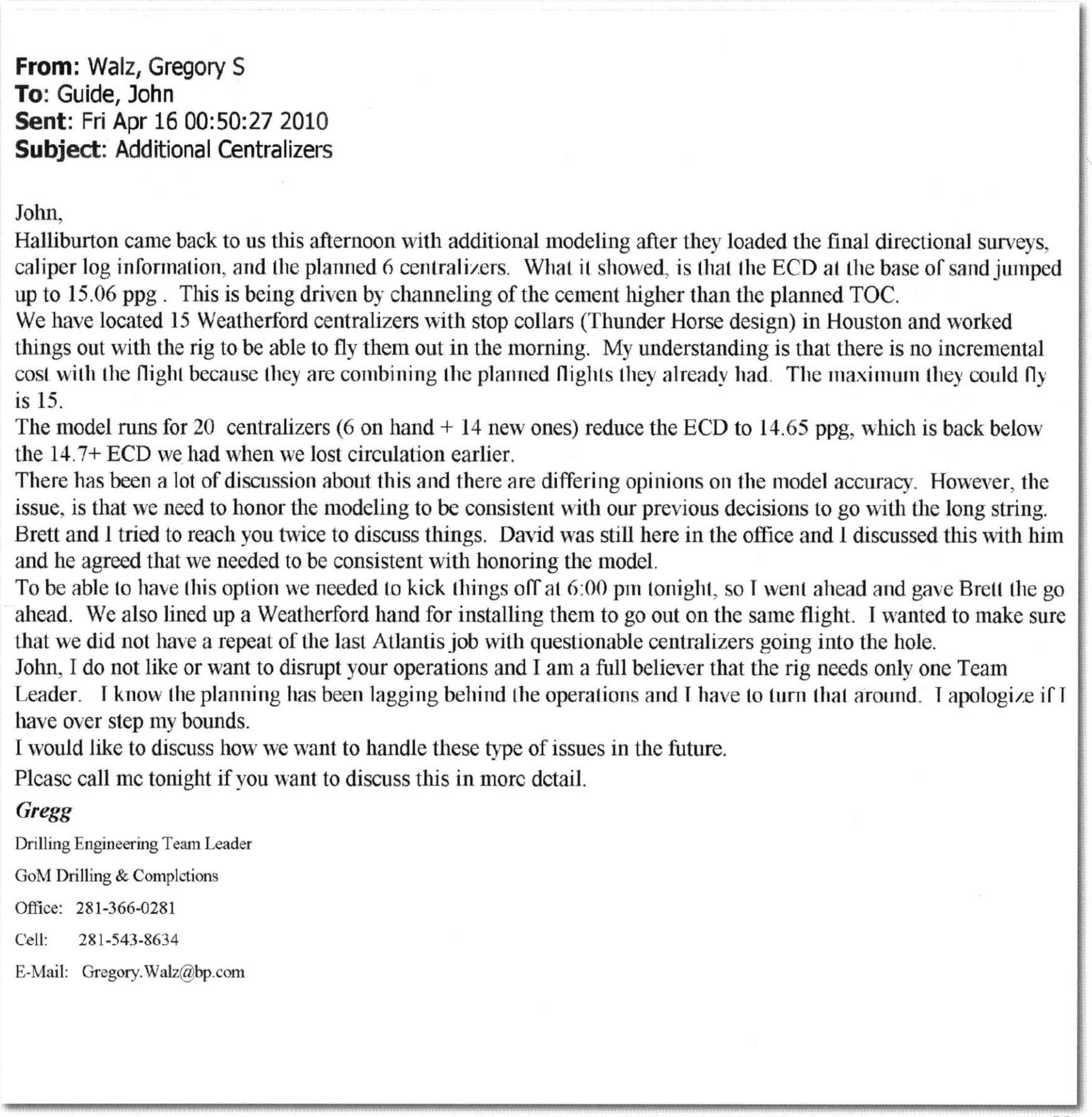

From: Walz, Gregory S
To: Guide, John
Sent: Fri Apr 16 00:50:27 2010
Subject: Additional Centralizers

John,

Halliburton came back to us this afternoon with additional modeling after they loaded the final directional surveys, caliper log information, and the planned 6 centralizers. What it showed, is that the ECD at the base of sand jumped up to 15.06 ppg . This is being driven by channeling of the cement higher than the planned TOC.

We have located 15 Weatherford centralizers with stop collars (Thunder Horse design) in Houston and worked things out with the rig to be able to fly them out in the morning. My understanding is that there is no incremental cost with the flight because they are combining the planned flights they already had. The maximum they could fly is 15.

The model runs for 20 centralizers (6 on hand + 14 new ones) reduce the ECD to 14.65 ppg, which is back below the 14.7+ ECD we had when we lost circulation earlier.

There has been a lot of discussion about this and there are differing opinions on the model accuracy. However, the issue, is that we need to honor the modeling to be consistent with our previous decisions to go with the long string. Brett and I tried to reach you twice to discuss things. David was still here in the office and I discussed this with him and he agreed that we needed to be consistent with honoring the model.

To be able to have this option we needed to kick things off at 6:00 pm tonight, so I went ahead and gave Brett the go ahead. We also lined up a Weatherford hand for installing them to go out on the same flight. I wanted to make sure that we did not have a repeat of the last Atlantis job with questionable centralizers going into the hole.

John, I do not like or want to disrupt your operations and I am a full believer that the rig needs only one Team Leader. I know the planning has been lagging behind the operations and I have to turn that around. I apologize if I have over step my bounds.

I would like to discuss how we want to handle these type of issues in the future.

Please call me tonight if you want to discuss this in more detail.

Gregg

Drilling Engineering Team Leader
GoM Drilling & Completions
Office: 281-366-0281
Cell: 281-543-8634
E-Mail: Gregory.Walz@bp.com

BP

Walz later explained his decision, as shown in Figure 4.3.14, to order the additional 15 centralizers t Guide in the following email, sent that night.[107]

Walz justified the decision to order additional centralizers because "we needed to be consistent with honoring the model." That model had convinced the team that a long string could be successfully cemented, so long as ECDs were kept in a low, narrow range. That model had also assumed that the centralizers would achieve a 70% standoff ratio.

The Macondo Team Decided Not to Install the Additional Centralizers

Sometime after 5 a.m. on April 16, a helicopter arrived at the *Deepwater Horizon*, carrying the 15 additional centralizers and Weatherford service technician Daniel Oldfather.[108] The helicopter did not, however, carry the stop collars and accessories that would be needed to secure the centralizers on the casing. Those had been shipped by boat and were scheduled to arrive by 4 p.m. (before the casing would be run).[109] Oldfather explained this to the rig crew when he landed.[110]

Figure 4.3.15. Centralizers delivered to the *Deepwater Horizon* on April 16, 2010.

BP

Morel was still visiting the rig at the time the helicopter landed. He examined the centralizers when they arrived. Like the other BP engineers, he had expected that the centralizers would have integrated stop collars. He now recognized that this was not the case.[111] Morel called Guide and told him that these were not the "one-piece" centralizers that he was expecting. Guide agreed they were not what he had planned on using either.[112] Morel took digital pictures of the centralizers and emailed them to Guide, telling him that "the centralizers do not have the stop [collars] on them."[113] However, Morel also told Guide that the centralizers could still be used because the boat carrying the collars would arrive in "plenty of time before needing them."[114]

After learning that the new centralizers had separate stop collars, Guide reversed Walz's decision to install them on the production casing in an email to him midday on April 16,[115] shown here in Figure 4.3.16.

Guide's email explained to Walz that the separate stop collars were prone to coming off the casing as it was being run into the well. Not only did this mean that the centralizers could slip away from their predetermined positions on the casing, but the centralizers could also get "hung up" against other parts of the well as the casing was being run. This could prevent the casing from being

Figure 4.3.16. John Guide April 16, 2010 email to Gregg Walz about centralizers.

From: Guide, John
Sent: Fri Apr 16 17:48:11 2010
To: Walz, Gregory S
Subject: Re: Additional Centralizers
Importance: Normal
Attachments: David Sims.vcf

I just found out the stop collars are not part of the centralizer as you stated. Also it will take 10 hrs to install them. We are adding 45 pieces that can come off as a last minute addition. I do not like this and as David approved in my absence I did not question but now I very concerned about using them

BP

lowered all the way to the bottom of the wellbore—a serious problem that would take significant time to fix.[116] Guide also noted that installing this type of centralizer would alone take 10 hours.[117] In a phone call with Walz, Guide weighed the risks of losses that fewer centralizers presented against the risk of a "last minute" addition of unfamiliar centralizers. There was no discussion at that point of stopping the job in order to procure the "correct" style of centralizers.[118] Instead, Guide told Walz and Sims he was reverting to the original plan. Sims agreed. Walz also accepted the reversal, saying, "I agree. This is not what I was envisioning," and apologized to the rest of the drilling team for the "miss-step" of ordering centralizers.[119]

During the same time period, Morel was attempting to position BP's six centralizers where they would be most effective, rather than place them at fixed intervals. As early as April 14, he had emailed Gagliano his suggested placement.[120] On April 15, when he mistakenly told Gagliano that it was "too late" to get more centralizers to the rig, he changed his recommendation, switching the position of two centralizers.[121] The next afternoon, the day BP reverted to the six centralizer plan, Morel changed the position of two other centralizers on his own "casing tally."[122] Morel supposedly based his recommendation on the caliper data and a wellbore image, though it is unclear precisely how he used them.[123]

Morel's placement of the centralizer subs was different than Gagliano's. Gagliano had assumed the centralizer subs would be evenly spaced apart while Morel placed them at irregular intervals.[124] It appeared that Morel expected Halliburton to run a new model based on his casing tally and centralizer placement. Morel's discussion with Cocales regarding the placement concluded, "We can argue this one out after we get the actual vs model data and see how it reacts."[125] As it turned out, BP never requested a model that reflected the actual centralizer placement, and Halliburton never ran one.

Neither Halliburton nor the BP engineering team appears to have considered that inadequate centralization might increase the chance of a blowout. Rather, they concluded that the worst-case result of using only six centralizers would be the need to conduct a remedial cement squeeze job.[126] As Cocales emailed Morel, "I would rather have to squeeze than get stuck above the WH [wellhead]. So Guide is right on the risk/reward equation."[127] In other words, Cocales preferred the increased risk of having to perform a remedial squeeze job to the increased risk of one or more of the 15 slip-on centralizers getting stuck in the well while the crew was running the production casing.

The BP team did not explicitly communicate its decision to use only the six centralizer subs on the rig to Halliburton or Weatherford.[128] When Gagliano eventually learned of the decision (from a Halliburton cementer aboard the rig), he asked BP to confirm it, and when he received no reply, he ran a new model on April 18.[129] It predicted poor centralization, "SEVERE" gas flow potential, and mud channeling. When Gagliano emailed the latest cement job procedures to the BP team at 9 p.m. that night, he attached this report.[130] He spoke with Walz the next morning (April 19) about the potential for channeling.[131] Walz in turn spoke with Guide about the issue.[132] BP nevertheless proceeded with its plan to run only six centralizers.

As BP has pointed out, Gagliano's April 18 model was based on several imperfect inputs. Notably, Gagliano assumed that BP would use seven centralizers, not six, and again, that BP would space them evenly along the casing, not place them in sections of the borehole where they might be especially effective.[133] Gagliano also utilized an incorrect pore pressure in the reservoir zone, which could influence the model's prediction of gas flow into the cement column.[134] It is unclear, however, whether eliminating these inaccuracies could have eliminated the channeling and gas flow predicted by the model. The use of fewer centralizers would decrease centralization, and the actual placement of two-thirds of the centralized joints was within 15 feet of the placement of the centralizers in the model.[135] In any case, the April 18 model was the most accurate model of the cementing process that existed before the blowout,[136] and it predicted that channeling would occur.[137] (As of 10 months after the blowout, Halliburton had still not produced modeling results that more accurately reflect Macondo conditions.)

BP began installing the casing at 3:30 a.m. on April 18 and finished at 1:30 p.m. on April 19.[138]

Float Collar Installation and Conversion at Macondo

Once the production casing string had been run, the crew turned to converting the valves in the float collar. Until this time, the float valves had been propped open by an auto-fill tube. Rig personnel needed to push the auto-fill tube down and out of place, thereby converting the float valves and allowing them to close (Figure 4.3.1). Once closed, the float valves would become one-way valves that would permit drilling mud and cement to flow down through the inside of the casing but would prevent "reverse flow" or "u-tubing."[139]

Shoe Track Length and Placement

The shoe track is the space between the float collar and the reamer shoe at the bottom of the casing. (A **reamer shoe** is a bullet-nosed, perforated piece of equipment that guides the casing toward the center of the hole as it is lowered into the well). At the end of the cement job, this space is filled with the "tail" portion of the cement that was pumped down the well. That tail cement may be contaminated by mud scraped from the casing by the top wiper plug. Indeed, one purpose of the shoe track is to contain contaminated tail cement.

A longer shoe track increases the volume for capturing contaminated tail cement, which in turn reduces the likelihood that such cement will flow into the annular space. A larger shoe track also dilutes the impact of any contamination in the tail cement. Morel suggested the shoe track at Macondo may not have been long enough but ultimately left the decision whether to extend the length up to the well site leaders on the rig.[140] According to Guide, BP also wanted to set the shoe track deeper in the well so that it was entirely below the hydrocarbon-bearing zone.[141] Ultimately, the shoe track was not below all of the hydrocarbon-bearing zones because the total depth of the well was shallower than planned due to problems of losing returns into the formation.[142]

Macondo Float Collar

The production casing at Macondo contained a Weatherford Flow-Activated Mid-Bore Auto-Fill Float Collar, which rig personnel had installed just above the 180-foot shoe track at the bottom of the casing string.[143]

The Weatherford float collar held two aluminum float valves set approximately 6 inches apart and propped open by an approximately 14-inch-long auto-fill tube (made out of phenolic resin).[144] As shown in Figure 4.3.17, the auto-fill tube allowed mud to flow up through the float valves while the casing string was run. Once the production casing had landed, however, the crew needed to push the tube out of the way to allow the float valves to close.

The float collar's auto-fill tube contains a 2-inch weighted ball, which is free to move within the tube but not out of it. At the top of the float assembly is a plastic cage that prevents the ball from escaping but allows mud to flow through. At the bottom is a phenolic resin collar that is less than 2 inches in diameter, which also allows mud, but not the ball, to flow through. When casing is being run, mud flowing up through the tube pushes the ball against the inside of the cage. When the casing lands, the ball falls to and plugs the bottom of the tube, leaving two small holes on the side of the tube as the only path through the tube for mud circulated down through the well.[145]

Figure 4.3.17. Auto-fill float collar.

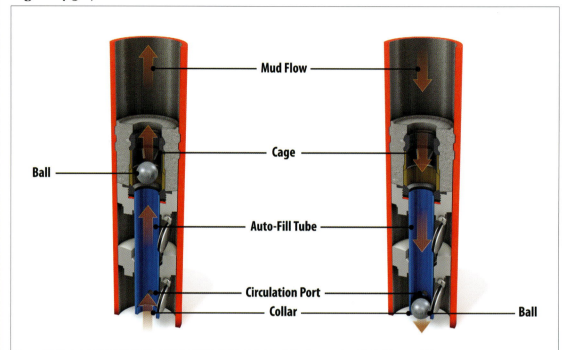

TrialGraphix

Flow while casing is being lowered (left) and flow during conversion (right).

The crew converts the float valves by pumping mud down through the tube, against the ball, and out the two holes in the side. As rig personnel increase the flow rate of mud, the constricted flow path creates a differential pressure against the auto-fill tube. Once the flow rate exceeds a certain threshold, the differential pressure should break four shear pins that hold the auto-fill tube in position and force the tube downward and out of the float collar assembly. With the auto-fill tube removed, the float valves spring shut, "converting" the float collar into a one-way valve system.[146]

According to calculations based on Weatherford's specifications, the Macondo float collar assembly would have converted at a flow rate of approximately 6 barrels per minute (bpm), which would have created a 500 to 700 pounds per square inch (psi) differential pressure across the auto-fill tube.[147] Achieving the requisite flow rate through the two small holes is the only way to convert the collar. Significantly, increasing pump pressure above 500 to 700 psi would not push the auto-fill tube through and convert the valves unless the flow through the two side holes exceeds the flow rate recommended by Weatherford.

Attempted Float Conversion at Macondo

Rig personnel prepared to convert the float collar at approximately 2:30 p.m. on April 19.[148] The crew turned on the pumps and began pumping mud down the well in an effort to establish circulation to convert the float equipment. Morel and BP well site leader Bob Kaluza oversaw the operation.

The crew ran into a problem. They could not establish circulation (and hence had a zero flow rate), suggesting that the float collar or shoe track was somehow plugged. The crew increased pump pressure nine times before finally establishing mud circulation. They increased pump pressure to 1,800 psi, then to 1,900 psi, but could not establish circulation.[149] Rig personnel then pressured up to 2,000 psi four times but still could not circulate. The crew then pressured up to 2,250 and then 2,500 psi and again failed to establish circulation.[150] The crew then made a ninth attempt to establish circulation, pressuring up to 2,750 psi, then 3,000 psi. At 3,142 psi, the pressure finally dropped and mud began circulating down through the float collar assembly.[151] Significantly, however, the crew never thereafter achieved sustained flow rates of 6 bpm, which were required for conversion of the float valves based on calculations using Weatherford specifications.

The rig crew sought advice from shore during these attempts to establish circulation. At 3:28 p.m., Hafle emailed a representative from Allamon, another equipment supplier, and asked for the specifications of the auto-fill float equipment. The Allamon representative responded and suggested "rocking the casing in 1000 psi increments up to 5,000 psi."[152] Morel called Clawson at Weatherford, reported that they could not break circulation, and asked how much pressure could be applied.[153] After checking with the Weatherford engineering department, Clawson called back Morel and told him they could increase pressure up to 6,800 psi.[154] However, he also told Morel that at 1,300 psi the ball would pass through the bottom of the auto-fill tube without converting the floats.[155] Morel called Guide onshore and received permission to increase pressure to 2,200 psi.[156] The crew pressured up to 2,250 and then 2,500 psi but still failed to establish circulation.[157] Guide later gave permission to increase pressure to 5,000 psi.[158]

Questions remained after establishing circulation. At 5:30 p.m. on April 19, Clawson of Weatherford emailed BP's Morel inquiring about progress.[159] Morel responded, "[W]e blew it at 3140, still not sure what we blew yet," indicating the rig crew did not know what they had dislodged with the amount of pressure applied.[160] Kaluza said, "I'm afraid we've blown something higher up in the casing string."[161] Hafle said, "Shifted at 3140 psi. Or we hope so."[162] Despite these uncertainties, the rig crew proceeded onward.

Low Pressure After Circulation Established

After establishing circulation, BP observed another anomaly. The pump pressure required to circulate mud through the well was significantly lower than expected.[163] As shown in Table 4.3.1, mud engineers from M-I SWACO had calculated that 370 psi would be required to circulate at 1 bpm and 570 psi at 4 bpm post-conversion. However, after the crew established circulation, it took only 137 psi to circulate at 1 bpm, which made Kaluza uncomfortable.[164] The crew increased circulation to 4 bpm, which required only 340 psi of pressure—230 psi less than M-I SWACO had predicted.

The low circulating pressure raised concern among personnel on the rig floor.[165] Kaluza spoke to Morel, who was on the rig.[166] Morel called Guide onshore, who agreed the pressures appeared low.[167] Cocales asked M-I SWACO to rerun its model to confirm that the original calculations had not been mistaken; M-I SWACO's models continued to predict substantially higher circulating pressures than actually observed.[168]

Guide and Kaluza instructed the crew to switch from pump 4 to pump 3 to see if changing pumps might change the circulation pressure.[169] They observed a slightly higher circulation pressure (396 psi at 4 bpm) after switching pumps, but this was still significantly lower than the expected pressure.[170]

Table 4.3.1. Low pressure observed after circulation established.

Circulation Rate	1 bpm	4 bpm
Pressures Observed	137 psi[171]	340 psi (on pump 4)[172] 396 psi (on pump 3)[173]
Pressures Modeled	370 psi[174]	570 psi[175]

At Guide's suggestion, the crew checked whether the Allamon **diverter** in the drill pipe might be leaking. The diverter is a valve opened during casing installation to allow drilling fluid flowing up inside the casing to flow into the annulus and back to the surface. At Macondo, the diverter was located in the drill pipe, above the wellhead at a final depth of 4,424 feet.[176] The test confirmed the diverter was closed.[177] Morel and Kaluza considered the possibility of a breach somewhere in the casing string.[178] However, they determined that a leak in the casing could not be fixed at the moment and, if present, would be revealed by later pressure tests (such as the positive pressure test).[179]

BP never resolved the low circulation pressure issue, concluding instead based on discussions with the rig crew that the pressure gauge was likely broken.[180] Morel and others felt comfortable proceeding because of the fact that the cement would be pressure tested later.[181] According to BP interview notes, Kaluza later described the low circulation pressure as an anomaly and said that after he had discussed it with Guide and well operations advisor Keith Daigle, Guide instructed Kaluza to begin pumping cement.[182]

Pre-Cementing Wellbore Conditioning at Macondo

Circulation After Landing the Long String

After converting the float valves, BP circulated mud again to clean the inside of the production casing string, remove any debris and cuttings dislodged by the casing installation, and condition the mud in the wellbore for cementing.

Planned Pre-Cement Circulation Volumes and Rates

An API recommendation from May 2010 was to circulate a minimum of 1.5 annular volumes or one casing volume after casing installation, whichever is greater.[183] Had this recommendation been in place at Macondo, this would have meant circulating 4,140 barrels (bbl) of drilling fluid. Halliburton recommends performing at least one full bottoms up circulation on a well before pumping a cement job.[184] This standard would have required BP to circulate 2,760 bbl of drilling fluid through the wellbore.[185]

Early BP drilling plans discussed pre-cementing circulation but did not call for a full bottoms up circulation. Omitting a full bottoms up is not unusual at deepwater wells because of the large mud volumes involved—circulating bottoms up could have taken as long as 12 hours at Macondo.[186] BP's September 2009 and January 2010 drilling programs called for circulating and conditioning 1.5 × pipe volume of drilling fluid "unless loss returns are experienced."[187] Although the plan did not specify which "pipe" volume it was referring to, circulation volumes are typically based on the volume of the casing used. The total long string casing and drill pipe volume at Macondo was 884 bbl, so it appears the plan called for the rig crew to circulate 1,326 bbl of mud before cementing.[188]

BP changed its plans in response to the April 9 lost circulation event, decreasing both the pre-cementing circulation volume and rate in order to reduce ECD. BP's April 12 plan thus called for circulating volume equal to one casing plus drill pipe capacity if hole conditions allowed, at a reduced rate of 8 bpm.[189] In its subsequent April 15 plan, BP further lowered the pump rate to "reduced rates (3 bpm) based on MI-SWACO models to keep ECD below 14.5 ppg."[190]

Figure 4.3.18. BP's pre-cementing mud circulation.

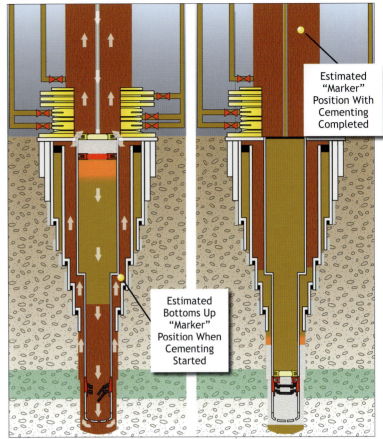

Estimated "Marker" Position With Cementing Completed

Estimated Bottoms Up "Marker" Position When Cementing Started

TrialGraphix

Even after receiving full returns during circulation on April 16, BP engineers remained concerned about lost returns during pre-cementing circulation.[191] They feared that circulating too extensively could damage the inside of the wellbore or instigate another lost returns event.[192] Onshore, Walz discussed whether to circulate full bottoms up with Gagliano late in the morning on April 19.[193] Afterward, Walz also spoke with Guide about circulation.[194] Ultimately, Guide recommended against circulating bottoms up because of concern over lost returns and gave approval to begin cementing.[195] On the rig, Halliburton cementing engineer Nathaniel Chaisson brought up the idea of circulating a full bottoms up but was told by a BP well site leader that a lower volume would be pumped.[196] Halliburton's April 18 cementing proposal lists reduced volumes, calling for 111 barrels at 1 bpm, followed by 150 barrels at 4 bpm for a total of 261 bbl.[197] Chaisson noted in the April 18 plan that the volumes and pump rates listed were "as per co. man,"[198] indicating that one of the BP well site leaders had provided it.

Pre-Cement Circulation Volumes and Rates

At approximately 4:18 p.m. on April 19, the rig crew re-established mud circulation after running the long string.[199] The rig crew then circulated a total of approximately 350 barrels of mud at rates up to 4 bpm before beginning the cementing process.[200] This figure exceeds the 261 bbl called for in the April 18 Halliburton cement job procedure[201] but is significantly lower than the 2,760 bbl required for a full bottoms up.[202]

Additional Circulation During Course of Cementing

BP has argued that the Chief Counsel's team must also take into account the additional mud volume circulated up the annulus from the bottom during the cement job itself in determining the total volume of mud circulated prior to the conclusion of the cement job. During the cement job, rig personnel pumped approximately 1,020 bbl of base oil, spacer, cement, and mud down into the well, which would have displaced an equal volume of mud.[203]

When combined with the pre-cementing circulation, this means that rig personnel pumped a total of 1,370 bbl of fluids (mud, spacer, and cement) down the well by the time cementing was complete.[204] This would have brought the bottomhole mud up into the riser to a depth of 4,250 feet below the ocean surface by the end of the cement job as shown in Figure 4.3.18. It would have taken a total of 2,760 bbl of circulation to bring the bottom mud all the way back to the rig.[205]

Table 4.3.2. Plans reduce pre-cement circulation volumes and rates.

Plan	Recommended Volume	Volume in Barrels	Recommended Circulation Rate
API RP 65, Part 2[206] (First edition)	1.5 annular volumes or one casing volume, whichever is greater	4,140 bbl (1.5 annular volumes)	
Full Bottoms Up		2,760 bbl[207]	
BP September 2009 Plan[208] and January 2010 Plan[209]	1.5 x pipe volume	1,325.73 bbl[210]	—
BP April 12 Plan[211]	1 casing and drill pipe capacity, if hole conditions allow	883.82 bbl[212]	~ 8 bpm
BP April 15 Plan[213]	1 casing and drill pipe capacity, if hole conditions allow	883.82 bbl[214]	3 bpm, based on M-I SWACO models to keep ECD below 14.5 ppg
April 18 Halliburton Cement Proposal[215]	—	111 bbl / 150 bbl per company man	1 bpm / 4 bpm
April 19 Actual Circulation		350 bbl	1-4 bpm

Cementing Process at Macondo

Halliburton's cementing team began pumping cement for the production casing on April 19.[216] In all, they pumped the following fluids down the well:

Table 4.3.3. Cementing volumes.

Material Pumped	Volume
Base oil	7 bbl[217]
Spacer fluid	72 bbl[218]
Unfoamed lead cement	5 bbl[219]
Foamed cement	39 bbl (Foamed to 48)[220]
Unfoamed tail cement	7 bbl[221]
Spacer	20 bbl[222]

After pumping these fluids, the cementing crew pumped mud into the drill pipe to push the cement down the well into position.[223]

Over the next three-and-a-half hours, the cement traveled down the drill pipe and into the well. During that time, rig personnel watched pump pressures at the rig for signs of cementing progress. Morel saw small pressure spikes suggesting that the top and bottom plugs had passed through the crossover joint in the long string.[224] Personnel on the rig agreed that the plugs bumped.[225] At 12:38 a.m. on April 20, Chaisson marked in his tally book that the plugs bumped at a pressure of 1,175 psi.[226]

Morel noted that the bottom plug landed 9 bbl ahead of plan.[227] This meant that the rig crew had to pump 9 bbl less fluid down the well than they planned before the bottom plug reached the float collar, potentially suggesting that the bottom plug had bypassed mud on its way down the well, and that the bypassed mud had contaminated the cement.

The top plug landed according to plan.[228] Chaisson watched the Sperry-Sun data[229] and estimated 100 psi of lift pressure before the top plug bumped.[230] Guide looked at the data from shore and thought it "easy" to see lift pressure.[231] Throughout cementing, the rig crew saw "full returns."[232]

BP and Halliburton declared the job a success based on the indirect indicators—lift pressure, bumping the plugs on time, and full returns. Chaisson sent an email to Gagliano at 5:45 a.m. saying, "We have completed the job and it went well."[233] He attached a detailed report stating that the job had been "pumped as planned" and that "full returns were observed throughout."[234] Just before leaving the rig, Morel emailed the rest of the BP team: "Just wanted to let everyone know the cement job went well. Pressures stayed low, but we had full returns the entire job, saw 80 psi lift pressure and landed out right on the calculated volume.... We should be coming out of the hole shortly."[235] Later, Morel followed up with an email saying "the Halliburton cement team...did a great job."[236] Sims congratulated Morel and the BP team, writing, "Great job guys!"[237]

The Float Check at Macondo

After cementing was complete, rig personnel conducted a **float check** to ensure the float valves had closed properly. Rig personnel began by pressuring up the system after bumping the top wiper plug.[238] They then released the pressure and monitored the system for pressure differentials and flow back from the well.[239] BP well site leader trainee Lee Lambert and Halliburton cementer Vincent Tabler opened a valve at the cementing unit to see how much mud flowed out of the well when they released the pressure. [240] (Some modest flow back is expected due to the compressibility of fluids during the pumping of the cement job.) Models had predicted 5 or 6 bbl of flow back.[241] The two men observed 5.5 bbl of flow, which tapered off to a "finger

Figure 4.3.19. Decision tree.

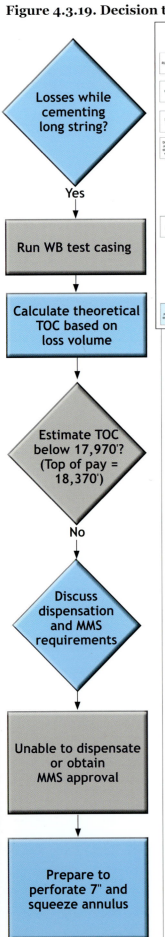

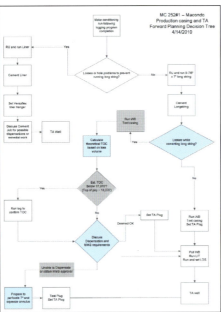

BP/TrialGraphix

tip" trickle.[242] Tabler testified they watched flow "until it was probably what we call a pencil stream," which stopped, started up again, and then stopped altogether.[243] The total flow at that point was close to the predicted flow,[244] and the two men concluded the float valves were holding.[245]

Cement Evaluation at Macondo

BP's Decision Tree for Cement Evaluation

BP's decision process for determining whether to run evaluation tools after the cement job focused on lost circulation concerns as shown in Figure 4.3.19. On April 15, Hafle developed a cementing decision tree that effectively reduced the decision process to a single question: "Losses while cementing long string?"[246] If the cementing crew reported losses while pumping the cement job, the decision tree stated that BP engineers would "Calculate theoretical [top of cement] based on loss volume." If that calculation estimated that TOC was *below* 17,970 feet that would mean that there was less than 100 feet of cement above the top of the pay zone—400 feet less than MMS requires.[247] In that situation, the decision tree required a "log to confirm" the TOC.

If the theoretical calculation predicted that TOC was *above* 17,970 feet, the decision tree stated that the Macondo team would discuss MMS requirements and consider seeking a dispensation. If unable to get dispensation or "obtain MMS approval," then BP would "perforate" the casing and "squeeze" the annulus to remediate the cement job. An operator would not normally run a cement evaluation log and plan to remediate cement before temporary abandonment operations; the Macondo team's explicit discussion of these contingencies illustrates how concerned they were about the possibility of cement losses.[248]

On April 15, Morel distributed a full plan for the temporary abandonment procedures at Macondo. The plan summarized the cement evaluation decision tree and provided further detail on the criteria for how to evaluate the cement job:[249]

1. If cement job **is not** successful: (no returns or lift pressure seen): set wear bushing / Run IBC-CBL log / Wait on decision to do remedial work (MMS and BP).

2. If cement job **is** successful (partial returns or lift pressure seen) **or** IBC-CBL log and required remedial work is completed.

The plan thus stated that the BP team would declare the cement job "successful" if it saw "partial returns" or "lift pressure." It anticipated that the team might need to run cement evaluation tools ("IBC-CBL log") but required doing so only if "no returns or lift pressure seen." Steps one and two were the only steps in the BP plan that contemplated cement evaluation: In step three, the crew would move on to the temporary abandonment phase of the well and begin to displace mud in the wellbore with seawater.

BP Ordered Cement Evaluation Services From Schlumberger

On the same day that Morel distributed the temporary abandonment procedures, BP well site leader Ronnie Sepulvado placed an order with Schlumberger for cement evaluation services.[250] Sepulvado did so to ensure that a cement evaluation team would be available on the rig if the cement job did not go as planned. The order included a "full suite of logs,"[251] including a cement bond log, isolation scanner, variable density log, and inclinometer survey.[252] Schlumberger planned to evaluate the annular cement from the float collar to about 500 feet above the expected TOC.[253] The total cost for the services would be about $128,000.[254]

On April 18 and 19, a team of technicians from Schlumberger flew out to the rig.[255] BP told the team that the cement evaluation log would be run only if there were lost returns.[256] The Schlumberger team waited for more than a day on the rig to see if BP needed their services.

BP Sent Schlumberger Home

At 7:30 a.m. on April 20, the Macondo team discussed the cement job during its daily morning phone call with its contractors. BP concluded during the call that the cement job had gone well enough that it could send home the Schlumberger technicians. According to Guide, "everyone involved with the job on the rig site was completely satisfied with the job."[257] Having seen lift pressure and no lost returns during the cement job, BP sent the Schlumberger team home and moved on to prepare the well for temporary abandonment. At approximately 11:15 a.m., the Schlumberger crew left the rig on a regularly scheduled BP helicopter flight.[258] Not running the cement log probably saved BP about eight hours of rig time.[259]

Technical Findings

The Primary Cement at Macondo Failed to Isolate Hydrocarbons

It is undisputed that the primary cement at Macondo failed to isolate hydrocarbons in the formation from the wellbore—that is, it did not accomplish zonal isolation.[260] If the cement had set properly in its intended location, the cement would have prevented hydrocarbons from flowing out of the formation and into the well. The cement would have been a stand-alone barrier that would have prevented a blowout even in the absence of any other barriers (such as closed blowout preventer rams, drilling mud, and cement plugs).

Although the Chief Counsel's team is certain that the Macondo cement failed, data currently available do not allow the team to determine precisely why. It may never be possible to make such a determination. Government investigators recovered samples of debris from the blowout that may be cement, but they have not currently determined whether it came from the well and, if so, from where within the well.[261] There are no plans to directly examine the annular cement currently remaining at Macondo for clues. Even if someone were to plan such an examination, the blowout and subsequent remedial efforts may have obscured or erased any clues that might otherwise have been discovered.

BP, Halliburton, and Transocean have each speculated about potential failure mechanisms. Based on information currently available, the Chief Counsel's team can conclude that most (if not all) of the cement pumped at Macondo flowed through the float valves and that most of the cement that rig personnel intended to place in the annular space around the production casing did in fact reach that location. (Chapter 4.1 discusses the remote possibility of a casing breach that would have affected cement placement.) Several events may have contributed to cement failure, either alone or in combination:

- cement in the annular space may have flowed back into the production casing due to u-tube pressure and failure to convert the float valves;
- drilling mud may have contaminated the cement in the shoe track and/or annular space badly enough to significantly slow cement setting time;
- cement in the annular space may not have displaced mud from the annular space properly, leaving channels of mud behind;
- cement in the shoe track may have flowed down into the rathole (the open section of wellbore below the reamer shoe), "swapping" places with drilling mud and increasing the potential for flow through the shoe track;
- cement slurry characteristics (such as retarder concentration, base slurry stability/ rheology, or foam instability) may have compromised the sealing characteristics of the cement (discussed in Chapter 4.4); and
- severe foam instability may have allowed nitrogen bubbles to break out of the slurry, with unpredictable consequences (also discussed in Chapter 4.4).

Any theory regarding the precise mechanisms of the Macondo cement failure must account for several issues that the Chief Counsel's team has identified. Most importantly, if our team is correct that hydrocarbon flow came through the shoe track and up the production casing, then the tail cement in the shoe track must have failed to block that flow. It would have taken only a relatively small amount of properly set cement in the shoe track to block that flow. This suggests one of three nonexclusive possibilities to the Chief Counsel's team.

Drilling mud contamination. The first is that enough drilling mud contaminated the shoe track to delay cement setting time so that the shoe track cement did not provide a competent flow barrier at the time of the blowout. This probably would have taken a significant amount of mud; testing by Chevron indicated that even with 25% mud contamination, the Macondo cement formulation would develop adequate compressive strength without serious delay.[262]

The mud in question could have been entrained in the cement flow during cement placement by, for instance, the wiping action of the plugs. If the plugs landed off-schedule (as post-blowout statements by Morel suggest), that would support this theory. Cementing experts emphasize that the shoe track is designed to prevent cement contaminated by plug bypass from entering the annular space. Shoe track cement should therefore properly be treated as one part of the overall cement barrier system and may not bar hydrocarbon flow on its own.

Drilling mud could also have "swapped" into the shoe track from the open hole section below the casing (sometimes called the rathole). The rathole volume was similar to the shoe track volume. Mud contamination could also have come from the annular space around the production casing if channeling or other phenomena caused contamination of that area and float equipment malfunctions allowed this material to flow back into the shoe track under u-tube pressure.

Gross nitrogen breakout. The second possibility is that the foamed middle section of the cement slurry was so unstable (as discussed in Chapter 4.4) that nitrogen gas bubbles in it "broke out" of suspension while the cement was flowing down the drill pipe and production casing. This could have left large gas-filled voids not only in the middle section of cement that was injected with nitrogen, but also in the tail cement (which became the shoe track cement). That tail cement should not otherwise have had nitrogen in it. A problem with this theory is that pumping data from the cement job do not show the sorts of gross anomalies that one would expect if cement and nitrogen flowed through the float collar separately.

Nitrogen breakout could also have occurred after the cement arrived at the bottom of the well. This might not have produced anomalies in the pumping job data but still could have compromised the quality of the set cement. As described in Chapter 4.4, unstable nitrogen foamed cement can be excessively porous and permeable once set. Hydrocarbons can flow through such cement.

Gross cement slurry failure. A final possibility is that the Macondo cement slurry was unstable even before being foamed with nitrogen. As Chapter 4.4 explains in greater detail, pre-blowout testing shows that the Macondo slurry had a very low yield point, and post-blowout testing shows that a cement slurry produced using the Macondo recipe had a tendency to settle as it set. It is possible that these problems compromised the quality of the Macondo cement job so that cement in the shoe track could not have prevented hydrocarbon flow. A problem with this theory is that it appears, based on available information, that the cap cement in the annulus above the pay zone set up properly and created a barrier to flow up the annulus.

Using Six Centralizers Increased the Risk of Cement Failure

Reduced pipe centralization increases the risk of poor mud displacement, the risk that mud channels will compromise zonal isolation, and the risk that hydrocarbons will migrate into and through the annular cement as it sets. Without a direct examination of the Macondo cement, the Chief Counsel's team cannot determine whether any of these things occurred, let alone whether they caused or contributed to the blowout. The team can only conclude that BP's engineering decision increased the risk of cementing failure.

The Chief Counsel's team cannot at this time accept Halliburton's conclusory assertion that the limited number of centralizers at Macondo caused inadequate mud displacement, channeling, and cement failure.[263] To support its view, Halliburton relies heavily on the results of the model that Gagliano produced on April 18.[264] But Gagliano produced the April 18 report using several assumptions that did not match the eventual Macondo conditions. Halliburton points out that Gagliano received these assumed figures from BP, but that it is irrelevant; because the April 18 modeling inputs were inaccurate, the modeling output was unreliable even if one were to assume that those models accurately predicted problems with a cement job.[265] (Halliburton personnel have argued that their model would still have predicted channeling even with corrected inputs. However, Halliburton has yet to provide the results of a corrected model to the Chief Counsel's team or the public. This leads the Chief Counsel's team to infer that the results are not favorable to Halliburton.)

The Chief Counsel's team also cannot accept BP's equally conclusory assertion that the decision to use only six centralizers "likely did not contribute to the cement's failure to isolate the main hydrocarbon zones...."[266] Chapter 4.1 explains that the Chief Counsel's team finds it likely that hydrocarbon flow came up the production casing through the shoe track. But even though insufficient centralization may not have directly affected the integrity of the cement in the shoe track, it very well could have damaged the integrity of the cement in the annular space around the pay zone. If that cement had worked properly, shoe track cement failure would have been irrelevant.

BP's technical guidance and early Macondo well plans called for more centralizers than were actually run and for centralizers to be used over a larger casing interval.[267] If BP believed that its engineers could reliably reduce the number of centralizers (and hence cost) by scrutinizing caliper logs and pinpointing the placement of centralizers, one would expect its guidance documents and well plans to describe this practice. And while BP has repeatedly questioned the accuracy of the Macondo cementing models and the value of Halliburton's model in general,[268] it offers little affirmative technical analysis of its own to support its claim that centralization was not an issue at Macondo. Moreover, *before* the Macondo blowout, BP engineers thought the model's predictions of channeling were sufficiently credible that they flew 15 more centralizers to the rig in response.

Limited Pre-Cementing Mud Circulation Increased the Risk of Cement Failure

BP's decision to circulate a limited volume of mud at a relatively low rate before cementing may have led to inadequate mud conditioning and wellbore preparation. BP's decision was perhaps an understandable response to its concerns about formation integrity and lost returns, but it also increased the risks of cementing failure.

BP has defended its decision not to circulate bottoms up before cementing. It has argued, among other things, that modern technologies can identify wellbore cleanliness problems without full mud circulation and that the Macondo team took other measures to prepare the wellbore for cementing. For instance, the team circulated bottoms up *before* running the production casing[269] and pumped additional spacer during the cementing process to remove debris from the well.[270] At the same time, BP cannot dispute that circulating bottoms up is a "best practice" specified by Halliburton and other cementing experts,[271] and that its team did not do so. Although circulating less mud may have reduced the particular risk of lost returns, it nevertheless increased other aspects of the risk for cement failure, as compared to completing a full bottoms up.

Low Cement Volume Increased the Risk of Cement Failure

The limited volume of cement used at Macondo increased the risk of cement failure. BP pumped only about 60 barrels of cement (after nitrogen foaming) at Macondo. While BP may have thought it necessary to pump a small amount of cement to reduce the risk of lost returns, this approach magnified three other risks.

First, it meant there would be less cement in the annular space above the hydrocarbon zones—less even than BP's technical guidance recommends.[272] Second, it increased the risk that placement errors would leave insufficient cement in the shoe track or in the annular space corresponding to the hydrocarbon zone. And third, it increased the detrimental effects of any mud contamination. Mud contamination may have been a particular problem at Macondo because the design called for a tapered long string casing. That casing design called for the top and bottom wiper plugs both to wipe mud from a relatively long length of casing and to wipe two different casing diameters.[273]

Before the blowout, BP's engineering team recognized that their design called for a low cement volume that would provide little room for error. [274] And since the blowout, BP has recognized that "small cement slurry volume" increased cementing difficulties at Macondo.[275]

Cementing Pump Rate Increased the Risk of Cement Failure

In concert with Halliburton, BP chose to pump the primary cement at a relatively low rate.[276] This low rate would have decreased the efficiency with which the cement would have displaced mud from the annular space, especially given Halliburton's predictions regarding the impact of a reduced number of centralizers.[277] This, in turn, would have increased the risk of mud-related cementing failures such as channeling, contamination, and gas flow.

Using a Reamer Shoe Instead of a Float Shoe May Have Increased the Risk of Cement Failure

BP could have decreased cementing risks using a float shoe. Like a reamer shoe, a float shoe is a rounded piece of equipment that attaches to the bottom of a casing string and helps to guide the string down. But unlike the reamer shoe, the float shoe includes a check valve that functions much like the valves in the float collar. That extra check valve serves as an extra line of defense against cement contamination and helps keep debris and contaminants away from the float collar's valves. The existence of the extra check valve also helps to ensure proper cement placement by preventing cement from flowing back up the casing. Industry engineers often install float shoes where they are concerned about cement contamination.[278] While cement contamination was (or should have been) a concern at Macondo, BP chose not to install a float shoe on its production casing.

Rathole Issues Could Potentially Have Increased the Risk of Cement Failure

BP chose not to take precautions against **rathole** swapping. The rathole, again, is the open section of wellbore below the end of the production casing. As described above, mud in this portion of the wellbore can swap places with cement in the shoe track if the mud is less dense than the tail cement. This can contaminate the cement in the shoe track or potentially create a flow path through the cement in the shoe track.

One common precaution to guard against this phenomenon is to pump a small volume of dense mud into the rathole. If this mud is more dense than the cement, it will tend to stay in place rather than swap places with the cement. Although early BP plans called for this procedure,[279] the engineers eventually chose not to do it because the volume was small and improper placement could cause ECD concerns.[280] They reasoned that this created relatively small risks: the density differential between the mud and tail cement was not large, and the rathole volume was relatively low.[281] Halliburton personnel admitted after the blowout that rathole swapping could create a problem, but they had not considered the issue before pumping the job.[282]

Rig Personnel May Not Have Converted the Float Valves

Although rig personnel and BP concluded that they successfully converted the float valves, the Chief Counsel's team finds that the float valves at Macondo may not have actually converted.[283] Unconverted float valves could have compromised the bottomhole cement job at Macondo.

Rig Personnel Never Pumped Mud at the Rates Weatherford Specified to Convert the Float Collar

Planning documents and pumping data show that rig personnel never pumped mud down the well at sustained rates high enough to ensure float valve conversion. While well plans specified mud circulation rates that would have converted the float valves, actual rates never exceeded 4.3 bpm—significantly less than the 6 bpm required to convert the equipment:

Table 4.3.4

	Flow Rate Needed to Convert	Differential Pressure Needed to Convert
BP September 2009 Plan[284] and BP January 27, 2010 Final Drilling Program[285]	12 bpm maximum	~ 600 psi
BP April 12, 2010 Drilling Plan[286] and BP April 15, 2010 Drilling Plan[287]	8 bpm minimum	~ 500 to 700 psi per Weatherford recommendation
Weatherford Manufacturer Recommendation[288] Adjusted for 14.1 ppg Mud Weight[289]	6 bpm[290]	600 psi[291]
April 19 actual[292] steady flow rate never exceeds 4.3 bpm,[293] which would result in a differential pressure of approximately 328 psi[294]		

BP contracted Stress Engineering Services, a third-party engineering firm, to conduct post-blowout testing on float collars similar to those used at Macondo.[295] On the basis of this testing, BP asserts that temporary surge flow rates caused by sudden pressure changes in the well would have converted the float equipment.[296] BP contends that there were two potential surge-inducing events. The first was the sudden drop in pressure from 3,142 psi once mud circulation began.[297] The second was during the cement job when the bottom plug burst at 2,392 psi.[298] The Stress Engineering analysis shows that the Macondo float valves may have converted because of pressure-induced surge flows. But if this in fact happened, it was by happenstance, not design. More importantly, without having pumped mud consistently through the float collar at Weatherford-prescribed rates, BP personnel had no sound basis for concluding that the float valves had converted. And the later float check that they performed was not a reliable indicator that the float collar had sealed.[299] BP's own report agrees.[300]

Although rig personnel deemed the Macondo float check to be a success, the check was actually inconclusive because of the small density differential between the cement and drilling mud in the well. Halliburton's April 18 model predicted 38 psi of differential pressure.[301] (The Chief Counsel's team's calculations based on actual volumes pumped indicate a u-tube pressure of about 56 psi—an inconsequential difference.[302]) A Weatherford representative confirmed that 38 psi of differential pressure is "pretty tiny,"[303] and other experts agree that it would be hard to detect.[304] The small u-tube pressure would also have meant that any cement backflow may have been too small and gradual for rig personnel to detect in the time that they monitored for flow.

The Drop From 3,142 psi May Have Been Due to a Clogged Reamer Shoe or a Failure of the Float Collar System

Rig personnel assumed that the sudden drop in pump pressure from 3,142 psi indicated that they had converted the float collar. If the float collar did not actually convert, then something else must have caused this pressure drop. The Chief Counsel's team has identified two possible explanations.

The Reamer Shoe May Have Been Clogged

The first possibility is that the unexpected pressure increases and sudden pressure drop may have been caused by a clog in the reamer shoe that eventually cleared in response to elevated pump pressure.

Drilling mud pumped down the Macondo production casing and through the float collar assembly had to exit the bottom of the casing through three 1⅝-inch holes ("circulation ports") at the

Figure 4.3.20. Clogged reamer shoe.

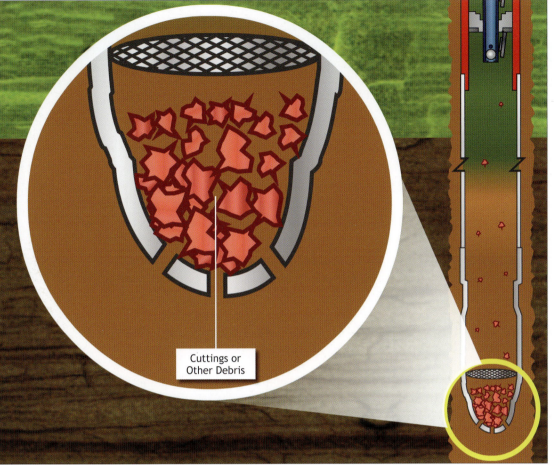

Cuttings or
Other Debris

TrialGraphix

Figure 4.3.21. Ball forced through tube.

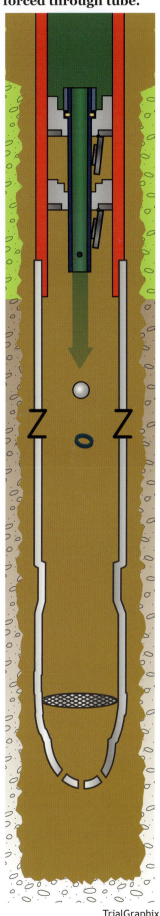

TrialGraphix

bottom of the reamer shoe.[305] Debris and/or cuttings may have plugged these holes during the course of casing installation as shown in Figure 4.3.20. This could explain why the rig crew was unable initially to establish mud circulation after landing the production casing. It could also explain why the pressure dropped suddenly from 3,142 psi—that pressure may have been sufficient to clear a clog in the reamer shoe to allow mud to flow again.

After the blowout, at least two BP personnel identified a clogged reamer shoe as a factor that may have complicated the float conversion process. Morel told BP investigators soon after the blowout that he believed the reamer shoe may have been plugged.[306] Sepulvado, who was onshore at the time of the blowout, similarly told the Chief Counsel's team that the only reason such high pressures would have been needed was because differential pressure was not getting to the ball,[307] which may have been caused by a clogged reamer shoe.[308] Besides interfering with float conversion, a clogged reamer shoe could have complicated cementing by altering cement flow out of the reamer shoe.

The Ball May Have Been Forced Through the Auto-Fill Tube

A second possibility, shown in Figure 4.3.21, is that the sudden pressure drop may have been caused when pump pressure forced the ball inside the auto-fill tube through the end of the auto-fill tube. The collar that would normally have retained the ball within the auto-fill tube was held in place with brass pins. It is possible those pins and the collar failed, allowing the ball to pass through.[309] This would have left the auto-fill tube in place between the float valves and created a path for flow in either direction.

If the ports in the bottom of the auto-fill tube were clogged, the rig pumps may have placed enough force on the collar to shear the brass pins instead of the pins holding the auto-fill tube in place. Clawson informed Morel on April 19 that it would only take 1,300 psi of pressure to force the ball through the collar without converting the float valves.[310] It is not apparent whether Morel considered or informed others of this possibility.[311]

Unconverted Float Valves Would Have Increased the Risk of Cement Failure

If rig personnel never converted the float valves at Macondo, it would have left an open flow path through the float collar assembly. That flow path may have allowed cement to flow back into the casing from the annular space outside the casing, which would in turn have left less cement in the annular space. This flow would also have: (1) increased the potential for contamination of the shoe track cement with mud; (2) brought foamed cement from the annulus into the shoe track (which should have contained only unfoamed tail cement); and (3) allowed any nitrogen that broke out of the foamed cement to compromise the shoe track cement. The open flow path would also have made it easier for any hydrocarbons that bypassed the cement to flow through the float collar assembly.[312]

Properly Converted Float Equipment Is Not a Reliable Barrier to Hydrocarbon Flow

The Chief Counsel's team does not believe that even properly converted float valves would have constituted a reliable physical barrier to hydrocarbon flow. While BP's internal investigation report appears to state that float valves could be a barrier,[313] several senior BP personnel disagreed with that statement.[314] Weatherford does not consider float equipment a barrier to hydrocarbon flow and instead provides the equipment only to prevent backflow of cement.[315] The API similarly states only that "float equipment is used to prevent the cement from flowing back into the casing when pumping is stopped"[316] and does not include float equipment among its list of subsurface mechanical barriers.[317]

Management Findings

BP's Management Processes Did Not Force the Macondo Team to Identify and Evaluate All Cementing Risks and Then Consider Their Combined Impact

BP engineers failed to fully appreciate the cementing challenge they faced at Macondo. Every deepwater cement job presents a technical challenge, but the Macondo cement job involved an unusual number of risk factors. Several were inherent in the conditions at the well. BP and Halliburton created several others during the course of the design and execution of the primary cement job. The list includes:

- narrow pore pressure/fracture gradient;
- use of nitrogen foamed cement;
- use of long string casing design;
- short shoe track;
- limited number of centralizers;
- uncertainty regarding float conversion;

- limited pre-cementing mud circulation;

- decision not to spot heavy mud in rathole;

- low cement volume;

- low cement flow rate;

- no cement evaluation log before temporary abandonment; and

- temporary abandonment procedures that would severely underbalance the well and place greater stress than normal on the cement job.

BP engineers certainly recognized some of these risk factors and even tried to address some of them. For instance, the team asked Halliburton to use additional spacer during the cement job to compensate for the limited pre-cementing circulation.[318] But it does not appear that any one person on BP's team—whether in Houston or on the rig—ever identified all of the risk factors. Nor does it appear that BP ever communicated the above risks to its other contractors, primarily the Transocean rig crew. For instance, Transocean was never aware that Halliburton had recommended more than the six centralizers that were used.[319]

More importantly, there is no indication that BP's team ever reviewed the *combined impact* of these risk factors or tried to assess the overall likelihood that the cement job would succeed, either on their own or in consultation with Halliburton. Rather, BP appeared to treat risk factors as surmountable and then forgettable. For instance, after Guide had decided to use only six centralizers despite the risk of channeling, one BP engineer wrote to another team member, "But who cares, it's done, end of story, [we] will probably be fine and we'll get a good cement job."[320] Reviewing the aggregate effect of risk factors may not even have led BP to change any of its design decisions. But if done properly, it may have led BP engineers to mitigate the overall risk in ways that could have prevented the blowout. Indeed, a major oil company representative stated that the risk factors at Macondo were so significant that his organization would not have counted the Macondo cement job as a barrier to annular flow outside the production casing even after a successful negative pressure test.[321]

A closely related issue is that once BP's engineering team properly identified a risk, it often examined and addressed the risk without a full appreciation of other risks its *response* might create. For instance, BP's team focused almost exclusively on the risk of lost returns in designing its cementing program. BP engineers may well have been right to view this as the largest individual risk they faced. But they failed to consider the secondary impacts of their numerous responses to that risk, which included reducing pre-cementing circulation, cement volume, and cement flow rate. Those responses may have increased the overall likelihood of cement failure even as they decreased the potential for lost returns.[322]

BP Did Not Properly Manage Design Changes and Procedural Modifications

Impact of Changes to Its Mud Circulation Plan

BP's engineering team does not seem to have recognized that late-stage changes to mud circulation plans might impact float collar conversion. Before the early April lost circulation event, the team intended to circulate fluids at 8 bpm—a rate that would have converted the float valves. But the BP team later reduced the planned circulation rate to 4 bpm because of ECD concerns—a rate that would not have converted the float valves according to the manufacturer's specifications. The April 15 drilling plan highlights the disjoint: It simultaneously calls for circulation rates of *at least* 8 bpm to convert the float equipment but recommends circulating mud at 3 bpm "to keep ECD below 14.5 ppg."[323] Circulating at 8 bpm would clearly exceed that ECD threshold, and an independent expert found this inconsistency irreconcilable.[324]

If BP had recognized that lowering planned circulation rates could impact float collar conversion, it could have solved the problem easily. Weatherford can readily produce float collars that convert at different flow rates—changing the conversion flow rate can be as simple as changing the number of shear pins or the size of the holes in the bottom of the auto-fill tube. BP could therefore have used a different float collar assembly that would have converted at the lower flow rates it planned. Its engineering team does not appear to have considered this possibility or the internal inconsistency in its drilling plan.

Centralizer Sub Procurement

By January 2010, BP's well plan had called for at least 11 centralizers for its final production casing string. Weatherford, BP's centralizer supplier, recommends that its clients notify it of equipment needs four to six weeks in advance.[325] But BP engineers waited until the last day of March to begin the process of ordering centralizers, leaving themselves less than three weeks of lead time. If BP had ordered centralizers earlier, Weatherford personnel would have had ample lead time to manufacture more centralizer "subs" to meet BP's request,[326] and BP's team would not have been forced to decide whether to use slip-on centralizers.

When BP eventually ordered centralizers from Weatherford, the engineer who made the request only asked for a range of "7-10" centralizers rather than the 11 centralizers that BP's January 2010 plan specified. It appears that BP engineers relied on their own estimates of centralizer needs given well conditions, but it is unclear why those conditions would have been any different than when the original well plan was designed.[327] When Weatherford responded that it had only six centralizers in stock, BP's team viewed this as sufficient even though it was less than the number the engineer requested and about half the number called for in the well plan. There is no indication that BP's team even asked whether additional centralizer subs could be manufactured in time, nor is there any evidence that BP attempted to secure acceptable equipment from other suppliers besides Weatherford.[328]

Managing equipment procurement is a key part of safe and efficient offshore drilling. By failing to plan centralizer procurement properly, BP's engineering team forced itself to choose between using only a few centralizer subs, adding slip-on centralizers that its team believed posed mechanical risks, or incurring costs by waiting for Weatherford to manufacture additional subs at the last minute.

Decision Not to Run Additional Slip-On Centralizers

BP also mismanaged its engineering response to Halliburton's advice to add centralizers. First, BP and Halliburton could have considered centralizer availability during the mid-April design review that led them to determine they could cement a long string without exceeding ECD thresholds. Instead, they simply assumed optimal centralization without examining whether they had the materials on hand to achieve it.

Once Gagliano advised BP's team that additional centralizers would be needed to avoid channeling, the team responded by procuring 15 additional centralizers immediately. The immediate response reflects appropriate levels of concern, but also highlights the problems with making complex design changes at the last minute. The engineering team believed that it was ordering slip-on centralizers with integrated stop collars even though a Weatherford representative sent the team specifications that showed otherwise. It appears that BP's team did not review these specifications carefully, perhaps because of time pressure. Careful review here would have avoided last-minute decision making on April 16.[329] The decision to send these additional centralizers prompted Guide to complain to his supervisor Sims the next day:

> David, over the past four days there has been so many last minute changes to the operation that the WSL's have finally come to their wits end. The quote is "flying by the seat of our pants." More over we have made a special boat or helicopter run everyday. Everybody wants to do the right thing, but this huge level of paranoia from engineering leadership is driving chaos.... The operation is not going to succeed if we continue in this manner.[330]

After the centralizers were delivered, BP made its final decision not to use them without careful engineering review. After Guide found out the type of centralizers Weatherford had provided, he argued that they should not be used because of recent problems that BP had experienced with the design.[331] (Guide mentioned time and cost concerns as well.) But Guide and the rest of the BP team appear to have been motivated by personal experience rather than any disciplined analysis. Notably, they did not consult the Weatherford centralizer technician that they had flown to the rig, who could have provided valuable input on the relative risks of centralizer hang-up.[332] It is not even clear whether BP believes *now* that its Macondo team should or should not have used the centralizers; the Bly report states that the team "erroneously believed that they had received the wrong centralizers."[333]

BP also did not examine whether the mechanical risks of running additional centralizers outweighed the cementing risks of *not* using them. BP's team could easily have asked Gagliano to run a new model to predict the impact of using only six centralizers and could have provided up-to-date wellbore and well design data to improve the accuracy of those predictions. The team also could have consulted its in-house cementing expert Cunningham.[334] BP could have asked Halliburton to incorporate Morel's irregular placement of centralizers into its model, rather than simply relying on Morel's apparent ad hoc analysis to determine their placement. It did none of these things.[335] BP's engineering team may have been motivated by skepticism of Halliburton's modeling,[336] but this was the only analytical tool the team had at the time.

Having made a last-minute decision to use fewer centralizers than planned, BP's team should have recognized that decision would increase the risks, first, of lost returns (by increasing ECD), and second, of overall cementing failure. Instead, the team appears to have viewed its centralizer decision-making process as a "miss-step"[337] that had little significance after it occurred. Had BP at least noted the risks of using fewer centralizers than it had planned, its rig personnel and contractors might have been better prepared for the events that followed.

Communication of Centralizer Decisions Hampered Risk Identification and Management

Once BP decided not to run the additional centralizers, it made no effort to inform its contractors of its decision. Weatherford's technician only learned that the centralizers would not be used by asking about the issue hours after the installation should have occurred.[338] When he did learn of it, the technician was concerned enough to call his supervisor—he had never been on an installation job that had been canceled.[339] But neither he nor anyone else at Weatherford expressed concerns to BP. Instead, the technician's supervisor instructed him to defer completely, stating: "Third party, we do what the company man requests."[340]

Gagliano only learned about the decision from Tabler, who in turn learned it from Chaisson, who in turn learned of the decision by happenstance.[341] Gagliano stated that he was "frustrated,"[342] and emailed BP's team to confirm the decision and to ask if he should rerun his models, but nobody ever responded to him.[343] Gagliano eventually updated the cement model on his own, but his model lacked up-to-date information from BP, and he sent it only after the casing run had begun. A prompt response from BP to Gagliano might have improved the Macondo team's appreciation of the risks they faced.

Use and Management of Modeling Results

BP engineers mismanaged their use of Halliburton's computer cementing models.

It is unclear why BP did not review Halliburton's modeling results more carefully and continually update Halliburton's data after April 14. Industry experts say that it is not uncommon for operators to depart from cementing rules of thumb (such as full bottoms up) in reliance on favorable modeling predictions. But operators who do so should continually update such models to ensure that their departures do not cause cementing problems. At Macondo, BP appears to have done little after April 14 to ensure that Halliburton was using up-to-the-minute data. BP provided Halliburton a caliper log but not updated information about reservoir pressure and centralizer placement. Instead, it appears that BP's engineering leadership paid little attention to refining the model once it produced results they found favorable.

BP's willingness to disregard Halliburton's April 18 modeling predictions is especially questionable given the degree to which BP relied on the model's earlier predictions. On April 14, BP relied almost exclusively on a Halliburton model to conclude that it could successfully cement a long string casing. At this time, BP engineers knew that the model was based on incomplete data. BP then disregarded the April 18 predictions even though the concerns it identified were similar to those that motivated more serious analysis on April 14. BP's apparent skepticism of the value of the April 18 results is hard to square with its near-total reliance on the April 14 results.

BP Did Not Adequately Evaluate the Significance of Float Conversion Difficulties

BP's management and review of the float collar conversion process were inadequate. As explained above, BP should have secured different float equipment once it modified its planned circulation rates. BP also mismanaged its evaluation of the float conversion process on the rig. BP rig personnel properly consulted their shore-based engineering team after encountering difficulties when converting the float collar. But after reinitiating circulation at much higher pressures than expected, BP's team appears to have assumed the float valves converted. If the team had instead reviewed the data carefully, it would have recognized that it had not yet circulated mud in excess of 4.3 bpm and might have increased circulation to ensure conversion.

Making matters worse, BP and Transocean personnel then tried to explain away concerns about lower-than-predicted circulation pressures by blaming a faulty pressure gauge. BP has since pointed out that the circulating pressures predicted by M-I SWACO were erroneous and that the circulation pressure observed was actually what should have been expected. But rig personnel believed at the time that M-I SWACO's predictions were accurate, and yet there is no evidence that they took steps to confirm the gauge was actually faulty or tried to replace it.[344]

If BP or Transocean had adequately considered the possibility that the float valves did not convert, they could have undertaken efforts to mitigate the potential risks. For instance, one standard industry tactic to address float valve failure is to add pressure inside the casing system after pumping cement and to thereby counterbalance any u-tube pressure that might otherwise induce flow back through open float valves.[345]

BP Focused Excessively on Full Returns as an Indicator of Cementing Success

The Macondo team's approach to cement evaluation at Macondo was flawed. Because the team focused its attention so heavily on the risk of lost returns, it overemphasized the significance of full returns as an indicator of cementing success.

Receiving full returns showed that cement had not flowed into the weakened formation but provided little or no information about: (1) the precise location where the cement had ended up; (2) whether channeling had occurred; (3) whether the cement had been contaminated;[346] or (4) whether the foamed cement had remained stable. Similarly, reports of on-time top plug arrival indicated, at most, only one thing for certain: The cement flowed through the float collar. (Morel's report that the bottom plug bumped early may suggest that mud contaminated the cement during job placement.) Accordingly, BP's technical guidance documents do not list reports of full returns or on-time plug bumping as indicators of zonal isolation.[347]

BP engineers also considered lift pressure a positive indication. Company technical guidance documents state that lift pressure can provide a coarse indication of TOC (if not zonal isolation) but that it "is unlikely to provide a sufficiently accurate estimate" of TOC when "cement and mud weights are very similar,"[348] as they were at Macondo. While one BP engineer stated that lift pressure was "easy" to see at Macondo,[349] another admitted after the blowout that it was not a valid confirmation of good cement placement.[350] Industry experts who reviewed the data after the fact were also skeptical. The Chief Counsel's team spoke with several experts who agreed that the roughly 100 psi pressure increase that rig personnel observed at Macondo after the bottom plug landed was too low to be a reliable indication that cement had turned the corner and flowed up into the annulus.[351] One described 100 psi of lift pressure as "nearly unreadable."[352] That relatively small pressure increase might have been caused by cement "turning the corner" into the annulus, but it might also have been caused by friction from cement flow.[353]

Better management would have encouraged the BP team to question the overall value of its pressure and volume indicators. BP's own report appears to agree. It states:

> A formal risk assessment might have enabled the BP Macondo well team to identify further mitigation options to address risks such as the possibility of channeling; this may have included the running of a cement evaluation log.... Improved technical assurance, risk management and management of change by the BP Macondo well team could have raised awareness of the challenges of achieving zonal isolation and led to additional mitigation steps.[354]

Rather than aiding decision making, the Macondo team's cementing decision tree reinforced the flaws in its analytic approach. Proper risk management in a complex engineering project requires a constant awareness of risks and potential risks. The decision tree instead encourages a simplified linear approach in which complex risks (such as the risk of failed cementing) can be forgotten or ignored on the basis of simple and incomplete indicators (such as partial returns or lift pressure).

Most Operators Would Not Have Run a Cement Evaluation Log in This Situation, but BP Should Have Run One Here, in Part Because of Its Chosen Temporary Abandonment Procedures

At least some personnel appear to have believed that the Macondo team was planning to run a cement bond log no matter what. On April 20, a BP completions engineer emailed Morel to ask for cement bond log data. When Morel responded "No CBL," the completions engineer wrote "Can you explain why? I thought y'all were planning to run one."[355]

A cement evaluation log would have provided more direct and reliable information about the cement job than pressure and volume indicators on which BP relied. While most operators would not have run a cement evaluation log until the completion phase, BP should have run one here[356] for at least two reasons. First, BP engineers recognized or should have recognized that this was a "finesse" cement job that presented higher-than-average risks.[357] Full returns would not identify if channeling had occurred; a cement bond log could.[358] Second, BP's temporary abandonment procedures would force the rig crew to rely on this finesse cement job as the sole hydrocarbon barrier in the Macondo wellbore. Alternatively, BP should have sought other means for addressing the risk of unsuccessful cementing.

Halliburton Did Not Adequately Inform BP of Cementing Risks or Suggest Design Alternatives

Halliburton did not provide BP the full benefit of its corporate cementing expertise. Since the blowout, senior Halliburton personnel have repeatedly and forcefully emphasized the complexity and difficulty of the Macondo cement job and the limitations of indicators such as full returns.[359] But Halliburton's personnel did not raise all of these concerns before the blowout, let alone emphasize them with the same force.

It appears that Gagliano mentioned the possibility of cement channeling to individual BP engineers on April 15 and then again later on April 19.[360] But he did not flag the concern in his emails or express serious reservations. Gagliano told Congressional investigators that he "recommended to BP that they use 21 centralizers" but admitted that he "did not think there would be a well control issue."[361]

Gagliano also testified that he would have recommended that BP perform a cement bond log given the reduction in the number of centralizers but did not do so because "we do not recommend running a [cement] bond log"[362] and, anyway, he "was never asked."[363] Although Gagliano was present when BP discussed criteria for the cement bond log, he never told anyone full returns alone could not identify channeling.[364] Moreover, the only risk factor that Halliburton identified during the design process was the relatively low number of centralizers. Halliburton did not discuss any other risk factors or recommend other design changes that might have mitigated those risks. Halliburton personnel were aware that BP's design called for a low cement volume and a low cement flow rate. They also knew of the decision not to circulate bottoms up, the float valve conversion difficulties, and the low post-conversion circulating pressures. [365] But they never raised concerns about these risk factors, let alone offered BP an independent assessment of the overall likelihood of success of the cement job.

The format of Halliburton's modeling reports exacerbated communication difficulties. After the blowout, Halliburton personnel argued that the reports included predictions of channeling and gas flow that BP engineers should have heeded.[366] Halliburton could have highlighted these warnings—along with overall assessments of cementing success—in a simple summary early in

Figure 4.3.22. Page 23 of Halliburton's April 18, 2010 OptiCem™ report.

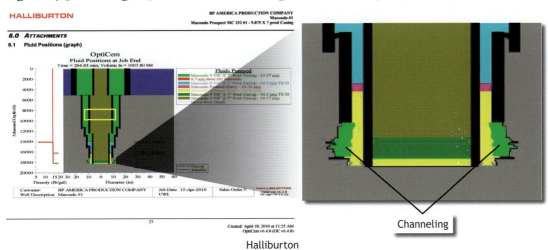

Halliburton

Halliburton personnel explained the green areas as predicted channeling.

the report. Instead, the reports presented information in an obscure and unnecessarily technical manner. (For instance, as shown in Figure 4.3.22, the reports present channeling predictions only as unexplained jagged lines in a well diagram).[367] As a result, BP engineers reviewed the predictions in a cursory fashion, if at all.[368]

Halliburton missed another opportunity to communicate its concerns when it reported the overall success of its cement job. Chaisson expressed complete satisfaction with the cement job in his post-job report but later clarified that "[cementing] was successful on the surface. As far as being successful downhole, actually if it were successful at getting zonal isolation, I cannot be sure of that."[369] Halliburton explains the difference between its pre-blowout reports and its post-blowout skepticism by suggesting that it is BP's responsibility as the operator to evaluate the significance of cementing indicators and BP's responsibility to mitigate risks at the well. Whether that turns out to be true as a legal matter, Halliburton could have helped avoid the blowout if it had highlighted the risks of the cement job and the limitations of the few cementing indicators it had reviewed.⬥

Chapter 4.4 | Foamed Cement Stability

BP and Halliburton chose to cement the final Macondo production casing into place using nitrogen foamed cement. That technology offered several advantages at Macondo, but it also posed a risk: An improperly designed or incorrectly pumped nitrogen foamed cement slurry can be unstable and lead to a failed primary cement job. Data from pre- and post-job laboratory testing lead the Chief Counsel's team to conclude that the foamed cement slurry pumped at Macondo was very likely unstable. The Chief Counsel's team finds that Halliburton failed to review properly the results of its own pre-job tests, and that a proper review would have led Halliburton to redesign the cement slurry system. The Chief Counsel's team also finds that BP inadequately supervised the cement design and testing process.

Foamed Cement

Cementing personnel create **nitrogen foamed cement** by injecting inert nitrogen gas into a base cement slurry. This produces a slurry that contains fine nitrogen bubbles. If the system is properly designed, the bubbles will remain evenly dispersed in the slurry as it cures, and the set cement will retain the bubbles in the same form.

Foamed cement offers two principal technical advantages. First, the nitrogen bubbles in the foamed cement slurry make the overall cement mixture less dense than the base cement slurry. Second, cementing personnel can adjust the density of the foamed cement slurry in response to well conditions by adjusting the rate at which they inject the nitrogen into the base cement slurry. Whereas a base cement slurry typically weighs about 15 pounds per gallon (ppg), foamed cement can weigh as little as 5 ppg.[1] All other things being equal, a low-density column of cement in the annular space around a well casing will exert less hydrostatic pressure on the formation than a high-density column of cement. As a result, using a low-density foamed cement can reduce the risk of formation breakdown. Such a breakdown may result in the loss of cement into the formation, compromising zonal isolation and reducing the productivity of the well over the long term.[2]

Risks of Unstable Foamed Cement

A foamed cement system must exhibit good foam **stability**.[3] A stable nitrogen foamed cement slurry will retain the nitrogen bubbles internally and maintain its design density as the cement cures. The result is hardened set cement that has tiny, evenly dispersed, and unconnected nitrogen bubbles throughout. If the foam does not remain stable as the cement cures, the small nitrogen bubbles may coalesce into larger ones, potentially rendering the hardened cement porous and permeable to fluids and gases, including hydrocarbons.[4] If the instability is particularly severe, the nitrogen can **break out** of the cement, with unpredictable consequences.[5] While technical authorities do not appear to have definitively determined the effects of pumping unstable foamed cement downhole, they uniformly agree that only stable foamed cement designs should be used.[6]

Foamed Cement Testing

When designing a nitrogen foamed cement system, it is critical to test the stability of the foamed slurry.[7] The American Petroleum Institute (API) has published recommended procedures for conducting **foam stability tests**.[8] The technician mixes a volume of base cement slurry with air (not nitrogen) in a sealed blender to generate a foamed slurry of the same density that will be used in the field (see Figure 4.4.1). The laboratory may then conduct foam stability tests using one of two methods.

Figure 4.4.1. Foam testing apparatus.

Five-blade blender.

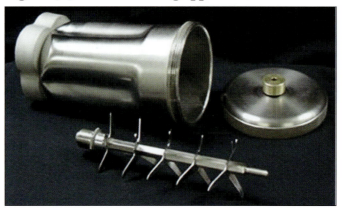

API RP 10b-4

The first method involves pouring a sample of the foamed cement into a graduated cylinder (see Figure 4.4.2). After two hours, the technician visually examines the foamed slurry for signs of instability, such as large coalescing bubbles or cement density variations caused by nitrogen bubble migration or escape.

The second method involves pouring the foamed cement into a plastic cylinder, sealing it, and then allowing it to cure and set (see Figure 4.4.3). The technician then removes the solid cement sample from the cylinder and measures the density of solid cement at the top, middle, and bottom of the sample. If there are density variations from top to bottom, or if the densities are equal to one another but significantly higher than the target density, the foamed cement is deemed unstable.

Figures 4.4.2 and 4.4.3. Foam testing apparatus.

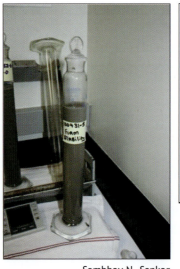

Sambhav N. Sankar

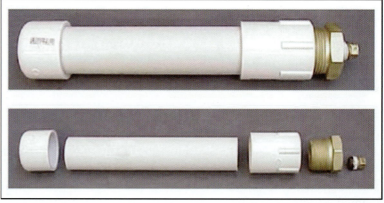

API RP 10b-4

Left: Graduated cylinder for unset foam test.
Above: Curing mold for set cement tests.

The API lists five signs of foamed slurry instability in the laboratory:[9]

- more than a trace of free fluid;
- bubble breakout noted by large bubbles on the top of the sample;
- excessive gap at the top of the specimen;
- visual signs of density segregation as indicated by streaking or light to dark color change from top to bottom; and
- large variations in density from sample top to bottom.

None of these criteria is quantitative. All rely to some degree on the judgment of laboratory personnel or cementing experts.

Foamed Cement at Macondo

Decision to Use Foamed Cement

BP and Halliburton planned from the very beginning to use foamed cement technology for at least some of the cementing work at Macondo. It is common to use foamed cement on the first few casing strings in a deepwater well because shallow formations are often too weak to withstand the hydrostatic and dynamic pumping forces exerted by a heavier, normal-density cement slurry. (The *Marianas* crew and Halliburton cemented at least two of Macondo's early casing strings with foamed cement.)[10]

Operators use foamed cement less frequently for deeper casing strings and in applications for which synthetic oil-based mud is being used as a drilling fluid. While at least one operator—Shell—often uses foamed cement in deepwater Gulf of Mexico production casings, BP appears to have had relatively little experience with using the technology for this purpose.[11]

To cement the final long string production casing at Macondo, Halliburton and BP began planning as early as February 2010 to start with a base slurry having a density of 16.7 ppg and then to add enough nitrogen to reduce the density to 14.5 ppg. It appears that BP drilling engineer Brian Morel first raised the idea of using foamed cement technology for the production casing. He suggested the idea because using foamed cement might provide long-term strength benefits over the life of the well.[12] Halliburton cementing engineer Jesse Gagliano agreed that foamed cement would be useful at Macondo.[13] But an internal BP cementing expert cautioned Morel as early as March 8 that:

> Foaming cement after swapping to [oil-based drilling mud] presents some significant stability challenges for foam, as the base oil in the mud destabilizes most foaming surfactants and will result in N_2 [nitrogen] breakout if contamination occurs. This drives the need for a lot of attention to the spacer programs and often results in non-foamed cap slurries being placed on top of the foamed slurry to mitigate breakout.[14]

The early April lost returns problems appear to have further solidified the decision to use nitrogen foamed cement. According to BP and Halliburton's calculations, using the lighter foamed cement would reduce the risk of fracturing the formation at the well and thereby reduce the risk of losing returns during the cementing process.

Pre-Blowout Cement Testing

When the *Deepwater Horizon* arrived at Macondo to replace the *Marianas*, it had on board a large quantity of cement dry blend that Halliburton had originally designed for use at Kodiak #2, the previous BP well the *Horizon* crew had drilled.[15] Gagliano had designed the primary features of that blend in late 2009.[16]

Dry Blend. The term **dry blend** refers to the combination of dry cement components that are blended together onshore for use on the rig. The Macondo dry blend included Portland cement, two different grades of silica powder, potassium chloride, a proprietary antisettling agent, and a proprietary flow-enhancing additive. The rig cementing team added water, two liquid chemical additives, and a glass fiber material to the dry blend to produce the base slurry.

On February 10, Gagliano instructed technicians in Halliburton's Broussard, Louisiana, laboratory to conduct pilot tests on a cement slurry recipe based on this dry blend. The slurry recipe specified the amount of water and the type and quantity of liquid chemical additives that should be mixed with the dry blend to produce the cement slurry. If the dry blend had been unsuitable—either because of its original design or because it had degraded during storage—then Halliburton could have delivered a new dry blend to the rig for use at Macondo.

Foamed Cement Pilot Testing

Gagliano's February 10 pilot cement design listed the precise amount of liquid retarder, surfactant, and fresh water that the laboratory should add to the dry blend to produce a cement slurry for testing. The "recipe" that Halliburton tested in February was identical to the recipe that it eventually used at Macondo, with one exception: The February recipe included roughly twice the amount of liquid chemical "retarder" that Halliburton eventually used (0.20 gallons per sack (gal/sack) vs. the final 0.09 gal/sack) and correspondingly less water. (Adding retarder extends the setting time of cement.)[17] The laboratory used the dry blend from the *Deepwater Horizon* but used local tap water and stock liquid chemicals rather than water and liquid chemicals from the rig.

The Broussard laboratory conducted several tests in February, including two separate foam stability tests.[18] Both foam stability tests were "set" slurry tests, in which personnel poured foamed cement into a cylinder, allowed it to cure for 48 hours, and then examined the density of the top and bottom of the set cement cylinder.

Laboratory personnel appear to have conducted the first February foam stability test on or about February 13. The top and bottom of this sample weighed 16.8 ppg and 17.6 ppg, respectively. These measurements indicated serious instability because they differ significantly from each other, and they are both higher than the target density of 14.5 ppg. *The test measurements showed either that: (1) The lab personnel were unable to generate a proper foamed slurry; (2) gas bubbles migrated within the foamed slurry; (3) gas escaped from the slurry before it could set; or (4) some combination of these things occurred.*

Laboratory personnel appear to have conducted a second February foam stability test on or about February 17. The top and bottom of this sample weighed 15.9 ppg and 15.9 ppg, respectively.

While these two measurements were identical, the data still indicated serious instability because both measurements were significantly higher than the target density of 14.5 ppg. Again, nitrogen gas must have escaped from the tested slurry before it could cure, or the lab personnel had been unable to generate a proper foamed slurry.

These two February 2010 lab tests should have caused Halliburton technical personnel to conclude that the foamed cement Halliburton was planning to pump at Macondo was likely unstable.

Three other facts about the February tests are worth noting. First, laboratory personnel did not condition the cement before conducting the February 13 foam stability test but conditioned the cement for two hours before conducting the February 17 test. Second, rheology test results showed that the yield point of the base slurry was quite low. This can be an independent warning that the base slurry may be unstable and that a foamed slurry prepared from that base slurry may also be unstable.[19] Third, time-lapse strength testing showed that the pilot cement recipe set extremely slowly, suggesting that the recipe included too much retarder.

Halliburton did not report any of the February pilot testing data to BP until March 8.[20] On that date, Gagliano attached an official data report of the February test results to an email in which he discussed his recommended plan for cementing one of the Macondo casing strings.

The official data report included only the results of the February 17 foam stability test, in which the top and bottom portions of the set cement both weighed 15.9 ppg. (The official laboratory reports list the results in terms of specific gravity (SG) rather than pounds per gallon.) Because the top and bottom weights matched, the test did not demonstrate density segregation, but the test was still a clear failure because both weights were significantly higher than the target density.

For some unexplained reason, Halliburton's official data report to BP *incorrectly* stated that laboratory personnel had not conditioned the cement prior to the February 17 foam stability test.

Apparently, Halliburton did no further testing of the proposed Macondo cement slurry until April 2010, as the final production casing planning was under way.

April 13 Pre-Job Testing

On April 1, Morel sent an email to Gagliano, BP senior drilling engineer Mark Hafle, BP operations engineer Brett Cocales, and Quang Nguyen of Halliburton requesting that Halliburton begin testing cement for the final production casing cement job. Morel wrote, "This is an important job and we need to have the data well in advance to make the correct decisions on this job."[21] Gagliano responded on the same day with an email stating that he had already run the February pilot tests, and that he would run further tests "[o]nce I get samples from the rig sent into the lab" and once he had the latest data on the downhole temperatures at the well.[22] Gagliano attached the same official laboratory report that he had sent on March 8.

Gagliano appears to have first ordered additional testing on April 12.[23] This time, the laboratory tested samples of dry blend, additives, and water from the rig, and used a design recipe that was nearly identical to the one that Halliburton eventually pumped. (The tested recipe contained slightly less retarder than the pumped recipe—0.08 gal/sack instead of 0.09 gal/sack.) According to Gagliano, the main goal of this test was to determine how much retarder the recipe should use.[24]

It appears that the laboratory performed a foam stability test on this recipe on or about April 13 and conditioned the cement slurry for 1.5 hours at 180 degrees before conducting the test.[25] They finished the test on or about April 15. After curing, the top and bottom of the set cement sample weighed 15.7 ppg and 15.1 ppg, respectively.

This April 13 test result, just a week before the blowout, indicated serious instability.[26]

On April 17, Gagliano sent an email to Morel, Cocales, and BP drilling engineer team leader Gregg Walz and attached two official laboratory reports.[27] The data reports included results from various tests on cement slurry recipes with two slightly different retarder concentrations: 0.08 gal/sack and 0.09 gal/sack. BP and Halliburton had discussed increasing the retarder concentration in order to compensate for the fact that they planned to pump the cement at a low rate. The slow pumping rate would translate to increased cement travel time, which would in turn raise the risk of premature cement thickening.

Neither data report included the results of the April 13 foam stability test (or any other foam stability test). *Gagliano did not otherwise alert BP to the foam stability test results.* Gagliano's cover email discussed the data from recently completed thickening time tests, presumably because this measured the cement characteristic that would vary depending on retarder concentration. Gagliano also stated that he had not yet obtained compressive strength results for the final cement recipe that BP planned to use—which included slightly more retarder.

Morel complained to Hafle that Gagliano had started the compressive strength tests later than he should have. Morel asked Hafle if Morel would be "out of line" by sending the following message to BP wells team leader John Guide and Walz:

> I need help next week dealing with Jesse. I asked for these lab tests to be completed multiple times early last week and Jesse still waited until the last minute as he has done throughout this well. This doesn't give us enough time to tweak the slurry to meet our needs.... As a team we requested that [Gagliano] run another test with 9 gals on Wednesday, I know the first [compressive strength] test had issues, but I do not understand what took so long to get it underway and why a new one wasn't put on right away. There is no excuse for this as the cement and chemicals we are running has been on location for weeks.[28]

Hafle agreed that Morel's concerns were reasonable and that BP should ask Halliburton to replace Gagliano soon (a request that BP appears to have made earlier as well).[29] Morel and Hafle conveyed their concerns to Walz, Cocales, and Guide, and on April 18, Walz responded that he and Guide would be meeting soon with Halliburton.[30]

Meanwhile, on April 17, Morel responded directly to Gagliano's email. Morel wrote:

> I would prefer the extra pump time with the added risk of having issues with the nitrogen. What are your thoughts? There isn't a compressive strength development yet, so it's hard to ensure we will get what we need until it[']s done.[31]

Morel thus told Gagliano that he would prefer to alter the cement slurry recipe to include more retarder to increase the thickening time (or "pump time") of the cement. In the same email, he appears to have recognized that adding more retarder would potentially increase the risk of nitrogen foam instability.

Laboratory personnel appear to have conducted a second April foam stability test on or about April 18.[32] They used the same amount of retarder (0.08 gal/sack) but conditioned the cement at 180 degrees for three hours—the longest period yet. The top and bottom of the set cement sample weighed 15.0 ppg and 15.0 ppg, respectively.

While these numbers are the same as each other, they are both 0.5 ppg higher than the target of 14.5 ppg. This means one of two things. First, laboratory personnel may have generated a foamed cement slurry that *initially* weighed 15.0 ppg and retained that density throughout the test. If this was the case, however, the laboratory documents should at least have noted the difficulty; API standards state that if laboratory procedures generate a foamed slurry density that is above the design density, "it will be difficult to obtain the proper foamed cement density in the field, and the slurry should be redesigned."[33]

Second, laboratory personnel may have generated a foamed slurry of 14.5 ppg, but some nitrogen gas may have escaped from the slurry as it set, making the slurry more dense. Because the change from 14.5 to 15.0 ppg is not indisputably "large" within the meaning of API testing criteria, this might suggest that the foamed cement was stable. Halliburton appears to contend that this is what happened and argues that the April 18 test shows that its cement slurry was stable.

Internal documents provided by Halliburton do not clarify which of these two things happened.

Availability of April 18 Test Results

The documents also do not establish conclusively *when* Halliburton completed its April 18 foam stability testing. Handwritten notes in the documents suggest that laboratory personnel began the test at 2:15 a.m. on April 18,[34] and Halliburton has confirmed this time in correspondence to the Chief Counsel.[35] Halliburton at one point stated publicly that the test took 48 hours to complete.[36] If that were true, the test results would not have been complete until at least 2:15 a.m. on April 20, which would have been after the time Transocean's rig crew and Halliburton's cementing personnel *finished* pumping the primary cement job at 12:35 a.m. on April 20.[37]

Six months after the blowout, and after the Chief Counsel's team publicly questioned the stability of the Macondo cement design and the timing of lab testing, Halliburton still had not determined whether its personnel had completed the April 18 foam stability test before pumping the Macondo job.[38] Finally, eight months after the blowout, Halliburton informed the Chief Counsel that it had "learned more about the specific facts surrounding the cement lab testing," including that "the second April foam stability test was finished before the final cement job started."[39] In the words of its counsel:

> Halliburton can now demonstrate that an email notification was sent to Jesse Gagliano on April 19, 2010, at approximately 4:14 pm indicating that all tests associated with the final cement job were then "finished in lab," more than three hours before the cement job commenced. Attached to this letter is a copy of a spreadsheet containing the "web log" data referenced above and explained further in Halliburton's January 7th letter to you. This constitutes objective evidence...that the foam stability test was run in less than 48 hours and that the test was completed prior to the final cement job.[40]

Halliburton contended that the "finished" notification "would not have been generated had the foam stability test failed or been incomplete."[41]

The Chief Counsel's team cannot accept or reject Halliburton's contentions based on these statements by its counsel. While Halliburton did provide a one-page spreadsheet that it views as "objective evidence" of the timing of its test, the Chief Counsel's team cannot decipher the document (displayed as Figure 4.4.4) without the aid of Halliburton personnel.

Halliburton flatly refused to produce any witness who could explain this document (or any of the other timing and testing issues discussed above) in a transcribed interview.

Figure 4.4.4. Halliburton evidence of test times.

IP Address	UserID	Time Stamp	Web Server Log Information
		18.04.2010 09:52:03	From: noreply@halliburton.com; To: jesse.gagliano@halliburton.com; Subject of Email: Daily Summary Report
34.34.133.22	HBAM242	2010-04-18 20:44:50	GET /pls/viking/labdb_report.stepTwo?p_trid=73909 HTTP/1.1
34.34.133.22	HBAM242	2010-04-18 20:44:58	GET /pls/viking/labdb_report.stepThree?p_trid=73909&p_tsid=151852 HTTP/1.1
34.34.133.23	-	2010-04-19 02:49:26	GET /osso_login_success?urlc=v1.4~5E653963CE68A4ED43251CE43A205D9B7086D803CCD13504E93F7EDF2732CFF6B5E3D3C3D127BCED52C2S007DEE121C4867F0A323F67857B056890280CDFC276BB16694F32B2E517C34927EESF3421FAE4DDD7B19F9A62217A75D98E86E0458774B5442DEACD45A94060FC6F84ED7CA753A3110216F883F68CCC6E1DC36A7D7EE6EA99B9941A2F46D0A007F83612A1C80AF89CF3E1CD36079B34349877C6EF0801CA307D846FD6B91B87BD9D93CCD005E52554B7B66A7D354B877ED5A7C45D8242D7584E66100764AE13369CB2D5E9B0324C30262DDA8A5ED681B9F2771A64FE5C0F1DD6ADC47C0C73E9508698E6A779C20B90D5626F110209015A07D32D95E542D497194FD79B82645EA26D68BC1B73F1E4528ACE1558AC30B311F1619B3CA6025ACF3EAD1D0B10747619CAF240468D01B991ECB5CB14C3E93EE3C484795C6DA60084EF081A661C18DF8C195B21FEA9DCD749C07D3CFD43C29A9165898D409359D739095C534F3CAB83DBBF84D HTTP/1.1
		19.04.2010 09:52:08	From: noreply@halliburton.com; To: jesse.gagliano@halliburton.com; Subject of Email: Daily Summary Report
		19.04.2010 16:04:13	From: noreply@halliburton.com; To: jesse.gagliano@halliburton.com; Subject of Email: Test Status Changed (US-73909/2)
34.34.133.23	HX11269	2010-04-19 16:13:27	GET /pls/viking/labdb_test.testresults?p_request_id=73909&p_slurry_id=150924&p_test_id=43&p_request_test_id=806072 HTTP/1.1
34.34.133.23	HX11269	2010-04-19 16:13:28	GET /pls/viking/labdb_test.testresults?p_request_id=73909&p_slurry_id=150924&p_test_id=43&p_request_test_id=813603 HTTP/1.1
34.34.133.23	HX11269	2010-04-19 16:14:25	GET /pls/viking/labdb_test.testresults?p_request_id=73909&p_slurry_id=150924&p_test_id=43&p_request_test_id=813603&p_message=Results%20successfully%20updated HTTP/1.1
34.34.133.23	HX11269	2010-04-19 16:14:32	GET /pls/viking/labdb_test.testresults?p_request_id=73909&p_slurry_id=150924&p_test_id=43&p_request_test_id=806072 HTTP/1.1
34.34.133.23	HX11269	2010-04-19 16:14:33	GET /pls/viking/labdb_test.testresults?p_request_id=73909&p_slurry_id=150924&p_test_id=43&p_request_test_id=813603 HTTP/1.1
		19.04.2010 16:14:43	From: noreply@halliburton.com; To: jesse.gagliano@halliburton.com; Subject of Email: Request 73909, Status: Finished in Lab
34.34.133.23	HX11269	2010-04-19 16:14:46	GET /pls/viking/labdb_test.testresults?p_request_id=73909&p_slurry_id=150924&p_test_id=43&p_request_test_id=813603 HTTP/1.1
34.34.133.23	HX11269	2010-04-19 16:14:47	GET /pls/viking/labdb_test.testresults?p_request_id=73909&p_slurry_id=150924&p_test_id=43&p_request_test_id=806072 HTTP/1.1
34.34.133.22	HX46076	2010-04-20 08:36:37	GET /pls/viking/labdb_test.testresults?p_request_id=73909&p_slurry_id=150924&p_test_id=43&p_request_test_id=806072 HTTP/1.1
34.34.133.23	HX46076	2010-04-20 08:36:37	GET /pls/viking/labdb_test.testresults?p_request_id=73909&p_slurry_id=150924&p_test_id=43&p_request_test_id=813603 HTTP/1.1
		20.04.2010 09:52:11	From: noreply@halliburton.com; To: jesse.gagliano@halliburton.com; Subject of Email: Daily Summary Report
34.34.133.22	-	2010-04-20 11:31:10	GET /osso_login_success?urlc=v1.4~78AECBB7DF3421DD9F6E1C845ED7C386DE726BE1B2FEFD0F1D9669DB7EED49633BB73DB640D6B484F56E44FA456B225B05EDFA2F96C9E5746825AB490FE2BC191E21939751490E4610FC302D5388AB16E487526D7CEBDFDCD3D36256E14B7BD9941406DB3169C961B56D01AAEEC3A0B77054D1B9CD7739644856C67D5FF4C6CD6A9C4E50A10E61076E13624C863709003F5A1CFA9C4F6E6B7ADA26F7F2137EB8639F3710D5E4B13A60D60B4F55E5472F673D6D0516F206C34815F3337CC9F1F482317526A47ED03BC5CD212B71B24A511513D26C63F2A697CD02682641532D96BFE33AAD34BA87A71395D802D528FF058447F4BDEC0C66E9BCA2CA64893EE4CF3E2FE2FC7B82B9A49CA1FF6B0A2C354851DF5729CB6A0E31CE58E2837390911D8F9EE1E1645D331D7403CD66B04F38A6ED5B251AE849E0C5CE19E0FA6F7DCBBC9B429F477A6BC7557574C8D20503333D5E9759002670D43DAB9B479C6BC6DA91D3F03065B3B5119 HTTP/1.1

Halliburton

Significant problems remain even if the Chief Counsel's team accepts Halliburton's assertions about when the April 18 test had been completed. While Halliburton argues that its computer system generated a notice that the April 18 test results were available before its personnel pumped the cement job, it has carefully avoided saying that any of its engineers actually *knew* that the results were available, let alone *reviewed* them, before pumping the job. Indeed, BP documents show that Halliburton first reported the April 18 result to BP on April 26, six days after the blowout.[42] And while Halliburton contends that the "finished" notification meant that the April 18 foam stability test did not fail by its standards, it refuses to identify those standards, let alone the person who actually applied them.

Halliburton presumably would not deny this information to the Chief Counsel if it were favorable to the company.

Post-Blowout Cement Testing

Testing by BP

BP's internal investigation raised several questions about Halliburton's cement slurry design and pre-job testing procedures.[43] BP asserted that the final April 18 foam stability test "indicated foam instability based on the foamed cement weight of 15 ppg."[44]

BP also commissioned third-party testing by CSI Laboratories, an independent cement consulting company.[45] CSI could not conduct these tests on the actual materials that had been used at the Macondo well because those materials sank into the ocean with the rig.

CSI also could not conduct these tests using the precise off-the-shelf ingredients specified by the cement slurry recipe because Halliburton refused to provide its proprietary additives to CSI. CSI therefore developed a model slurry to mimic the characteristics of the slurry used at Macondo. CSI prepared the model slurry by mixing commercially available cement and additives according to the final Macondo cement recipe. To replace proprietary Halliburton additives, CSI used third-party chemicals that served similar purposes (for example, using a commercially available third-party retarder instead of Halliburton's proprietary SCR-1000 retarder). Despite these differences, BP's investigation team asserted that the model slurry was "sufficiently similar to support certain conclusions concerning the slurries actually used in the Macondo well."[46]

CSI reported that foamed cement generated from the model slurry was unstable under several test conditions. Based in large part on this analysis, BP's investigation team concluded in its report that "the nitrified foamed cement slurry used in the Macondo well probably experienced nitrogen breakout, nitrogen migration and incorrect cement density."[47]

Testing by Chevron and Chief Counsel's Team

The Chief Counsel's team conducted its own independent tests of cement slurry stability on behalf of the Commission.

The Chief Counsel's team worked with an independent expert and cement experts from Chevron to conduct these tests.[48] Halliburton recognized that Chevron's laboratory personnel were highly qualified for this work; Chevron maintains a state-of-the-art cement testing facility in Houston, Texas, and employs a staff of cement experts to supervise cement design and testing for its oil wells. Halliburton also agreed to supply the Chief Counsel's team with off-the-shelf cement and additive materials of the same kind used at the Macondo well. Although these materials did not come from the specific batches used at the Macondo well, they are in all other ways identical in composition to the slurry pumped there.

Halliburton refused to provide the Chief Counsel's team with full details of the methods and protocols that its laboratory used to conduct its February and April cement tests. *Most notably, Halliburton refused to provide any information on whether and how its staff had conditioned the cement before conducting the foam stability tests.* (At the time Chevron conducted its tests, Halliburton had not yet produced any internal laboratory documents to the Commission staff. Halliburton later provided some internal documents that disclosed conditioning times.) When the Chief Counsel's team sought input from BP and other parties regarding these and other issues, Halliburton demanded that the team refrain from doing so.[49] The Chief Counsel's team agreed to honor Halliburton's request by working solely with Chevron experts and an independent expert to develop protocols for testing Halliburton's cement materials.

Chevron conducted numerous tests on the Commission's behalf. Chevron's laboratory report states that many of its results "were in reasonable agreement" with results reported by Halliburton. However, Chevron's staff did not obtain foam stability test results that agreed with Halliburton's. Instead, Chevron's report stated that its staff was "unable to generate stable foam with any of the tests" that they conducted to examine foam stability.[50] Chevron's testing strongly suggests that the foamed cement slurry actually used at Macondo was unstable. Appendix D is Chevron's letter to the Chief Counsel's team that accompanied its report.

Technical Findings

The Foamed Cement Slurry Used at Macondo Was Very Likely Unstable

Of all the tests done so far to evaluate the stability of the Macondo foamed cement slurry design, only one (the April 18 Halliburton pre-job test) even arguably suggests that the design would be stable.

Even the April 18 test result predicts only borderline stability. Industry experts believe that the three-hour high-temperature conditioning regimen for this test biased it in favor of success. Several have stated that cement laboratories should not condition a slurry sample *at all* before running foam stability tests, let alone at such elevated temperatures.[51] They reason that during field cementing operations, crews do not usually mix or circulate the base slurry before foaming it with nitrogen. Halliburton explained that its laboratory personnel derived the conditioning time from pumping time,[52] and then contended in writing that there is "sound operational basis" for conditioning cement in a laboratory prior to foam stability testing.[53] But when the Chief Counsel's team asked Halliburton to provide "[a]ny scientific study or other document" supporting the latter statement,[54] Halliburton cited only one thing: API Recommended Practice 10b-2.[55] Section 15 of that document states, "The cement slurry is conditioned to simulate dynamic placement in a wellbore." But this document discusses methods for testing the static stability of *unfoamed* cement slurries. By contrast, API's practice recommendations for testing *foamed* cement do not mention pre-test conditioning at all.

Halliburton also declined to provide any information that would help the Chief Counsel's team determine whether lab personnel had difficulty generating a proper density foamed slurry sample on April 18, which might account for the 15.0 ppg density of that sample.

Indeed, Halliburton repeatedly flatly refused Chief Counsel's personal requests for documents or recorded testimony regarding many otherwise unsupported assertions from Halliburton's lawyers. For example, Halliburton's lawyers have consistently asserted that the April 18 foam stability test produced passing results. Commission staff requested "any document specifying or prescribing the conditioning time...test duration, or success criteria" for this and other tests, and requested the opportunity to conduct and transcribe interviews with Gagliano, his supervisors, and any "individual or individuals competent to testify regarding standard Halliburton laboratory practices."[56] Halliburton produced no documents and provided no witnesses. It noted that it had allowed the Chief Counsel to interview Gagliano and a Halliburton cement expert early in the investigation—*before* the Chief Counsel had learned of the failed February and April tests and *before* the Chief Counsel's testing had identified concerns with the Macondo cement slurry recipe. Halliburton then stated:

> [H]alliburton is compelled to view these requested "interviews" as being more in the nature of adversarial depositions designed to defend the [Chief Counsel's] preliminary conclusions as opposed to furthering an objective evaluation of what occurred. Given Staff's apparent shift in purpose, Halliburton respectfully declines to make such witnesses available.

In contrast to the April 18 test, 12 other stability tests—three by Halliburton and nine by Chevron—clearly predict that the foamed cement slurry design would be unstable. One can debate the significance of these tests individually. For instance, the February Halliburton tests predicted severe instability but were performed with a recipe containing more retarder, which can potentially reduce slurry viscosity and make it more unstable.[57] And one can also debate how well laboratory testing approximates field conditions.[58] However, the sheer number of failed foam stability tests combined with other indicia of instability (discussed below) lead the Chief Counsel's team to conclude that the foamed cement slurry used at Macondo was very likely unstable.

The Commission-sponsored tests further suggest that the Halliburton base slurry was unstable even *before* being foamed with nitrogen. Chevron's lab report notes that its personnel observed base slurry "settling" in six of the nine tests it performed. The base slurry also consistently showed a very low yield point, which can be a warning that the slurry will be unstable before and after foaming. Base slurry instability also could have severely compromised the bottomhole cement job at Macondo.

The Chief Counsel's team notes that Halliburton's Broussard laboratory did retain a small sample (1.5 gallons) of dry blend material from the *Deepwater Horizon*. This material was left over from Halliburton's April pre-job testing process. At the time of this writing, the federal government had taken custody of the material and was holding it pending laboratory testing. Industry experts have informed the Chief Counsel's team, however, that the dry blend material has probably chemically degraded by now to the point where any laboratory testing results would be inconclusive. If this is the case, Halliburton's four pre-blowout tests and the Commission's nine post-blowout tests are the most probative information regarding the performance of the Macondo cement slurry.

Halliburton May Not Have Reviewed the April 18 Test Results Before Beginning the Cement Job

Currently available data lead the Chief Counsel's team to conclude that Halliburton did not fully review its April 18 foam stability tests before pumping the Macondo cement job. While Halliburton states that its personnel completed the test at approximately 4:14 p.m. on April 19, it has provided neither documentary nor testimonial evidence to show that its personnel actually reviewed that data before pumping the job or communicated it to anyone at BP.

Once again, the Chief Counsel repeatedly offered Halliburton opportunities to produce witnesses with relevant knowledge to be examined by the Chief Counsel. Halliburton consistently refused to support its lawyers' assertions with sworn testimony or additional documentation.

Even if Halliburton did review final test results before pumping the cement job, it did not transmit those results to BP until April 26—six days after the blowout.[59] On that date, Jesse Gagliano sent BP an official laboratory data report containing the results of the second April foam stability test. Halliburton never sent BP the results of the April 13 foam stability test.

Halliburton Should Have Redesigned the Slurry Before Pumping It

Halliburton personnel should have redesigned the Macondo slurry before pumping it. Richard Vargo, a Halliburton cementing expert who testified at the Commission's hearings on November

8, 2010, appears to agree. He testified: "I don't think at this point I would choose to run this slurry."[60]

Table 4.4.1 summarizes Halliburton's internal laboratory data concerning the stability of the Macondo cement slurry.

Table 4.4.1

Test ID	Apparent Date	Target Density in ppg	Top Density in ppg	Bottom Density in ppg	Retarder Concentration in gal/sack	Conditioning Time in Hours	Stable?	Available Before Job?	Sent to BP Before Job?
65112/1	Feb. 13	14.5	16.8	17.6	0.20	0:00	Unstable	Yes	No
65112/3	Feb. 17	14.5	15.9	15.9	0.20	2:00*	Unstable	Yes	Yes
73909/1	Apr. 13	14.5	15.7	15.1	0.08	1:30	Unstable	Yes	No
73909/1	Apr. 18	14.5	15.0	15.0	0.09	3:00	Arguable	Uncertain	No

* Reported to BP as 0:00

Halliburton personnel should have redesigned the cement slurry design after receiving the February pilot test results. Both of the February foam stability tests clearly indicated that the pilot cement design was severely unstable.

Halliburton has repeatedly argued that these pilot tests do not reliably predict the stability of the cement system used during the Macondo cement job. Specifically, Halliburton notes that the final cement design was different and that the final well conditions differed from BP and Halliburton's assumptions in February.[61]

These facts are irrelevant to the question of whether Halliburton should have redesigned its slurry. The pilot test results showed that Halliburton's then-current design would be unstable under BP's then-available predictions of well conditions.[62] This should have led Halliburton to inform BP of the problem and to redesign the slurry as necessary. Instead, the Chief Counsel's team has found nothing to suggest that Halliburton personnel seriously considered the issue.

Halliburton missed another clear warning in April. The April 13 foam stability test data should again have prompted Halliburton to inform BP of stability problems and to redesign the slurry immediately. Halliburton personnel have since testified that they would not use a slurry that generated such test results.[63]

Halliburton contends that its laboratory personnel conducted the April 13 test improperly and that the results are therefore "irrelevant."[64] Halliburton cites a laboratory document to support this conclusion, but the Chief Counsel's team and an independent cementing expert were unable to confirm the conclusion merely by reviewing that document. The Chief Counsel asked Halliburton to provide witness testimony to support this assertion, but Halliburton declined.

Even if Halliburton personnel did conduct the April 13 test improperly, this is again irrelevant to the question of whether Halliburton should have redesigned the slurry. As of April 15, the *only* data Halliburton had in hand predicted that the Macondo slurry design would be unstable, and Halliburton had very little time before it would have to pump the cement job. Under the circumstances, Halliburton should have immediately redesigned the slurry and immediately retested the new design. It appears that some Halliburton personnel recognized the problem and responded by rerunning the test two days later with additional conditioning time, perhaps hoping for a more favorable result. But that response was wholly inadequate given how soon the job was to be pumped and the fact that the April 13 test results were consistent with the two earlier February test results. On April 15 or shortly thereafter, Halliburton should have immediately alerted BP to the stability problem and immediately begun redesigning the Macondo slurry.

The Chief Counsel's team is not certain why Halliburton chose not to redesign its slurry. There are at least two possible explanations. One is that the Halliburton personnel who were responsible for approving or recommending the design were unaware of the foam stability test results or their importance. The other is that those personnel *were* aware of the results but did not consider them sufficiently problematic.[65]

Management Findings

Halliburton Mismanaged Its Cement Design and Slurry Testing Process

The number and magnitude of errors that Halliburton personnel made while developing the Macondo foamed cement slurry point to clear management problems at that company. In addition to the errors described above, the Chief Counsel's team believes that Halliburton personnel:

- began pumping the Macondo job without carefully reviewing laboratory foam stability data and without solid evidence that the foamed cement design would be stable;

- reported foam stability data to BP selectively, choosing in February not to report the more unfavorable February 13 test, and choosing in April not to report the more unfavorable April 15 test result (although Halliburton contends these results were erroneous);

- selected the pre-test conditioning time informally, choosing different conditioning times (ranging from no time to three hours) in each of the four foam stability tests without any stated explanation;

- assumed, without apparent scientific basis, that conditioning the base slurry before foaming was scientifically equivalent to foaming the cement then pumping it down the well; and

- recommended a cement design without conducting any formal internal review of that design. Notably, the only design element that Halliburton manipulated between February and April was retarder concentration, even though BP's well design changed significantly during that period and even though bottomhole well conditions were unknown in February. Halliburton has provided no evidence that a supervisor or senior technical expert ever reviewed the final cement slurry design.

To date, Halliburton has not provided any documents or testimony to suggest that established company rules or guidelines prohibited its personnel from doing any of these things. And if such guidelines did exist, it appears that Halliburton failed to enforce them on the Macondo job.

Halliburton's Lab Report Format Complicated Data Evaluation

Halliburton's lab reports to BP were highly technical. As with its modeling runs, discussed in Chapter 4.3, Halliburton did not provide a summary of results, an overall assessment of slurry design, or even reference values for any of the laboratory data it provided to BP. Halliburton could have improved the value of the reports by, for instance, inserting its criteria for a successful foam stability test alongside the reported foam stability data. This would not only have helped BP personnel understand the significance of relatively obscure numerical data, but might also have helped Halliburton personnel do so as well.

BP Did Not Adequately Supervise Halliburton's Work

BP technical guidance documents for cementing emphasize the importance of timely cement testing,[66] and BP Macondo team members themselves recognized that timely cement testing was important.[67] The team also expressed internal concern well before the blowout that Jesse Gagliano was not providing "quality work"[68] and was not "cutting it"[69] by waiting too long to start important tests. They had already asked Halliburton to reassign Gagliano, and Halliburton had apparently agreed to do so.[70] But while BP engineers discussed "how to handle Jesse's interim performance" by email on the very day of the blowout,[71] they did not double-check his work or supervise him more closely pending his replacement.

In particular, although BP personnel recognized the "significant stability challenges" of using foamed cement for the Macondo production casing,[72] and that changes to the retarder concentration in the cement design might increase the risks of foam instability,[73] BP does not appear to have insisted that Halliburton complete its foam stability tests—let alone report the results to BP for review—before ordering primary cementing to begin.[74] When asked why, a BP representative said, "I think we didn't appreciate the importance of the foam stability tests."[75]

BP also did not adequately supervise the slurry design process or review earlier test results.[76] BP documents show that its engineers questioned Gagliano's slurry recipes in other instances.[77] But the Chief Counsel's team found nothing to suggest that BP questioned the Macondo slurry recipe, even after the slurry failed to perform properly during the cement job for the 16-inch casing string. (A BP engineer explained that Halliburton dismissed the failure as the result of cement contamination and noted that this is a typical response for any cementing contractor.)[78] While the Macondo team consulted its in-house cementing expert on other issues, they did not ask him to review the foamed slurry recipe.[79] The expert raised several concerns as soon as he reviewed the recipe after the incident—among other things, he expressed surprise that the slurry design did not include a fluid loss additive and did include a defoamer additive.[80]

BP's failures are especially troubling because it had previously identified several relevant areas for concern during a 2007 audit of Halliburton's capabilities. In that year, BP hired Cemtech Consulting to review a Halliburton foamed cement job on the Na Kika project in the Gulf of Mexico.[81] Cemtech's report identified several issues that mirror problems at Macondo. For instance, Cemtech observed that Halliburton's initial foamed slurry design at Na Kika "had tendencies to stratify" (that is, was unstable) and required redesign. Cemtech also made broader observations such as:

- "The HES [Halliburton] Fluids Center chemists and senior lab technicians do a very good job of testing cement slurries, but they do not have a lot of experience evaluating data or assisting the engineer on ways to improve the cementing program."
- "COMMUNICATION and DATA TRANSFER/DOCUMENTATION could be improved to help avoid unnecessary delays or errors in the slurry design testing, data reporting, and evaluation of the cement program."
- "Lab reports could be improved! They are difficult to evaluate; often incomplete; and are submitted WITHOUT supporting lab charts and DATA to validate the test results. LAB DATA SHOULD BE MANDATORY!"

It does not appear that BP pressed Halliburton or its own Gulf of Mexico engineering teams to improve in these areas. ♦

Chapter 4.5 | Temporary Abandonment

BP developed a temporary abandonment procedure for Macondo that unnecessarily introduced significant risks into the operation. BP disagrees with this finding and argues instead that the specific procedure it used at Macondo was necessary under the circumstances.1 The Chief Counsel's team disagrees. BP could have avoided the additional risks created by the procedure by making a few simple changes.

Temporary Abandonment

Temporary abandonment refers to the procedures that a rig crew uses to secure a well so that a rig can safely remove its blowout preventer (BOP) and riser from the well and leave the well site. BP planned to have the *Deepwater Horizon* temporarily abandon the Macondo well after the rig finished its drilling operations so that another rig could later move to the Macondo site and complete the well construction process. (That rig would perforate the casing and install equipment to collect hydrocarbons.)

Many operators divide operations in this way to save costs; deepwater drilling work requires a large and expensive rig like the *Horizon*, but completion work can be done by a smaller and less expensive rig.

There does not appear to be any standard industry procedure for temporary abandonment. Instead, different operators perform the process differently based on their internal technical guidance, the design preferences of individual engineers, the capabilities of individual rigs, and the needs of particular wells.

At the time of the Macondo incident, MMS regulations did impose some important requirements on operators that wished to temporarily abandon a well. The regulations specified that the operator must set "a retrievable or a permanent-type bridge plug or a cement plug at least 100 feet long in the inner-most casing" and that the top of the plug must be "no more than 1,000 feet below the mud line"[2] (as discussed in Chapter 6). Operators typically refer to this plug as a **surface plug** to distinguish it from other plugs that may be set deeper in the well. Despite the name, surface plugs are not set at the surface or even at the very top of the well.

Figure 4.5.1. Planned configuration after temporary abandonment.

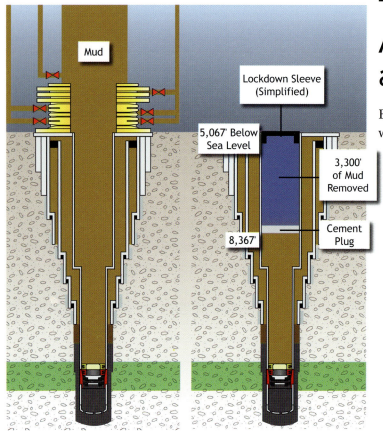

Mud

Lockdown Sleeve (Simplified)

5,067' Below Sea Level

3,300' of Mud Removed

8,367'

Cement Plug

TrialGraphix

After finishing cementing the production casing (left), the rig crew began temporary abandonment procedures that would have allowed the *Deepwater Horizon* to remove its riser and BOP from the well and move on to another job (right). The blowout occurred before the rig crew set the cement plug and lockdown sleeve.

Temporary Abandonment at Macondo

BP's temporary abandonment procedure for the Macondo well had the following basic sequence:

- run the drill pipe into the well to 8,367 feet below sea level (3,300 feet below the mudline);

- displace 3,300 feet of mud in the well with seawater, lifting the mud above the BOP and into the riser;

- perform a negative pressure test to assess the integrity of the well (including the bottomhole cement) and ensure that outside fluids (such as hydrocarbons) are not leaking into the well;

- displace the mud in the riser with seawater;

- set the surface cement plug at 8,367 feet below sea level; and

- set the lockdown sleeve (LDS) in the wellhead to lock the production casing in place.

This procedure is notable in at least two respects. First, it called for rig personnel to set a surface plug deep in the well, 3,000 feet below the mudline. (BP requested and obtained authorization to depart from MMS regulations in order to do this.) Second, the procedure called for rig personnel to displace the wellbore and riser to seawater before setting the surface plug.

After the incident, the BP Macondo team uniformly explained that it developed its particular temporary abandonment procedure in order to set a lockdown sleeve during temporary abandonment and to do so as the last step in the process.[3] The lockdown sleeve decision triggered a cascade of derivative decisions regarding the temporary abandonment procedure that are summarized here and described in greater detail below.

- BP engineers decided to set the lockdown sleeve during temporary abandonment because the *Deepwater Horizon* could do that job more quickly and efficiently than a completion rig.

- Having decided to set the lockdown sleeve during temporary abandonment, BP engineers wanted to ensure that other temporary abandonment operations would not damage the sleeve. To address this concern, they decided to set the sleeve last.[4]

Figure 4.5.2. Lockdown sleeve.

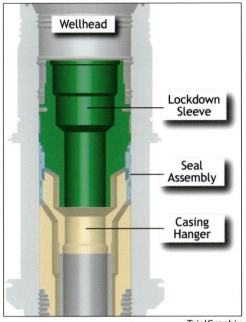

TrialGraphix

BP's desire to set a lockdown sleeve during temporary abandonment drove the development of its temporary abandonment procedure. The lockdown sleeve locks down the casing hanger and seal assembly.

- Deciding to set the sleeve last then drove BP's decision to set its "surface" cement plug unusually deep in the well. The process of setting the Macondo lockdown sleeve would require the rig crew to press (or pull) down on the sleeve with 100,000 pounds of force. The Macondo team chose to generate that force by hanging close to 3,000 feet of drill pipe below the lockdown sleeve.[5] In order to leave room for that length of drill pipe, BP needed to set the surface cement plug even farther down, from 3,000 to 3,300 feet below the mudline.[6]

- Deciding to set the cement plug deep in the well in turn led BP engineers to decide to remove a great deal of drilling mud from the well during temporary abandonment. The Macondo team believed that cement plugs set up better in seawater than in mud.[7] To set the deep cement plug in seawater, the team instructed the rig crew to replace 3,300 feet of mud in the well with seawater before setting the plug.[8]

Lockdown Sleeve. BP planned to set a **lockdown sleeve** during its temporary abandonment procedure at Macondo. A lockdown sleeve is a piece of equipment that is installed in the wellhead to guard against uplift forces that may be generated during the production of hydrocarbons at a well. The sleeve locks the production casing hanger and seal assembly to the high-pressure wellhead housing so that the forces generated during hydrocarbon production do not lift the casing hanger and seal assembly out of place.

Operators do not normally set lockdown sleeves during temporary abandonment.[9] They normally set lockdown sleeves later in the life of a well.[10] BP decided to set the lockdown sleeve during temporary abandonment because it believed that a drilling rig, such as the *Marianas* or *Deepwater Horizon*, could do this job more quickly and at a lower cost than a completion rig.

This series of design decisions ultimately led BP to instruct the *Deepwater Horizon* crew to replace 8,367 feet of drilling mud from the riser and well with lighter seawater *before* setting *any* additional mechanical barriers in the well, such as the surface cement plug.

Decision to Set Lockdown Sleeve During Temporary Abandonment

Lockdown sleeves need not be set during temporary abandonment. Indeed, the Macondo team originally planned to leave the job for a completion rig.[11]

Figure 4.5.3. BP subsea wells organization.

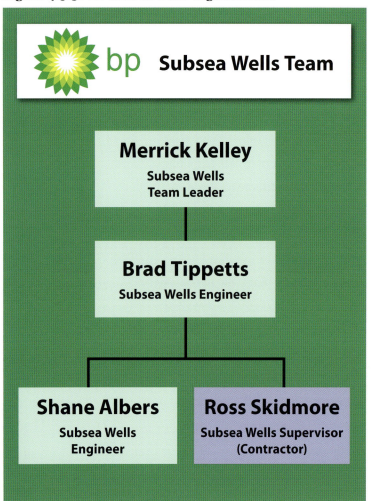

TrialGraphix

BP decided to set the lockdown sleeve during temporary abandonment because it believed that a drilling rig could do this job more quickly and at lower cost than a completion rig. As Chapter 3 discusses, BP began drilling Macondo with Transocean's *Marianas* rig. BP's subsea wells team (Figure 4.5.3) accordingly developed a lockdown sleeve setting procedure in October 2009 for the *Marianas*.[12] They reviewed the procedure on November 10, 2009, with Dril-Quip representative Barry Patterson.[13] Two days later, BP subsea wells engineer Brad Tippetts sent a request to Patterson for the information necessary to develop a final lockdown sleeve setting procedure.[14] Patterson included BP drilling engineer Brian Morel in this initial November conversation, but it does not appear that Morel participated or responded.[15]

After BP decided that the *Deepwater Horizon* would replace the hurricane-damaged *Marianas,* BP engineers developed a revised drilling program. On December 31, BP subsea wells team leader Merrick Kelley checked in to ask if the Macondo engineering team still planned to install the lockdown sleeve as part of its new drilling program.[16] Senior drilling engineer Mark Hafle said no: "We do not plan on installing lock down sleeve with the Horizon."[17]

Kelley responded by noting the time (and hence money) that BP could save by setting the lockdown sleeve with the *Horizon.* He explained that setting the lockdown sleeve during temporary abandonment "saves an incremental 5.5 days of rig time on the back side" and, with it, more than $2 million.[18] (Doing the job with a completion rig would take seven days, whereas the *Horizon* could do the job in 1.5 days during temporary abandonment.[19]) Hafle discussed the issue with BP drilling and completions operations manager David Sims,[20] and the Macondo team eventually decided to set the lockdown sleeve using the *Horizon.*

The Macondo team also considered an **open water** lockdown sleeve installation, in which a boat would set the lockdown sleeve using ROVs.[21] The open water installation process would save $120,000 in additional costs over having the *Horizon* do the installation.[22] But it also presented a greater risk of damaging the lockdown sleeve.[23] Kelley therefore recommended against it: "At the end of the day it boils down to the amount of risk we are willing to take to potentially save $120,000 by using a boat. To be honest and frank with you, performing this operation from the rig is the easiest and simplest way I know to install a[n] LDS.... For my money, it is just the right thing to do...."[24]

Ultimately, the Macondo team decided to set the lockdown sleeve with the *Horizon* during temporary abandonment.[25]

Development of the Lockdown Sleeve Setting Procedure

Finalizing the procedure for setting the lockdown sleeve was a necessary first step in developing the overall temporary abandonment procedure. The Macondo team did not finalize its lockdown sleeve setting procedure until very late in the drilling process. Indeed, as late as mid-April, the Macondo team was still reconsidering its decision to have the *Horizon* set a lockdown sleeve at all.

On April 8, 2010, Patterson again sent Morel the information about setting the lockdown sleeve that Morel had first received five months earlier.[26] Morel reviewed the procedure later that day.[27] Four days later, on April 12, BP well site leader Murry Sepulvado asked Morel via email for the temporary abandonment procedures (among other things), saying that rig personnel were "in the dark and nearing the end of logging operations."[28] Morel emailed BP subsea wells engineers Shane Albers and Tippetts to ask for a lockdown sleeve running procedure: "I need a procedure this morning, do you have one available?"[29] Tippetts responded five minutes later by attaching the detailed lockdown sleeve setting procedure that the subsea team had originally written for the *Marianas*. Tippetts said, "this should do for now," but noted that Albers was modifying the procedure "slightly" for the *Horizon* and that Albers "will send out the updated version later today."[30] Morel told Sepulvado, "I will have you something this morning."[31]

Later in the afternoon of April 12, Morel asked Kelley via email when BP would be setting a lockdown sleeve at Isabela, another BP well.[32] Morel knew that BP planned to set the Isabela lockdown sleeve using open water installation tools. Morel's question therefore suggests that he (and perhaps the Macondo team) was still considering another option for setting the lockdown sleeve—namely, using the open water tools that BP would use at Isabela instead of using the *Horizon*. But late that night, Kelley advised Morel and Hafle against that approach. Kelley said that the subsea team would not make it a priority to "combine the Isabela and Macondo lock down sleeve jobs." Kelley also warned that others in BP might challenge a decision to use open water tools to set the lockdown sleeve in order to save just 24 hours of rig time.[33]

Morel did not send out a final updated procedure on April 12. Instead, after the close of business on April 13, Morel sent BP wells team leader John Guide the *Marianas* procedure, with the caveat that the subsea wells engineers "are updating for the Horizon, but mostly will remain the same."[34] A little less than an hour later, at 6:50 p.m. on April 13, Albers sent Morel the final updated procedure.[35]

Numerous Last-Minute Changes During the Final Development of the Temporary Abandonment Procedure

In the nine days before BP began the temporary abandonment of the Macondo well, the company went through at least four different versions of temporary abandonment procedures.[36] Each version switched the order of several key steps.

April 12 Well Plan

In response to the April 12 prodding from Murry Sepulvado, Morel circulated a draft plan for upcoming operations at Macondo later that day.[37] The draft plan included temporary

ı. Multiple last-minute revisions to the temporary abandonment procedure.

12 Well Plan	April 14 Morel Email	April 15 Well Plan/ April 16 MMS Permit	April 20 Ops Note	April 20 Actual Procedure
Set lockdown sleeve	Run in hole to 8,367'	Negative pressure test to seawater gradient (with base oil to wellhead)	Trip in hole to 8,367'	Trip in hole to 8,367'
Run in hole to 6,000'	Set 300' cement plug in mud **Barrier**	Run in hole to 8,367'	Displace mud with seawater from 8,367' to above wellhead (BOP)	Displace mud with seawater from 8,367' to above wellhead (BOP)
Displace mud in well and riser from 6,000' with seawater	Negative pressure test with base oil to wellhead	Displace mud in well and riser from 8,367' with seawater	Negative pressure test with seawater to depth 8,367' rather than with base oil to wellhead	Negative pressure test with seawater to depth 8,367' rather than with base oil to wellhead
Set 300' cement plug in seawater **Barrier**	Displace mud in well and riser from 6,000' with seawater	Monitor well for 30 minutes/conduct second negative pressure test	Displace mud in riser with seawater	Displace mud in riser with seawater **BLOWOUT**
	Set lockdown sleeve	Set 300' cement plug in seawater **Barrier**	Set 300' cement plug in seawater **Barrier**	
		Set lockdown sleeve	Set lockdown sleeve	

TrialGraphix

abandonment procedures that instructed the rig crew to set the lockdown sleeve *first* and then to set a surface cement plug in seawater. The plug would be set just 933 feet below the mudline.[38]

Morel's draft did not include a negative pressure test. After reviewing it, well site leader Ronnie Sepulvado reminded Morel that he needed to include a negative pressure test.[39]

April 14 Morel Email

Two days later, Morel sent out a procedure that was different in several important respects.[40]

First, the new procedure stated that BP would set the cement plug first and *then* set the lockdown sleeve.

Second, Morel changed the depth of the cement plug in order to create the clearance necessary to set the lockdown sleeve. Morel moved the cement plug from 933 feet below the mudline to 3,300 feet below the mudline.

Third, Morel changed the procedure so that the rig crew would set the surface cement plug in *drilling mud instead of seawater*.

Fourth, Morel included a negative pressure test. Morel's procedure instructed the rig crew to perform the test "with base oil in kill/choke line to the wellhead."[41] Using **base oil** for a negative pressure test is a normal industry practice. Filling the choke or kill lines with base oil can simulate the pressure effects of displacing drilling mud in the riser and some portion of the wellbore with seawater *without* actually displacing any mud. This is because base oil is lighter than seawater. Morel presumably included this step to account for the new procedure to displace a large amount of mud from the wellbore before setting the surface cement plug. (Interestingly, the procedure called for the negative pressure test to be done after the cement plug had been set,[42] so that the test would examine the quality of the cement in the surface plug rather than the bottomhole cement.)

April 15 Well Plan and April 16 MMS-Approved Procedure

By April 15, with the approval of Guide and drilling engineering team leader Gregg Walz, Morel changed the plan again in at least two important respects.[43]

First, Morel's new plan required rig personnel to conduct a negative pressure test before setting the surface cement plug, so that the test would check the integrity of the bottomhole cement.[44]

Second, the new plan called for the rig crew to displace the riser to seawater immediately after conducting the negative pressure test.[45] Morel apparently made this change because one of the well site leaders had asked to set the cement plug in seawater.[46]

The Macondo team clearly recognized that its plan called for an unusually deep cement plug. Morel included an alternative plan with a shallower plug in the event that MMS did not approve the deep plug.[47]

Morel and Hafle worked together to develop an application for an MMS permit allowing the team to use the "deep plug" option. As part of that application, filed on April 16, Morel listed BP's planned temporary abandonment procedure and included a negative pressure test (even though MMS regulations did not require a negative pressure test, as discussed in Chapter 6). That test would now be conducted "with [the] kill line"—yet another change in the procedure.[48] MMS approved the permit application—and with it, BP's plan to use a deep plug—in less than 90 minutes.[49]

The language in BP's April 16 permit application describing the negative pressure test and displacement procedure was unclear. Some have said that the language, like that in the April 15 well plan, required BP to conduct its negative pressure test *before* displacing mud in the well with seawater.[50] Others have said (after the blowout) that the only sensible time to do the negative pressure test would have been *after* the rig crew displaced the mud beneath the wellhead with seawater to the depth of the cement plug.[51] This argument may be important; if the former interpretation is correct, the rig crew did not adhere to the approved MMS procedure.[52] In any event, the debate highlights the lack of specificity in the permitted language.

After MMS approved the temporary abandonment procedure, Morel realized there was a problem. By planning to set its surface cement plug very deep in the well *and* set it in seawater, BP would be severely underbalancing the well during temporary abandonment. BP could not generate enough differential pressure to simulate those conditions merely by pumping base oil through the kill line down to the wellhead. Accordingly, the base oil negative pressure test procedure would not constitute a proper negative pressure test of the system.[53]

The solution, as the drilling team saw it, was to conduct two negative pressure tests. The rig would conduct the first test as planned, with base oil to the wellhead *before* displacement to 8,367 feet. They would conduct the second test *after* that displacement.[54]

April 20 "Rig Call" and Morel "Ops Note"

The Macondo team had still not resolved the negative pressure test procedures even during the 7:30 a.m. "rig call" between the rig crew and shoreside personnel on April 20—the day of the blowout. The rig crew asked wells team leader Guide how they were supposed to run the negative pressure test. Guide responded that he would confer with the engineers onshore and get back to them.[55]

Guide decided that the crew would conduct only one negative pressure test. There would be no "first" test using base oil in the kill line. Instead, there would be a single test midway through the displacement at 8,367 feet. It is difficult to determine whether there was significant disagreement with this decision. Hafle stated that there was "some discussion but [that] John Guide [was] hard to argue with" and that "Walz was in discussion but didn't argue with John."[56] Morel (who was visiting the rig) stated that the well site leaders did not have strong opinions either way.[57] According to Guide, however, there was never any plan to perform more than one negative pressure test.[58]

Three hours after the rig call, Morel sent an "Ops Note" to the shoreside team and well site leaders. The Ops Note reflected the Macondo team's final changes to the temporary abandonment procedure.[59] The first time the rig crew saw the procedure was during the 11 a.m. pre-tour meeting on April 20.[60]

Whereas BP's April 16 submission to MMS may have stated that rig personnel would conduct the negative pressure test before displacement, the April 20 Ops Note directed the crew to conduct the negative pressure test midway through the displacement process.[61] The rig crew would first displace mud with seawater from beneath the wellhead to 8,367 feet. The crew would then conduct the negative pressure test on the kill line. After the test, the crew would displace the mud remaining in the riser and then set the cement plug.[62] Like the other procedures, the Ops Note lacked basic information about how the negative pressure test was to be conducted.[63]

The Macondo team apparently recognized that conducting a negative pressure test midway through displacement (rather than before displacement) was different from the procedure MMS had approved. But BP decided not to notify MMS of the change or seek further MMS approval.[64] According to members of the Macondo team, such notification and further approval were unnecessary because conducting the negative pressure test during displacement would be a more rigorous test than conducting it beforehand.[65] This explanation is called into question by the fact that BP did seek MMS approval before making a similar change in a negative pressure test procedure during temporary abandonment operations in 2006.[66]

According to BP well site leader Bob Kaluza, Hafle called him on the afternoon of April 20 to discuss the Ops Note. Hafle had been away on vacation while the rest of the shoreside team had put together the procedures in the Ops Note. Reviewing it, Hafle was concerned that the Ops Note procedure was different than the procedure MMS had approved. Kaluza woke up Morel. Morel explained that the rest of the shoreside team had decided to "deviate" from the procedure in the MMS-approved permit, which called for conducting the negative pressure test before displacement. "The team in town wanted to do something different," Kaluza later explained according to notes of BP's post-blowout interviews. "They decided we could do the displacement and negative test together – don't know why – maybe trying to save time.... Anytime you get behind, they try to speed up."[67]

It is impossible to know whether the changes to the negative pressure test procedure (including elimination of a second negative pressure test at a different depth) contributed to the blowout. As Chapter 4.6 explains in detail, personnel on the *Deepwater Horizon* missed clear warning signals from the negative pressure test they did conduct. Conducting an earlier version of the test may have removed one of the factors confounding successful interpretation of the test and eliminated the crew's erroneous explanation for the warning signals they observed.[68] And conducting a second test at a different depth might have given the rig crew another opportunity to recognize those signals.

Technical Findings

BP's Temporary Abandonment Procedure Created Significant Risks

BP's design decisions had significant consequences and increased the risks associated with the temporary abandonment at Macondo in several important ways.

First, the procedures created a severe hydrostatic underbalance in the well. By requiring the rig crew to remove so much mud from the wellbore during temporary abandonment, BP's procedures greatly reduced the balancing pressure that the mud column in the wellbore exerted on the hydrocarbons below. This increased stress on the bottomhole cement.[69] While temporarily abandoning a deepwater well typically involves placing some amount of stress on the bottomhole cement, BP's procedures stressed the cement more than usual[70]—to an extent never before seen by many in the industry.[71]

Second, the procedures led the rig crew to conduct riser displacement operations with *only one physical barrier in the well* (the bottomhole cement) and only one backup barrier (the BOP).[72] That backup barrier, in turn, was highly dependent on well control monitoring. As a result, BP's temporary abandonment procedure placed a high premium on kick detection and response during the displacement.[73] Unless the rig crew recognized a kick, they could not activate the BOP in time for it to function as a barrier.

Third, and as a result, the procedures placed a high premium on the integrity of the bottomhole cement and the negative pressure test that evaluated it.[74] Rig personnel could not rely on the bottomhole cement as a barrier until it had been verified, and the only procedure BP planned to use to verify the cement's integrity was the negative pressure test.

BP Did Not Need to Set a Lockdown Sleeve as the Last Step in Temporary Abandonment

As explained above, BP made many of its procedural decisions regarding temporary abandonment based on its decision to set a lockdown sleeve during the temporary abandonment phase of the well. BP did not need to set a lockdown sleeve during the temporary abandonment phase. The fact that BP nevertheless chose to do so is not problematic in itself. Indeed, locking down the casing earlier rather than later can increase safety by mitigating against potential uplift forces during drilling and abandonment (explained in Chapter 4.1). But BP increased overall risks by deciding to set the lockdown sleeve last in the temporary abandonment sequence.

A lockdown sleeve need not be set last in the temporary abandonment sequence. It can be set in mud prior to displacement and setting of the surface plug.[75] This is commonly done in the industry,[76] and BP engineers considered doing it this way at Macondo.[77]

Outer Lock Ring. Setting a lockdown sleeve before temporary abandonment can reduce the risk that underbalancing a well might lift the production casing out of place in the wellhead. Another mechanism for locking a production casing in place is an **outer lock ring**. Rig personnel can install an outer lock ring when they first set the casing in place. While this was not a common practice at the time of the Macondo incident,[78] some industry experts have recommended that it become standard.[79]

Indeed, the Macondo team initially planned to set the lockdown sleeve in mud, before setting a shallow surface cement plug in seawater. In a March 3 email, Hafle stated that the team would "set the plug after [lockdown sleeve] installation"; with no plug in the way, they could easily "supply the correct weight for installation."[80] On April 8, Morel checked with Dril-Quip representative Barry Patterson to make sure the lockdown sleeve procedure was compatible with "100,000 lbs air weight in 14.0 ppg mud."[81] On April 12, Morel emailed Tippetts to confirm that the plan was "to still have mud in the riser and wellbore when we set the LDS."[82] Subsea well supervisor Ross Skidmore preferred to set the lockdown sleeve in mud because the hole would be in its cleanest state at that point.[83]

As described above, by April 14, BP had changed its plan so that it would run the lockdown sleeve last, after setting a surface plug and displacing the riser to seawater.[84] When Skidmore heard about the change, he approached one of the BP drilling engineers on the rig and expressed his preference to set the lockdown sleeve in mud; the engineer indicated the decision had come from personnel onshore and was final.[85]

BP Did Not Need 3,000 Feet of Drill Pipe Below the Wellhead to Achieve the 100,000 Pounds Necessary to Set the Lockdown Sleeve

BP did not need to use 3,000 feet of drill pipe in order to generate the 100,000 pounds of downward force necessary to set the lockdown sleeve. Instead, BP could have instructed the rig crew to hang a much shorter length of pipe that included **drill collars** (a heavier type of drill pipe). Because drill collars are much heavier than other drill pipe, the crew could have used a much shorter length of them to generate the same downward force. BP could also have instructed the rig crew to generate some of the setting force using weight pushing down from *above* the running tool instead of hanging below it.[86] Using these methods, BP could have set the lockdown sleeve in place without requiring 3,000 feet of clearance beneath the sleeve, as called for in its final plan.[87]

BP engineers were well aware that they did not have to set the lockdown sleeve using 3,000 feet of hanging drill pipe. BP had previously set a lockdown sleeve with the same running procedures and weight requirement (100,000 pounds) at another well in the Gulf of Mexico, in Mississippi Canyon Block 129.[88] BP used drill collars at that well to generate the required setting force[89] and was thus able to set its surface plug only 1,600 feet below the mudline.[90] Similarly, BP set a lockdown sleeve with an even greater force requirement (125,000 to 135,000 pounds) in Mississippi Canyon Block 777.[91] There again, BP used drill collars to generate the required setting force and set a surface plug 1,500 feet below the mudline.[92] Such depths were more typical for pre-lockdown sleeve plugs.[93]

At one point, the Macondo lockdown sleeve was supposed to be set in much the same manner.[94] As far back as November 12, 2009, the Macondo team had planned to run drill collars beneath the lockdown sleeve in order to achieve the necessary setting weight.[95] That was still the plan on February 3 when the lockdown sleeve setting procedure was submitted for inclusion in the Macondo well planning spreadsheet.[96] But by March 2, Hafle had told Tippetts, "Here's the final plan.... We will *not* be using any drill collars. The rig has 5-1/2" [heavyweight drill pipe] and we will rent additional 5-1/2" [heavyweight drill pipe] to have 100k buoyed weight below" the lockdown sleeve.[97]

Despite Hafle's email, BP obtained drill collars and had them on the rig by April 17.[98] As late as April 12, Walz mentioned using drill collars to set the lockdown sleeve in an email to Morel,[99] and Morel included them in the April 12 drilling program.[100] The last final updated procedure that Albers sent to Morel on April 13 also included drill collars.[101] But by the time drill collars arrived on the rig, Morel had changed the procedures to specify a deep surface plug, 3,000 feet below the mudline, which suggests that he had not envisioned using drill collars to set the lockdown sleeve.[102] According to BP wells team leader Guide, the team changed the plan because the rig already had heavyweight drill pipe "racked back" and ready to run into the well.[103] In order to use drill collars at that point, the rig would need to make up each piece of pipe individually, which would take time and add to the general risk of personal injury.[104]

Figure 4.5.5. Bridge plug.

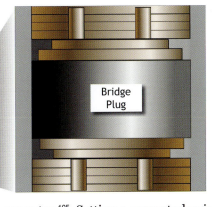

Cement Plug

Bridge Plug

TrialGraphix

BP could have used a mechanical plug in lieu of or in addition to a cement plug.

BP Could Have Set Its Surface Cement Plug in Mud Instead of Seawater

BP did not have to displace mud from the well and riser in order to set a cement plug; it could have set the cement plug in drilling mud instead.

Surface cement plugs can be set in mud just as they can be set in seawater.[105] Setting a cement plug in mud can present a risk of contamination and certain other chemical complexities.[106] But contamination issues can exist with cement plugs set in seawater as well,[107] and the complexities can be managed with proper cement slurry design and the use of spacer.[108] In order to help ensure that cement plugs set in drilling mud are secure, engineers also use **mechanical retainers** or **bridge plugs**—metal and rubber devices that fit into the casing and hold the cement,[109] as shown in Figure 4.5.5. The mechanical plug then serves as an additional barrier, apart from the cement it helps to set.[110]

BP generally, and the Macondo team specifically, were familiar with these options.[111] When an earlier surface cement plug at Macondo failed to set up, Morel and another BP engineer involved with the earlier plug discussed how "the biggest single factor for plug success is having a good base."[112] The engineers discussed how they could design that base by several means, including by contrasting fluid densities (lighter cement on heavier drilling fluid) and by using mechanical devices (retainers and bridge plugs).[113] Another engineer involved with the earlier plug commented, "We need to get better at setting plugs regardless of the method."[114]

BP representatives have acknowledged that surface cement plugs can be set in mud[115] and that doing so is not a mistake.[116] Indeed, BP has set surface cement plugs in mud before[117] and apparently considered doing so at Macondo as late as April 14.[118] BP has also frequently made use of mechanical devices for surface plugs, including both drillable and retrievable bridge plugs.[119]

In fact, BP engineers affirmatively considered running a mechanical plug at Macondo—specifically, a Baker Hughes model GT retrievable bridge plug.[120] The GT plug was much more expensive than a cement plug, but Morel preferred it (at least initially) because of its greater reliability. In an email to Hafle and others, he noted: "If Baker's GT plug wasn't available, we would either set a cement plug in its place or a Halliburton Fast Drill plug. Both are much cheaper options, but leave us with potential issues during the completions. They could potentially cost us more as well, because extra rig time might be involved with removing these type of plugs."[121]

BP engineers planned at various points to use a GT plug at Macondo.[122] The Macondo team would have rented that plug pursuant to a long-term GT plug rental contract that BP was arranging with Baker Hughes for several wells at the same time.[123] Because the BP personnel arranging the contract believed there was a "high probability of a long term installation of this plug at Macondo," they affirmatively committed to the rental.[124] BP initiated rental of the Macondo plugs on April 6.[125] The company paid $42,902 to Baker Hughes to make up, test, and keep a primary and backup GT plug on standby.[126]

On April 9, a Baker Hughes representative emailed Morel and Hafle to ask for an update on whether BP had decided to use the standby plug or not.[127] Morel responded with additional details but still no final decision: "If we need it, the rig will probably want to call it out next weekend or early the following week (18-19th of April). I will keep you informed."[128] Morel explained that the Macondo team would not commit to using the GT plug until it had decided if production casing was required.[129] But by April 12, two days before finalizing the decision to run production casing, the Macondo team decided to use a plain cement surface plug.[130] When the Baker Hughes representative emailed the two BP engineers again on April 19 to ask if they would need the plug he had kept on standby "since early April,"[131] Hafle responded, "We will be setting a cement plug instead."[132] Baker Hughes stopped the rental.[133]

It is not clear why the Macondo team chose to set a plain cement plug. Morel told one engineer that the reason was cost:[134] "Plan is to set a cement plug instead of running the GT plug as it doesn't cost us anything to leave it in the hole."[135] Morel told another set of engineers (the completion engineers) that the reason was risk: The "GT plug poses risks leaving it in the wellbore for an unknown amount of time."[136]*

BP Could Have Planned a Safer Temporary Abandonment Procedure Even Without Changing Its Design Assumptions

Even assuming that BP truly had to set the lockdown sleeve last and set its surface cement plug

---

* Some members of the Macondo team were concerned that leaving a mechanical plug in the well for an indefinite period of time might present complications during re-entry and completion. Retrievable plugs left in the wellbore for too long can corrode and become difficult to retrieve. Drillable plugs (like cement plugs) can produce debris when drilled out. Nevertheless, BP appears to have addressed or accepted these complications in other wells where the company set mechanical plugs. Indeed, a BP completion engineer reacted to Morel's email with wonderment: "I am curious about what risks he speaks of with leaving GT plugs in place for long periods. We had them in place at Dorado for a couple of years without problems."

deep in the well in seawater, BP could have taken at least three measures to mitigate the risk created by its unusual procedure. Each of these measures would have increased or improved the physical barriers in the wellbore during the displacement. While each would have taken some additional time,[137] they would have ensured that the cement job at the bottom of the well was not the only barrier physically in place during the displacement.

BP Could Have Retained Hydrostatic Overbalance

BP still could have retained hydrostatic overbalance even with the removal of 3,300 feet of mud from the wellbore. To do so, they could have replaced the mud at the bottom of the wellbore with heavier "kill weight" mud.[138] BP engineers should have been familiar with this concept,[139] and it is a common industry practice.[140] In doing so, they would have retained mud as a physical barrier in the wellbore during the displacement.[141†]

BP Could Have Set Intermediate Plugs

BP could have set additional plugs between the bottomhole cement and the surface plug.[142] BP engineers were familiar with this option, as the company had set multiple intermediate plugs (often including mechanical plugs) on previous wells.[143] Indeed, some in the industry treat the setting of intermediate plugs as standard practice.[144] But it appears that the Macondo team never considered it.[145] Setting intermediate mechanical or cement plugs would have increased the number of physical barriers in the wellbore during the displacement.

BP Could Have Conducted the Displacement (of Both the Wellbore and the Riser) With the BOP Closed

BP could have closed an annular preventer (or variable bore ram) before beginning the displacement and, in various configurations, then displaced the casing and riser using the drill pipe and choke, kill, and boost lines.[146] This would have been considered a particularly conservative approach in the industry, and unnecessary for most wells.[147] But the unusually deep cement plug and the uncertain nature of the bottomhole cement job at Macondo warranted extra caution.[148] Indeed, since the blowout, the industry appears to be moving in the direction of making this practice more prevalent.[149] Closing the BOP before the displacement would have eliminated the BOP's dependence on human monitoring and thereby converted it into a physical barrier in place during the displacement. The well would already have been shut in at the time of the kick, enabling the crew to more easily respond to and control the kick.

Management Findings

BP Failed to Develop Its Temporary Abandonment Procedure in a Timely Manner

The moment an operator designs a production well, it can (and should) develop a temporary abandonment procedure.[150] Even though BP planned Macondo as a production well from the start,[151] it did not include temporary abandonment procedures in its initial drilling program.[152]

---

† BP wells team leader John Guide suggested that for some wells underbalance is necessary because mud is simply not heavy enough to compensate for the loss of the riser. That was not true of the Macondo well. To be sure, if BP had insisted on using only one plug and setting that plug at 3,300 feet below the mudline, then replacing just the mud above that plug with kill weight mud would not have prevented underbalance. But BP could have set an intermediate plug deeper in the well (about 6,900 feet below the mudline), replaced the mud above that deeper plug with kill weight mud, and then set a surface plug higher up in the well. Therefore, BP could have left the Macondo well overbalanced by using a combination of kill weight mud and intermediate plugs.

As early as January 2010, the Macondo team planned to use the *Horizon* to install a lockdown sleeve and then temporarily abandon the well. But the company's January 2010 drilling program still did not include a temporary abandonment procedure.[153] By April 9, the Macondo team knew the total depth of the well.[154] At that point, they had enough information to design a temporary abandonment procedure specifically tailored to the final conditions at Macondo.[155] But three days later, on April 12, the well site leader was forced to ask the shoreside team for procedures himself, saying, "we are in the dark and nearing the end of logging operations."[156]

The Macondo drilling team did not begin developing a procedure in earnest until after this request. Perhaps because of the delays, the Macondo team changed its procedures repeatedly at the last minute, even up until the day the procedure was to begin (the day of the blowout). As Walz acknowledged in another context, "planning [was] lagging behind the operations."[157]

BP Changed Its Temporary Abandonment Procedure Repeatedly at the Last Minute Without Subjecting Those Changes to Any Formal Risk Assessment

BP's temporary abandonment procedures for Macondo changed at least four times over the last nine days before the blowout. This was an unusual number of changes so close to the procedure's execution.[158] BP also changed its lockdown sleeve setting procedures over time.

Several of BP's decisions—not using drill collars, not using a mechanical plug, setting the plug in seawater, setting the lockdown sleeve last—may have made sense in isolation. But the decisions also created risks, individually and especially in combination with the rest of the temporary abandonment operation. For instance, BP originally planned to install the lockdown sleeve at the beginning of the temporary abandonment. BP's decision to change plans and set the lockdown sleeve last triggered a cascade of other decisions that led it to severely underbalance the well while leaving the bottomhole cement as the lone physical barrier in place during displacement of the riser.

There is no evidence that BP conducted any formal risk analysis before making these changes or even after the procedure as a whole.[159] For example, on April 15, Morel (who was on the rig at the time) emailed the rest of the Macondo onshore engineers about setting a deep plug in seawater: "Recommendation out here is to displace to seawater at 8300' then set the cement plug. Does anyone have issues with this?"[160] The response, from Hafle, was simply: "Seems ok to me."[161] According to Guide, the team never discussed the risk of having such a deep surface plug.[162]

Post-incident interviews with the Macondo team confirm that it made significant procedural changes in a relatively casual manner. Walz admitted that there was "no structured approval process" and that "changes [were] made with email and verbal discussion."[163] Cocales stated that there was "no formal process on communicating changes to [the] well plan." Murry Sepulvado stated that it was not unusual to receive emails like the Ops Note containing procedural changes that had not been risk assessed through a formal process.[164] And according to Guide, such Ops Notes would not even flag whether changes had been made to the well plan.[165]

BP Allowed Equipment Availability to Drive Design and Procedure Decisions

BP inverted the normal process of well design in determining the depth of the surface cement plug, and the type and length of pipe to use in setting the lockdown sleeve.

Drilling engineers normally begin by considering their objective and the attendant risks and developing a well design and procedures that are efficient and safe. They then arrange for the equipment and materials necessary to execute the design.[166] BP did the opposite at Macondo. BP made decisions about what type of drill pipe to use (ordinary, heavyweight, or drill collars), and hence where to set its surface cement plug, based on the type of pipe available on the rig.[167] The *Deepwater Horizon* apparently already had heavyweight drill pipe "racked back" and ready to run into the well, which led the Macondo team to use that pipe instead of drill collars.[168]

BP's lockdown sleeve setting procedure underscored this logic: "To achieve 100,000 lbs of tail pipe weight drill collars & drill pipe will be used. The combination will depend on availability and will be determined while onsite." The caveat was repeated in step seven of the procedure, which stated "the decision on the pipe size & length will be made on the rig."[169]

BP Failed to Provide Written Standardized Guidance for Temporary Abandonment Procedures

BP had no consistent or standardized temporary abandonment procedure across its Gulf of Mexico operations.[170] Formal written guidance was minimal: The Drilling and Well Operations Practice manual and relevant Engineering Technical Practice (GP 10-36) mandated that, in each flow path, there should be two independent mechanical barriers isolating flow from the reservoir to the surface and that those barriers should be independently tested.[171] The documents did not specify the location of those barriers or the procedure by which they should be set. This left the Macondo engineers to determine such issues for themselves on an ad hoc basis. For example, when Hafle emailed the subsea engineers—"Can we set the plug after the LDS is in place?"—one subsea engineer wrote to another, "I do not know about setting the plug after the LDS. Do you? Could you ask someone around the office tomorrow about this to figure this out?"[172] Such uncertainty existed even with something as basic as regulatory requirements.[173] ◖

hapter 4.6 | Negative Pressure Test

he negative pressure test performed at Macondo showed repeatedly over a three-hour period that the well lacked integrity and that the cement had failed to seal off the hydrocarbons in the pay zone. BP well site leaders, in consultation with Transocean rig personnel, nevertheless mistakenly ncluded that the test had demonstrated well integrity and then proceeded to the xt phase of temporary abandonment.

e Chief Counsel's team finds that the lure to properly conduct and interpret e negative pressure test was a major ntributing factor to the blowout.

Vell Integrity Tests

er cementing the production casing, BP was nearly dy to **complete** the Macondo well and turn it into roducing well. (Completion refers to the process preparing the well for production and installing uipment to collect oil from the well.)

wever, BP only planned to use *Deepwater Horizon* drill the well, not to complete it. After installing production casing, BP planned to have the epwater Horizon leave Macondo for a different lling job elsewhere in the Gulf of Mexico. Another would perform the completion work at some determined time in the future.

e well would be temporarily "abandoned" during time between *Deepwater Horizon*'s departure d the completion rig's arrival. The *Deepwater Horizon* crew's last responsibility would be to secure well to ensure that nothing could leak in or out—to confirm the **well's integrity**—during that ervening time. It was during this **temporary abandonment** process, rather than during drilling, t the blowout occurred.

part of the temporary abandonment procedure, the rig crew conducted tests to check the well's egrity. If there were a leak in the system of cement, casing strings, and mechanical seals that nprised the well, these tests should have revealed it. The rig crew conducted three different tests: eal assembly test, a positive pressure test, and a negative pressure test. The tests each checked ferent parts of the well's integrity.

nificantly, however, the negative pressure test was the only one that tested the integrity of cement at the bottom of the well.[1] That cement is what the rig crew would rely on to isolate drocarbons in the pay zone and keep them from coming up the well.

sting this cement was thus critical to safety of everyone on the rig.

Figure 4.6.1. Well integrity tests.

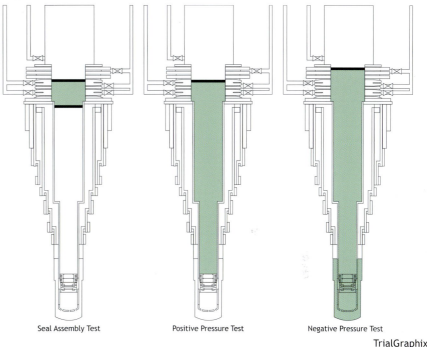

Seal Assembly Test　　Positive Pressure Test　　Negative Pressure Test

TrialGraphix

The rig crew conducted three pressure tests as part of the temporary abandonment procedure to verify the integrity of the well. From left to right: the seal assembly test, the positive pressure test, and the negative pressure test. Test regions are shown in green.

WELL INTEGRITY TESTS

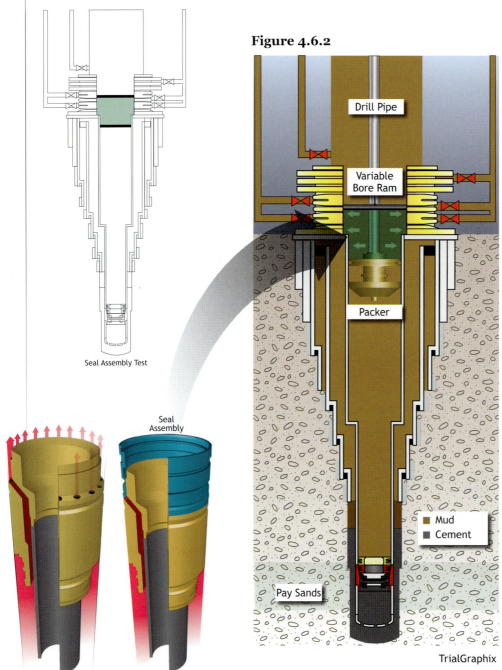

Figure 4.6.2

TrialGraphix

Seal Assembly Test

The **seal assembly test**, as its name implies, tests the casing hanger seal assembly. A long string production casing hangs from a casing hanger inside the wellhead. The casing hanger both supports the casing and seals off the annular space outside the top of the casing. After installing the casing, rig personnel conduct a test to determine that the casing hanger seal does not leak. To do so, the crew installs a plug, or **packer**, on the bottom of the drill pipe and lowers it beneath the seal assembly. The crew closes a variable bore ram of the blowout preventer (BOP) (above the seal assembly) around the drill pipe. This creates a small enclosed space inside the casing at the mudline. The rig crew then pumps additional fluid into this space, increasing the pressure. They then monitor the pressure for a predetermined time period. If the pressure remains constant, it means that the casing hanger seal is capable of containing high internal pressure. If the pressure drops, fluid is escaping through a leak. In the early morning hours of April 20, the rig crew performed two separate pressure tests on the seal assembly, both of which passed.[2]

Seal Assembly Test

Seal Assembly

TrialGraphix

The casing hanger, as described in Chapter 4.1, has flow passages that facilitate the flow of fluids during normal drilling operations. The seal assembly (blue) is fitted atop the casing hanger to halt annular flow after the primary cement job is complete. Together, the two bind the casing to the wellhead.

The seal assembly test checked the integrity of the interface between the casing and the wellhead. After lowering a packer into the well, the rig crew closed a variable bore ram around the BOP, sealing the space above and below the seal assembly. The rig crew then pumped fluid into this space, increasing the pressure inside it. If fluid did not leak out of the seal assembly, the pressure would remain constant.

igure 4.6.3

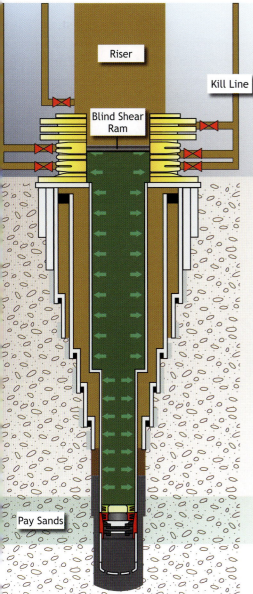

Pay Sands

TrialGraphix

he positive pressure test checks
ie integrity of the well by testing
hether the casing and wellhead seal
sembly can contain higher pressure
ian surrounds them. The *Deepwater*
orizon crew increased the pressure in
ie production casing string by pumping
uid into it through the kill line. If fluid
bes not leak out of the casing, the
ressure again remains constant.

Positive Pressure Test

Later that morning, the rig crew conducted a **positive pressure test**. A positive pressure test is like a seal assembly test, but over a larger area of the well. With the drill pipe pulled out of the well, the rig crew shuts the blind shear rams on the BOP to isolate the well from the riser. The crew then pumps additional fluid into the well below the BOP and monitors the pressure. If the pressure remains constant with the pumps shut off, that means that the casing, wellhead seal assembly, and BOP are containing internal pressure and are not leaking. Between 10:30 a.m. and noon, the crew conducted a positive pressure test to 250 pounds per square inch (psi) for five minutes and then a second to 2,700 psi for 30 minutes. In both instances, pressure inside the well remained constant over the test period.[3]

Because the seal assembly and positive pressure tests at Macondo appear to have been performed and interpreted correctly, this report does not explore them further.

Neither the seal assembly test nor the positive pressure test could check the integrity of the cement in the shoe track or in the annular space at the bottom of the production casing. The seal assembly test could not test anything below the packer. Similarly, the positive pressure test does not test anything below the wiper plug on top of the float collar.

The only test that was capable of testing the bottomhole cement, which was essential to preventing a blowout, was the negative pressure test.[4]

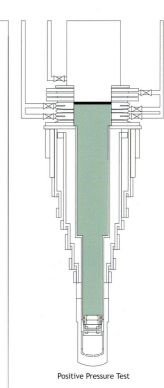

Positive Pressure Test

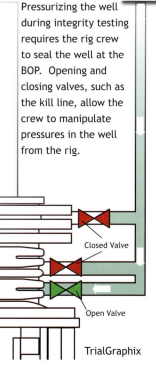

Pressurizing the well during integrity testing requires the rig crew to seal the well at the BOP. Opening and closing valves, such as the kill line, allow the crew to manipulate pressures in the well from the rig.

Figure 4.6.4

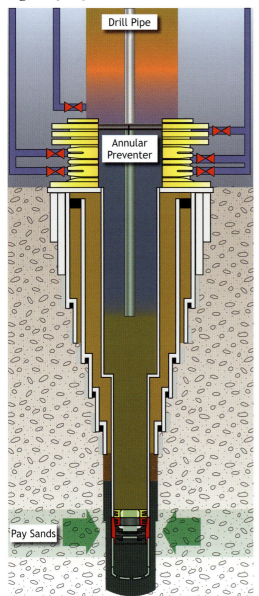

TrialGraphix

By moving mud from the production casing into the riser (displacement), the rig personnel reduced the pressure inside the well below the pressure outside the well (underbalancing). If there was good well integrity, the pressure inside the well would remain constant during the negative pressure test. If there was a leak of hydrocarbons into the well, the pressure in the well would rise (if the drill pipe or lines to the rig were closed) or fluid in the wellbore would be forced up and flow out at the rig (if the lines were open).

Negative Pressure Test

The **negative pressure test** is essentially the inverse of a positive pressure test. Rig personnel reduce the pressure inside the well below the pressure outside the well and then monitor the well to determine whether any hydrocarbons from the pay zones leak into the well from the formation outside it.

Whereas rig personnel identify a failed positive pressure test by observing diminishing internal pressure, they identify a failed negative pressure test when they observe *increasing* internal pressure while the well is shut in or *flow* from the well while it is open. In a successful negative pressure test, there should be no pressure increase inside the well and no flow from the well for a sustained period of time.[5] Increased pressure during this period indicates that the primary cement job at the bottom of the well has failed and hydrocarbons from the pay zone are entering the well.

The negative pressure test simulates the conditions rig personnel will create inside the well once they remove drilling mud from the riser (and from some portion of the well below the mudline) in order to temporarily abandon the well. Removing that mud removes pressure from inside the well.

Figure 4.6.5. End of cement to temporary abandonment.

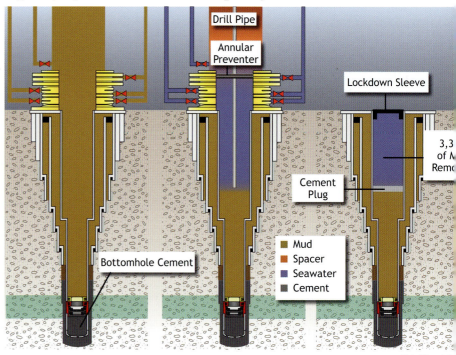

TrialGraphix

After the final casing string was cemented, heavy drilling mud filled the riser and the well (left). After the temporary abandonment planned for Macondo, the riser and its drilling mud would be removed. The drilling mud in the final casing string would be replaced with lighter seawater to a depth of over 8,000 feet below sea level (right). The removal of the hydrostatic pressure this drilling mud applied to the bottom of the well would increase the stress on the casing, seals, and cement. The negative pressure test simulated the conditions of temporary abandonment to confirm the integrity of the well in a controlled environment (middle).

The purpose of the negative pressure test is to make sure that when that pressure is removed, the casing, cement, and mechanical seals in the well will prevent high-pressure hydrocarbons or other fluids in the pay zone outside the well from leaking in. The test thus evaluates the integrity of the wellhead assembly, the casing, and the mechanical and cement seals in the well—indeed, it is the *only* pressure test that checks the integrity of the primary cement (see Figure 4.6.4).

For these reasons, both BP and Transocean have described the negative pressure test as critically important.[6]

Negative Pressure Test at Macondo

The negative pressure test at Macondo occurred in three separate phases over a five-hour period between approximately 3 and 8 p.m. on April 20.

First, the crew prepared to conduct the negative pressure test. To replicate conditions after temporary abandonment, the crew needed to "remove" the column of mud to a depth of 8,367 feet below sea level. In its place, the crew would "substitute" a column of seawater (see Figure 4.6.5). The crew accomplished this by pumping seawater (preceded by a buffer fluid known as **spacer** to separate it from the mud) down through a drill pipe lowered to that depth, illustrated in Figure 4.6.6. As they exited the stinger at the end of the drill pipe, the spacer and seawater would force—or **displace**—the surrounding mud up through the casing and into the riser. Once the seawater had displaced the mud and spacer into the riser above the BOP stack, the crew would close an annular preventer on the BOP around the drill pipe.

Closing the annular preventer would isolate the well below from the hydrostatic pressure exerted by the column of heavy drilling mud and spacer in the riser. At that point, the well would instead be subject to the lower hydrostatic pressure exerted by the lighter 8,367-foot column of seawater in the drill pipe. This would simulate the reduced hydrostatic pressure inside the well after temporary abandonment.

The next step was to conduct what became the first negative pressure test (the crew originally planned to conduct only one test). The crew would open a valve on the drill pipe at the rig and **bleed off** any pent-up pressure inside the drill pipe. In other words, the crew would allow fluids to flow out of the drill pipe until the flow stopped and the pressure in the pipe fell to 0 psi. The crew would then close—or **shut in**—the drill pipe and monitor the pressure inside it to see whether it remained at 0 psi or increased. This **drill pipe pressure** reflected the internal pressure of the well.

At Macondo, the crew had unexpected difficulty in bleeding the drill pipe pressure down to 0 psi. After each attempt, the crew would shut in the well, and the pressure would build back up. The rig crew attempted three times to bleed

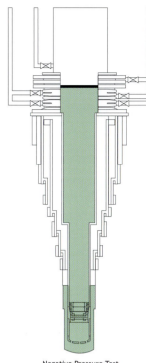

Negative Pressure Test

Figure 4.6.6. Preparations for the negative pressure test.

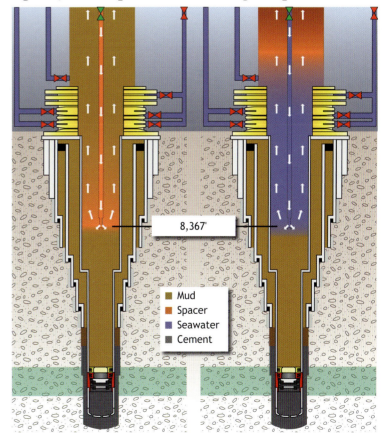

8,367'

	Mud
	Spacer
	Seawater
	Cement

TrialGraphix

To prepare for the negative pressure test, the rig crew needed to displace the mud in the drill pipe and casing string from a depth of 8,367 feet to above the BOP. The crew did so by pumping spacer fluid (left) and then seawater (right) down the drill pipe until the mud was above the BOP.

off the drill pipe pressure, but each time, the drill pipe pressure rose after being bled off. After the third attempt, drill pipe pressure rose from 0 to 1,400 psi as shown in Figure 4.6.7.

All parties now agree that this 1,400 psi pressure reading indicated that the well had failed the negative pressure test and that the cement job would not prevent hydrocarbons in the pay zones from entering the well.[7] The 1,400 psi pressure was the pressure of the hydrocarbon-bearing pay zone that was not properly sealed off by the primary cement.

The crew did not recognize that this first negative pressure test had identified a problem with the well—or if they did, they did not act upon that fact. Instead, they conducted a second test.

BP had submitted a permit modification to MMS stating that it would conduct the negative pressure test on the kill line rather than the drill pipe.[8] At least in part for this reason, BP well site leaders decided to follow up their first test on the drill pipe with a second negative pressure test in which they monitored pressure and flow on the kill line.[9] Rig personnel therefore opened the kill line, bled the pressure down to 0 psi, and monitored the line for 30 minutes. This time, there was no flow or pressure buildup in the kill line. The well site leaders and rig crew decided this was a successful negative pressure test and moved on to the next steps in the temporary abandonment procedure. But, as shown in Figure 4.6.8, although the pressure on the kill line may have stayed at 0 psi, drill pipe pressure remained at 1,400 psi.

Figure 4.6.7. First test failure.

Figure 4.6.8. Second test failure.

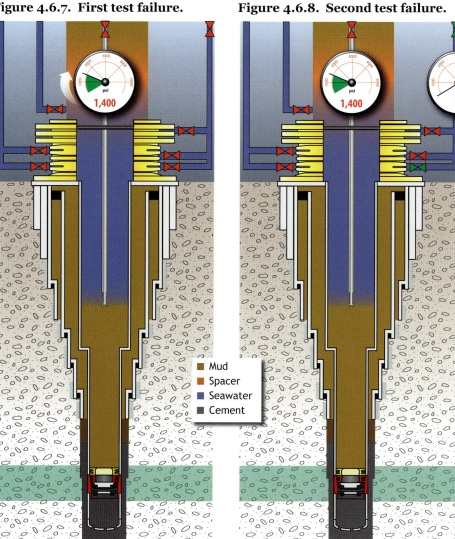

Mud
Spacer
Seawater
Cement

TrialGraphix

TrialGraphix

During the first negative pressure test, the crew repeatedly bled the drill pipe pressure down to 0 psi. However, more fluids bled than expected, and the drill pipe pressure repeatedly increased. After the last bleed, the drill pipe pressure rose from 0 to 1,400 psi, a clear failure.

During the second negative pressure test, the crew bled off the pressure in the kill line, rather than the drill pipe. The crew observed no excessive flow or pressure buildup on the kill line. The well site leaders and rig crew decided this was a successful test. But they had never accounted for the pressure on the drill pipe, which remained at 1,400 psi throughout the second test.

The well site leaders and rig crew never adequately accounted for that elevated pressure in the drill pipe.

The negative pressure test at Macondo "failed" in the sense that it did not show that the well had integrity. It was successful, however, in that it repeatedly and accurately identified a serious problem. All parties have since agreed that the 1,400 psi pressure reading on the drill pipe showed that hydrocarbons from the formation were entering the well from the pay zones and that the cement had failed to isolate or block off those pay zones. The larger question is why the men on the rig floor, who depended on this test to ensure well integrity, did not interpret the results of the negative pressure test correctly.

Answering this question is difficult because of the lack of consistent and detailed witness accounts. Some of the most valuable facts will never be known because many of the men involved in the test died in the rig explosion. The well site leaders involved in the test did survive but declined to speak to investigators about what happened (one citing his medical condition and the other invoking his Fifth Amendment rights).

However, the Chief Counsel's team did review notes taken by BP investigators who spoke with both well site leaders soon after the blowout. The Chief Counsel's team also had access to data records showing the pressures that the rig crew observed as well as testimony from witnesses who observed certain events in the drill shack that evening. The Chief Counsel's team based the following account on these information sources.

Preparations for the Negative Pressure Test

The rig crew began preparations for the negative pressure test at about 3 p.m. with a pre-job safety meeting. Because the crew would have to displace drilling mud to conduct the test, Leo Lindner, M-I SWACO's mud engineer, led the meeting. Well site leader Bob Kaluza was present for the meeting, though he left soon after it ended.[10] The meeting was held in or near the drill shack.

Shortly after 3 p.m., Transocean driller Dewey Revette pumped water to displace mud from three pipes, or "lines," that ran from the rig to the BOP stack: the **boost**, **choke**, and **kill lines** (see Figure 4.6.9).

Figure 4.6.9. Negative pressure test progress, 3 p.m. on April 20, 2010.

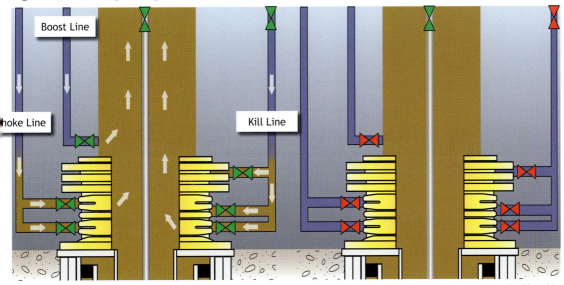

Boost Line

Choke Line

Kill Line

TrialGraphix

To begin preparations for the negative pressure test, the rig crew displaced the boost, choke, and kill lines with seawater. Seawater was pumped into the lines on the rig, forcing mud into and up the riser (left). After the lines were displaced, the crew closed the valves connecting them to the riser and BOP (right).

Rig personnel could use these lines to pump fluids into the well without pumping fluids through the drill pipe.[11]

The boost line was connected to the well immediately above the BOP. Rig personnel could pump fluids through it to accelerate the displacement of mud in the riser, literally "boosting" mud up toward the rig. The rig crew anticipated pumping seawater through the boost line later in the temporary abandonment process and prepared for doing so by displacing mud inside the line with seawater.

The choke and kill lines were connected to the BOP at various points on the stack. Rig personnel could use these lines to pump fluids in and out of the well even while certain BOP elements were fully sealed. These lines were therefore crucial to controlling kicks during drilling operations: After shutting the well in with the BOP, rig personnel could use them to "kill" the well (that is, overbalance it) with heavy mud and then "choke it off" by circulating hydrocarbons out. The rig crew could also use these lines instead of the drill pipe to conduct the negative pressure test. The men o[f] the *Deepwater Horizon* eventually did use the kill line for this purpose.[12]

Just before 4 p.m., the crew took its next preparatory step. They pumped seawater down the drill pipe to displace the drilling mud in the pipe and then continued pumping seawater until they displaced mud in the casing above 8,367 feet with seawater as shown in Figure 4.6.10.[13] Because mud is expensive and reusable, and because direct contact with seawater would contaminate it, the crew used spacer fluid as a buffer to separate the seawater from the mud. The crew's goal was to displace the heavy mud and spacer fluid entirely above the BOP.

Figure 4.6.10. 4 p.m.

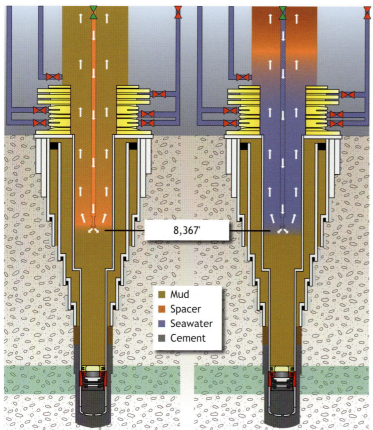

8,367'

Mud
Spacer
Seawater
Cement

TrialGraphix

The crew displaced the mud in the drill pipe and in the casing from 8,367 feet to above the BOP. The crew first pumped a spacer fluid down the drill pipe, which forced the mud out and up the casing and the riser (left). Following the spacer, the crew pumped seawater into the drill pipe. This forced the spacer and the mud up the casing. The crew's intent was to pump enough seawater to displace the spacer and mud above the BOP (right).

Use of Lost Circulation Material as Spacer

Operators commonly choose to use a spacer during displacement. However, BP chose to use a somewhat unusual *type* of spacer fluid at Macondo. BP chose to use a fluid composed of leftover **lost circulation materials** stored on the rig. As previously discussed, BP engineers had been concerned about the risk of further lost returns since the lost circulation event in early April. BP had asked M-I SWACO to make up at least two different batches, or "pills," of lost circulation material for that contingency—on[e] commercially known as Form-A-Set and the other as Form-A-Squeeze. BP decided to combine these materials for use as a spacer during displacement.

The combined spacer material that BP chose thus had two unusual characteristics. First, the material was denser than the drilling mud in the well and, at 16 pounds per gallon (ppg), much denser than 8.6 ppg seawater.[14] While using such a dense spacer would arguably assist in displacing mud down and out of the drill pipe, it could prove problematic as well. BP's plan called for the spacer to be pushed up through the wellbore and into the riser by the seawater flowing behind it. By using a spacer that was so much denser than the seawater, BP increased the risk that the spacer would instead flow downward *through* the seawater, potentially ending up *beneath* the BOP and confounding the negative pressure test.[15]

Second, the lost circulation materials that BP combined to create its spacer created a risk of clogging flow paths that could be critical to proper negative pressure testing. Much as blood clots to stop a bleeding wound, viscous lost circulation materials are designed to plug fractured formations to prevent mud from leaking out of a well. M-I SWACO therefore warned BP before the negative pressure test that spacer composed of lost circulation material could "set up" or congeal in "small restrictions" in tools on the drill pipe.[16]

The Chief Counsel's team found no evidence that anyone in the industry had ever used (or even tested) this type of spacer before, much less that anyone at BP or on the rig had done so.[17] There also appears to be no operational reason BP chose to use the lost circulation material as a spacer.[18] Rather, according to internal BP emails and the testimony of various witnesses, BP chose to use the lost circulation pills as a spacer in order to avoid having to dispose of the material as hazardous waste pursuant to the Resource Conservation and Recovery Act (RCRA).[19]

RCRA regulations would normally have required BP to treat and dispose of the two pills as hazardous waste. But BP and M-I SWACO reasoned that once the two pills had been circulated down through the well as a spacer they could be dumped overboard pursuant to RCRA's exemption for water-based drilling fluids.[20] This is what prompted BP to direct M-I SWACO to use the lost circulation material as a spacer.[21] This decision would save BP the cost of shipping the materials back to shore and disposing of them as hazardous waste.[22]

These disposal concerns also led BP to use an unusually large volume of spacer material at Macondo. Typically, 200 barrels of spacer are enough to provide an adequate buffer between mud and seawater.[23] BP chose to pump 454 barrels of its unusual combined spacer fluid at Macondo.[24]

Figure 4.6.11. 4:53 p.m.

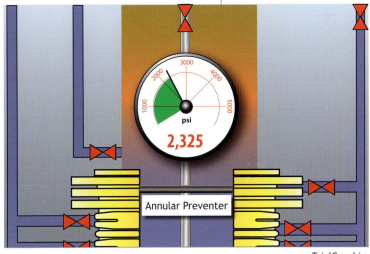

TrialGraphix

The crew closed the annular preventer around the drill pipe. The drill pipe pressure was approximately 700 psi higher than should have been expected, a sign that some spacer may have remained beneath the BOP.

Unlikely Displacement of All Spacer Above the BOP

After pumping 352 barrels of seawater behind the spacer, the crew closed the upper annular preventer, believing that they had displaced all of the spacer above the BOP.[25] BP's post-incident report calculates that the crew was correct, albeit by a slim margin of just 12 feet.[26] But that calculation is optimistic. It assumes that none of the heavy spacer fell back down through the much lighter seawater that was pushing it upward through the wellbore. Given the substantial density differential between the spacer and seawater and the substantial amount of time it took to displace 454 barrels of spacer, it is likely that at least some of the spacer fell backward through, or mixed with, the seawater on its way up the casing into the riser. Even putting aside that complication, Transocean and at least one independent expert have calculated that the tail end of the spacer did not end up above the BOP.[27]

Figure 4.6.12. 4:55 p.m.

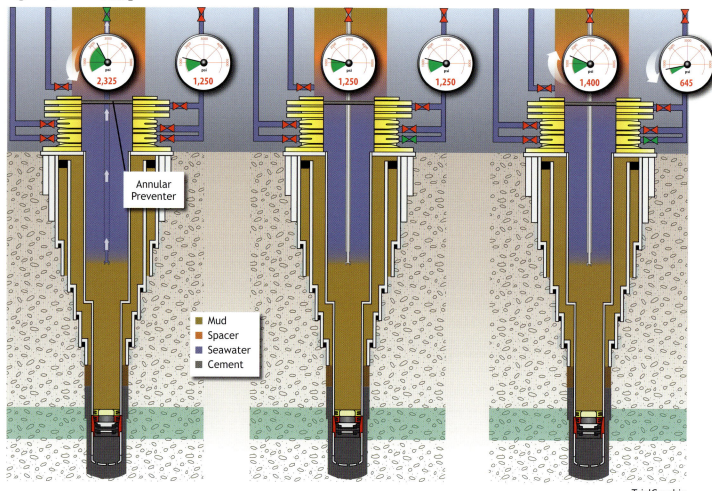

TrialGraphix

After the annular preventer was closed (left) the crew bled down the pressure in the drill pipe to equalize its pressure with the pressure in the kill line. Because both the drill pipe and the kill line go to the same vessel, when the valve connecting the kill line to the BOP is opened, the pressures should remain equal (middle). Instead, when the valve was opened, the pressures diverged (right).

Because the BOP and wellhead were a mile beneath the rig, the crew had no way of observing directly whether they had displaced all of the spacer above the annular preventer. But pressure readings on the drill pipe should have alerted them that something was amiss. When the crew first closed the annular preventer around the drill pipe (see Figure 4.6.11), the pressure on the drill pipe was approximately 700 psi higher than it should have been.[28] That anomaly should have merited further investigation because it could have indicated that spacer remained below the BOP. But it does not appear that anyone in the drill shack had ever calculated what the drill pipe pressure should have been.[29]

This higher-than-expected pressure was the first of many unrecognized and unheeded anomalous readings during the negative pressure test.

The rig crew next bled the drill pipe to 1,250 psi, in an effort to equalize pressure on the drill pipe with pressure on the kill line (which was 1,250 psi at the time, as shown in Figure 4.6.12).[30] Once the crew had bled the drill pipe pressure down to 1,250 psi, it opened a valve on the kill line at the BOP so that both the drill pipe and kill line were open to the well. At this point, the drill pipe and kill line should have behaved like two straws in the same glass of water: The pressure in both should have been a steady 1,250 psi. Instead, when rig personnel opened the valve, the drill pipe pressure jumped, and the kill line pressure dropped.[31]

Figure 4.6.13. 4:58 p.m.

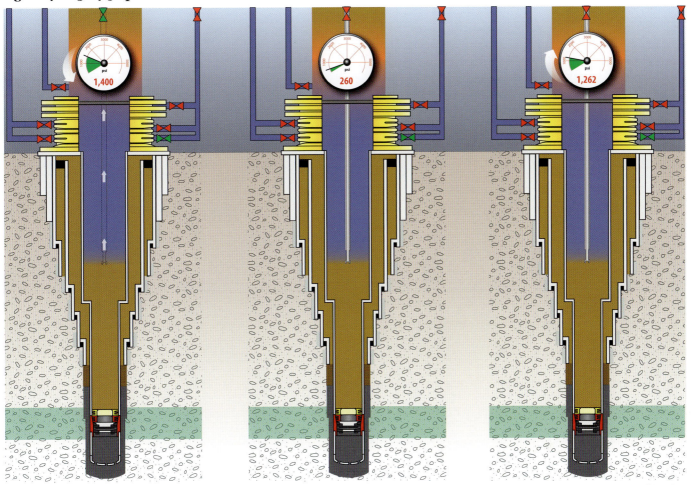

TrialGraphix

The crew began the negative pressure test by attempting to bleed the drill pipe pressure to 0 psi (left). However, the crew was unable to reduce pressure to below 260 psi (middle). This bleed returned an unknown amount of water to the rig. The crew shut in the drill pipe, and the pressure built up to 1,262 psi (right). In a successful negative pressure test, pressure does not build up.

This should have been another indication that spacer might have ended up beneath the BOP or that something else was amiss.[32] There is some evidence that the crew or well site leaders may have recognized a concern, but nobody appears to have acted upon it.[33] In what became a pattern, individuals on the rig did not take a simple precaution: They could have opened up the annular preventer, pumped more seawater into the well to ensure that all spacer had been displaced above the BOP, and begun the negative pressure test anew.[34] This would have taken time but also would have ensured that misplaced spacer did not confound the test results.

The First Negative Pressure Test

Just before 5 p.m., the crew opened a valve at the top of the drill pipe on the rig and attempted to bleed the drill pipe pressure down to 0 psi, as shown in Figure 4.6.13. The crew was unable to do so and could only reduce pressure to 260 psi.[35] It is not clear how many barrels of fluid the crew bled off at this point. Three witnesses have testified that 23 to 25 barrels were bled off; other accounts suggest it may have been more or less.[36]

The uncertainty over how much fluid flowed from the well during the bleed-off suggests that the well site leaders and crew failed to monitor the bleed-off volumes with requisite rigor. It does not appear that anyone had calculated ahead of time how many barrels should have flowed from the

Figure 4.6.14. 5:10 p.m.

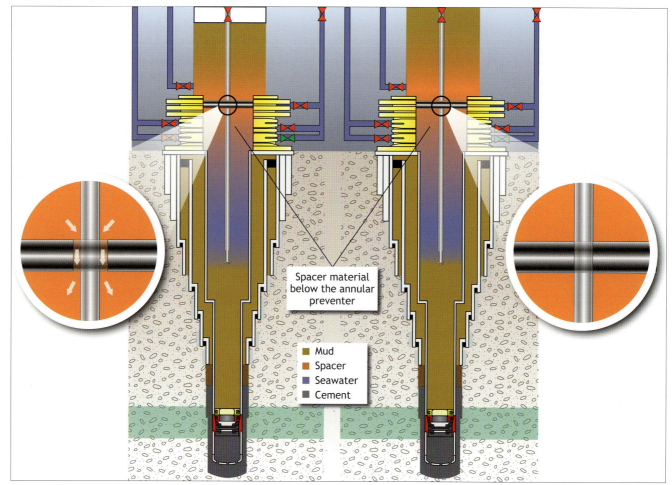

Spacer material below the annular preventer

Mud
Spacer
Seawater
Cement

TrialGraphix

The rig crew noticed that the fluid level in the riser was falling. Because the annular preventer was not sufficiently tight around the drill pipe, spacer fell beneath the BOP (left). In response, the rig crew tightened the seal of the annular preventer and refilled the riser (right), but did not circulate the spacer back above the BOP.

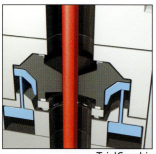

TrialGraphix

Annular Preventer. The annular preventer is a hard rubber donut that surrounds the drill pipe; when activated it expands and fills the space around the drill pipe, sealing the well below (see also Figure 2.9).

well during the bleed, even though such calculations would have been relatively straightforward.[37] After failing to bleed the pressure down to 0 psi, the crew closed the valve on the drill pipe, and the pressure built back up to 1,262 psi.[38]

These events indicated that the well was not behaving as a closed system. Something was entering the well, although the source of the material entering the well was indeterminate. If the well had been a closed system, the crew would have had no difficulty bleeding the drill pipe pressure down to 0 psi, and the well would have returned far less than 23 barrels of fluid during the bleed-off.[39] Also, the drill pipe pressure would not have increased.

As one independent expert has pointed out, this series of events actually constituted a failed negative pressure test, although the crew did not recognize that fact.[40]

At 5:10 p.m., the rig crew apparently noticed that the level of fluid in the riser was falling.[41] Spacer in the riser was leaking down through the annular preventer and into the well below the BOP.[42] Unlike many other indications, the crew could observe the fluid levels in the riser with their own eyes. When one rig crew member arrived on the rig floor, he saw others standing around the rotary table and using a flashlight to peer down into the riser to see how much fluid was missing.[43]

Around this time, the night crew began to gather at the drill shack in anticipation of the 6 p.m.

shift change. The night crew would include Transocean toolpusher Jason Anderson and M-I SWACO mud engineer Gordon Jones.

A group of visiting BP and Transocean executives also entered the drill shack as a part of a rig tour. They were escorted by Transocean offshore installation manager Jimmy Harrell and senior toolpusher Randy Ezell. The drill shack was so crowded with the shift relief and tour group that it was "standing room only,"[44] Transocean executive Daun Winslow recognized that the drilling team was confused about something. When the tour group left the drill shack, Winslow asked Harrell and Ezell to remain behind to assist.[45]

In response to the dropping levels of fluid in the riser, Harrell instructed the rig crew to tighten the seal of the annular preventer against the drill pipe as shown in Figure 4.6.14. Wyman Wheeler, the Transocean toolpusher on duty at the time, then topped off the riser with 20 to 25 barrels of mud, and the fluid level in the riser stayed steady.[46] The crew had thus identified and eliminated a leak in the well system that could have explained the anomalous pressure readings they had seen and their inability to bleed the drill pipe pressure to 0 psi.[47] By this time Kaluza returned to the rig floor.[48]

Despite clear evidence that spacer had probably leaked below the BOP, rig personnel again did nothing to ensure that they had fully displaced the spacer above the BOP and instead proceeded with the test.[49]

Having tightened the annular preventer, the crew once again tried to bleed the pressure in the drill pipe to 0 psi as shown in Figure 4.6.15. This time they were successful. According to witness accounts, 15 barrels of fluid were bled off from the drill pipe in the process.[50] Again, nobody had done any calculations to predict the returns. Those calculations would have predicted only three to five barrels of returns; the bleed-off process had produced more fluids than it should have.[51]

The crew shut in the drill pipe, but the pressure again built back up.[52] In this case, the pressure reached 773 psi and most likely would have gone higher had the crew not begun immediately bleeding it off.[53]

This second series of bleed-offs, excessive flows, and pressure buildups constituted another failed "negative pressure test" that the crew again did not recognize as such. With the annular preventer fully closed and sealed, the only explanation for the excessive returns and pressure increase would be that the primary cement job had failed to seal off the pay zone. Hydrocarbons were leaking from the formation into the well. Individuals involved in the test at this point should have recognized that the well lacked integrity.

Figure 4.6.15. 5:26 p.m.

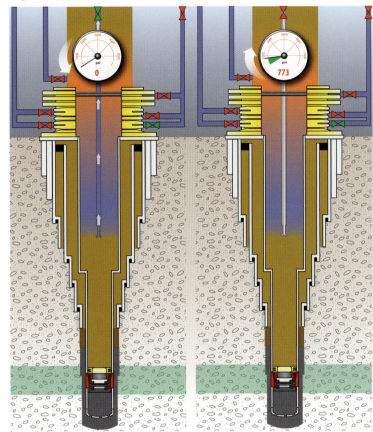

TrialGraphix

The rig crew attempted again, this time successfully, to bleed the drill pipe pressure down to 0 psi. Fifteen barrels of seawater were returned during this bleed (left). The drill pipe was shut in, but the drill pipe pressure rose to 773 psi. Fifteen barrels is a higher return than should have been expected, and the drill pipe pressure should not have built back up.

Figure 4.6.16. 5:53 p.m.

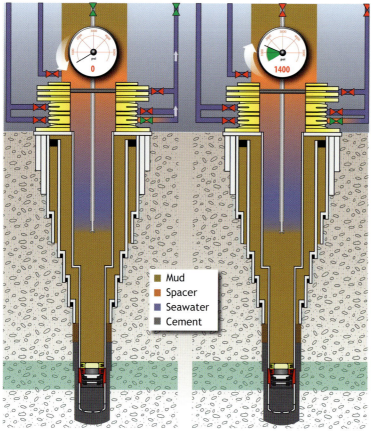

Mud
Spacer
Seawater
Cement

TrialGraphix

The rig crew bled drill pipe pressure down for a third time, this time through the kill line. Witnesses reported that three to 15 barrels were returned as the drill pipe reached 0 psi (left). When the drill pipe was shut, the drill pipe pressure rose to 1,400 psi. According to BP witnesses, the Transocean rig crew attributed this rise to a "bladder effect." A 15-barrel return would have been excessive, and the rise of drill pipe pressure to 1,400 psi was a clear sign that the negative pressure test had failed (right).

According to at least one witness, shortly before 6 p.m. Kaluza directed the crew to bleed down the drill pipe pressure by opening the kill line rather than the drill pipe. Because the kill line and drill pipe both led to the same place (again, like two straws in the same glass of water), bleeding pressure from the kill line would also cause drill pipe pressure to drop to 0 psi. It is not clear why Kaluza directed the crew to bleed down the drill pipe pressure by opening the kill line. The switch may be significant, however, as it suggests uncertainty about the pressure readings and flow observations. "Let's open the kill line and see what happens," Kaluza reportedly said.[54] Shortly afterward, Kaluza left the rig floor to speak with the Don Vidrine, the other BP well site leader whose shift was about to begin.[55]

Witnesses have provided differing estimates of the amount of seawater the crew bled from the kill line, ranging from three to 15 barrels.[56] Flows in the upper end of this range would have been more than expected—but once again, nobody calculated ahead of time what flows to expect. As the pressure on the drill pipe dropped almost to 0 psi,[57] the kill line continued to flow and spurt water until the crew closed the line's upper valve on the rig.[58] Over the next 30 to 40 minutes, the drill pipe pressure rose to 1,400 psi as shown in Figure 4.6.16.[59]

This was the clearest indication yet that the well lacked integrity. The 1,400 psi pressure buildup can only have been caused by hydrocarbons leaking into the well from the reservoir formation.

One expert described this test result as a "conclusive failure."[60] Later analysis has shown that 1,400 psi is approximately the reading that one would have expected reservoir hydrocarbon pressure to produce at the surface if there had been *no cement* at the bottom of the well during the negative pressure test.[61]

Kaluza returned to the rig floor with Vidrine, who would soon be relieving him.[62] While personnel at the rig had not treated earlier pressure readings and flow observations as problematic indications, the two well site leaders and other rig personnel did recognize that the rise in drill pipe pressure to 1,400 psi was a cause for concern.[63] According to witness accounts, Kaluza and Vidrine discussed the test in the drill shack together with Anderson, Revette, assistant driller Steve Curtis, and BP well site leader trainee Lee Lambert.[64] Because Kaluza's and Vidrine's accounts are only known through BP internal investigation notes, and the Transocean personnel principally involved did not survive (Ezell has stated that he did not take part in any such conversation and that he left the drill shack before the drill pipe pressure reached 1,400 psi), the details of the discussion are unclear. Transocean has challenged the accounts of the three BP witnesses, but those three accounts are consistent with each other, and at this point the Chief Counsel's team has no testimonial or documentary evidence that conflicts with them.

According to notes from BP's post-incident interviews of Kaluza and Vidrine, as well as testimony from Lambert, Anderson explained that the 1,400 psi pressure on the drill pipe was being caused by a "bladder effect" or "annular compressibility."[65] According to Lambert, Anderson explained that "heavier mud in the riser would push against the annular and transmit pressure into the wellbore, which in turn you would expect to see up the drill pipe," as illustrated in Figure 4.6.17.[66]

The Chief Counsel's team found no evidence to support this theory. Indeed, every industry expert that the Chief Counsel's team spoke with agreed that no such phenomenon exists. Even if it did exist, any pressure caused by this "bladder effect" would have disappeared after the rig crew bled off the drill pipe and kill line.[67]

Any "bladder effect" could not explain the 1,400 psi on the drill pipe.

Although there was a long discussion about the drill pipe pressure, it does not appear as though anyone in the discussion seriously challenged the bladder effect. According to BP witness accounts, Anderson explained that the pressure buildup after bleeding was not unusual. He told the well site leaders, "Bob and Don, this happens all the time."[68] Revette, the driller, apparently agreed that he had seen the bladder effect before.[69] Lambert testified that he asked about the phenomenon but accepted Anderson's explanation. On later reflection after the blowout, however, Lambert agreed that the explanation did not make sense.[70]

The conversation apparently turned to conducting another negative pressure test, this time on the kill line instead of the drill pipe. According to witness accounts, Vidrine insisted that the crew perform a new negative pressure test on the kill line because the latest permit that BP had submitted to MMS stated that BP would conduct the test on the kill line.[71] But it is unlikely that Vidrine made this decision solely because of the permit language; the rig crew had conducted the first test on the drill pipe without regard to the permit. Moreover, the BP team had already consciously deviated from the permit when it instructed the crew to conduct a combined displacement and negative pressure test–the permitted procedure did not specify such a step.[72] It appears instead that Vidrine insisted on a kill line test at least in part out of concern over the results of the negative pressure test on the drill pipe.[73] But again, neither Vidrine, Kaluza, nor the rig crew treated the test on the drill pipe as a failure. Instead, they chose to disregard it in favor of a new test on the kill line.

Figure 4.6.17

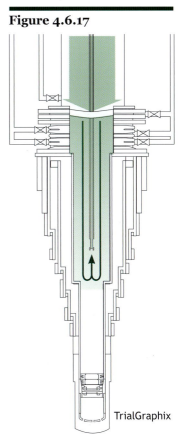

TrialGraphix

Bladder Effect. This figure represents the Chief Counsel's team's understanding of the "bladder effect" theory that supposedly explained the elevated pressure on the drill pipe. The "bladder effect" explanation contends that heavy fluids (mud and spacer) displaced to the riser were exerting force on the annular preventer from above, which in turn communicated pressure into the well. The Chief Counsel's team found no evidence to support this theory.

The Second Negative Pressure Test

Figure 4.6.18. 6:40 p.m.

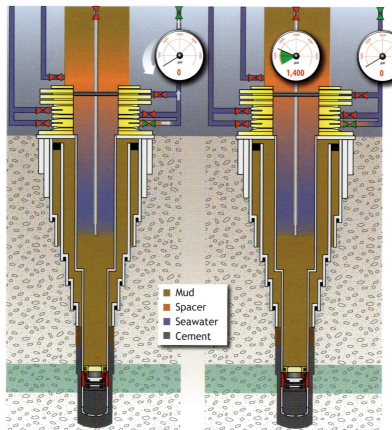

Mud
Spacer
Seawater
Cement

TrialGraphix

The rig crew conducted a negative pressure test on the kill line. The rig crew reduced the pressure on the kill line to 0 psi, bleeding an insignificant amount of water (left). No flow or pressure buildup was observed on the kill line, which on its own would have been a successful negative pressure test. However, the 1,400 psi on the drill pipe remained and was never properly accounted for (right).

Sometime after 6:40 p.m. on April 20, while the group in the drill shack continued to discuss the test, the crew moved the negative pressure test to the kill line at Vidrine's behest. The crew pumped a small amount of fluid into the kill line from the rig to ensure the kill line was full. They plumbed the kill line so that fluids could be bled off into the "mini trip tank" near the drill shack and then bled the pressure on the kill line down to 0 psi as shown in Figure 4.6.18. According to witness accounts, less than one barrel of seawater flowed from the kill line, an insignificant amount. Once that flow stopped, beginning at about 7:15 p.m., the crew monitored the kill line for 30 minutes and observed no additional flow or pressure buildup.[74]

The lack of pressure or flow on the kill line, on its own, would have meant a successful negative pressure test. *But the 1,400 psi on the drill pipe had never disappeared.*

The well site leaders and rig crew carried on their discussion about the test and whether the 1,400 psi on the drill pipe was acceptable. Vidrine later told BP interviewers that he continued talking about the 1,400 psi reading for so long that the rig crew found it "humorous."[75] Anderson and Revette apparently continued to explain the pressure as a "bladder effect." Kaluza's statements to BP investigators suggest that he was present for the discussion as well and that he too accepted the Transocean explanation. He justified his acceptance to the investigators by saying that if Anderson had seen this phenomenon so many times before it must be real.[76] In an email written after the blowout, Kaluza explained to BP management:

> Please consider this suggestion in the analysis about how this happened. I believe there is a bladder effect on the mud below an annular preventer as we discussed.... Due to a bladder effect, pressure can and will build below the annular bladder due to the differential pressure but can not flow – the bladder prevents flow, but we see differential pressure on the other side of the bladder.[77]

In the end, everyone apparently accepted that the negative pressure test on the kill line established that the primary cement job had successfully sealed off hydrocarbons in the pay zone.[78]

Transocean and BP have each contested their relative involvement in the negative pressure test and their relative legal responsibilities for interpreting it. The determination of legal responsibility is beyond the scope of this Report. However, experts and witnesses alike agree that industry practice requires the well site leader to make the final decision regarding whether the test has passed or failed.[79] There is also widespread agreement that the rig crew plays some role in interpreting tests, given their experience in running them and their authority to stop work if they recognize a safety concern.[80]

The Chief Counsel's team believes that the group of personnel involved in the Macondo negative pressure test—including Transocean drilling personnel and the two BP well site leaders—decided as a group that the test had succeeded.[81] It appears that the highly experienced Transocean crew[82] affirmatively advocated the view that the first and second negative pressure tests were acceptable once the "bladder effect" was considered, and the well site leaders eventually agreed. The long time spent conducting and discussing the tests shows a desire for consensus. It is possible, even likely, that this desire obscured the parties' responsibilities.

It does not at this time appear that either the BP well site leaders or the Transocean drilling crew ever sought guidance from others on the rig or onshore. For instance, based on available evidence, it does not appear that the BP well site leaders ever called the shoreside BP engineering team to ask for advice on interpreting or conducting the negative pressure test (Ezell also stated that nobody spoke with him regarding the test results).[83] BP did not require its well site leaders to obtain shoreside approval before directing the rig crew to begin temporary abandonment operations.[84] But the shoreside team had valuable expertise and experience. They could have answered questions about the test results, just as they often did regarding other drilling operations.[85] John Guide, BP's Houston-based wells team leader, later stated that given the pressure readings, he would have expected a call from the rig.[86]

Instead, Vidrine apparently deemed the test successful. No one disagreed,[87] and the rig crew moved on to begin displacing the remaining mud from the riser. Vidrine did speak to BP senior drilling engineer Mark Hafle by telephone shortly before 9 p.m., roughly an hour after the negative pressure test was finished. Hafle had called from Houston to see how operations were proceeding. Hafle had the transmitted Macondo drilling data up on his monitor. Vidrine told Hafle that there had been issues with the negative pressure test. He may specifically have told Hafle about the 1,400 psi seen on the drill pipe, and Hafle would have been able to see on his computer the recorded pressures from the test. But Vidrine explained that the test issues had been resolved.[88] The Chief Counsel's team has not seen any evidence of any further discussion of the test with BP personnel onshore.[89]

The second negative pressure test showed again that the well lacked integrity. The 1,400 psi reading from the drill pipe indicated that hydrocarbons were leaking into the well. The fact that the kill line pressure was 0 psi at this time suggests that something may have been blocking fluids from flowing through the kill line and transmitting pressure to the gauges on the rig. One possibility, alluded to earlier, is that the spacer below the BOP had migrated into the $3\frac{1}{16}$-inch

Figure 4.6.19. Spacer migration.

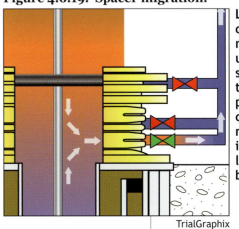

Leftover lost circulation material used as spacer for the negative pressure test could have migrated into the kill line during bleeds.

TrialGraphix

diameter kill line and clogged it.[90] It is also possible that rig crews accidentally closed a valve that should have been open. The kill line could also have been clogged by undisplaced mud in the kill line or by gas hydrates that solidified during the test (the same type of hydrates that complicated containment operations).[91] The exact reason may never be known.

Technical Findings

The Negative Pressure Test Showed That the Cement Failed

The pressure readings and flow indications during the negative pressure test were not ambiguous. In retrospect, BP, Transocean, independent experts, and other investigations all agree that this critical test showed that the cement had failed and there was a leak in the well.[92]

There were three instances in which pressure built up after being bled off, including the buildup experts have deemed a "conclusive failure" wherein pressure inside the drill pipe rose from 0 to 1,400 psi.[93] On at least one occasion, bleed-off procedures produced more flow than should have been expected. And while the rig crew observed no flow from the kill line during the second negative pressure test, the drill pipe pressure remained at 1,400 psi.

The test failure should have been clear even though the well site leaders and rig crew had complicated matters by using an untested spacer and by allowing the spacer to leak below the BOP during the test. The well site leaders and rig crew never should have accepted the test as a success or continued with displacement operations.

BP's Spacer Choice Complicated the Negative Pressure Test

BP's decision to use 454 barrels of a highly viscous spacer may have confounded the negative pressure test. All parties agree that at some point during the negative pressure test the spacer had leaked beneath the BOP and that the rig crew never circulated it out. That spacer may have migrated into and clogged the open kill line. If there had been a clear path through the kill line down to the wellhead, the rig crew would have observed the same 1,400 psi pressure inside the kill line that they saw on the drill pipe.[94] If that had happened, the crew might have recognized that the second negative pressure test had failed.

The Chief Counsel's team did not examine the legal significance of BP's decision to use lost circulation materials as spacer and then discharge them directly into the Gulf of Mexico. But the Chief Counsel's team does conclude that greater care should have been taken first in testing and then in monitoring the placement of this unusual spacer.

BP's own investigative report states that its team used the spacer because of a "perceived expediency."[95] Although BP had never used this material as a spacer before or tested it for such use, and although BP used twice as much spacer at Macondo as it had used at other similar jobs, the company did not undertake a risk analysis to consider the consequences of its decision.

BP thus did not consider the risk that a dense spacer made of lost circulation materials could be left beneath the BOP, potentially clogging crucial piping paths.

Rig Personnel Should Have Displaced All Spacer Above the BOP

Personnel involved in the test may have further confounded the negative pressure test by failing to set up the test as intended. They knew for at least two reasons that heavy spacer fluid had leaked beneath the BOP where it could potentially confuse test results. First, they observed that the pressure inside the drill pipe was 2,325 psi when the annular was closed at 5 p.m. Second, when they opened the kill line, they observed a drop in pressure on the kill line and a simultaneous jump in pressure on the drill pipe.

Despite these indicators, the individuals conducting the test did not try to correct the problem even after they decided to run a second negative pressure test. They could easily have circulated the spacer out of the wellbore to ensure that the test was set up as planned. They should have done so.[96]

Management Findings

Given the risk factors attending the bottomhole cement, individuals on the rig should have been particularly attentive to anomalous pressure readings. Instead, it appears they began with the assumption that the cement job had been successful and kept running tests and proposing explanations until they convinced themselves that their assumption was correct. The fact that experienced well site leaders and members of the rig crew believed that the Macondo negative pressure test established well integrity demonstrates serious management failures.

There Were No Established Procedures or Training for Conducting or Interpreting the Negative Pressure Test

Lack of Standard Procedures

Neither BP nor Transocean had pre-established standard procedures for conducting a negative pressure test.[97] While BP required negative pressure tests under certain conditions, one of its employees admitted that the tests "could be different on every single rig depending on what the [well] team agreed to."[98] Transocean likewise required negative pressure tests but did not have set procedures.[99] For example, the crew of the *Marianas* had done the immediately preceding negative pressure test at Macondo in a different way than the *Deepwater Horizon* crew did the April 20 test.[100] Partly because Transocean rigs conducted tests differently (in part because different rigs have different equipment), Kaluza and BP drilling engineer Brian Morel both spoke with an M-I SWACO engineer on April 20 to ask how the rig had previously conducted the negative pressure test.[101]

Unfortunately, the lack of standard test procedures is unsurprising. In April 2010, MMS regulators did not require operators even to conduct negative pressure tests, let alone spell out how such tests were to be performed.[102] (The Chief Counsel's team notes that some wells need not be negative tested.[103]) Nor had the oil and gas industry developed standard practices for negative pressure tests.[104] An independent expert admitted that he had to consult an academic text to find a description of a negative pressure test procedure.[105]

The recent regulatory proposal to require negative pressure test information in permit applications to MMS may trigger companies and the industry to establish standard negative pressure test procedures[106] (discussed further in Chapter 6). A negative pressure test procedure ought to include the depth of mud displacement, the volumes of fluids to be pumped into the well, the pressures and fluid returns to be expected during the test, and criteria for determining whether the negative pressure test passes or fails. The procedure should also include explicit instructions for diagnosing and addressing problematic or anomalous test readings.

Lack of Training at Macondo

BP well site leaders displayed troubling unfamiliarities with negative pressure test theory and practice. Neither Kaluza nor Vidrine calculated expected pressures or volumes before running the negative pressure test even though other BP well site leaders routinely do so.[107] Vidrine, Kaluza, and Morel all described the criteria for a successful test in terms of "flow or no-flow," which ignores the importance of monitoring pressures in the well.[108] Both well site leaders apparently accepted the "bladder effect" explanation, and Kaluza continued defending the theory and describing the Macondo test results as "rock solid" a week after the blowout.[109] These are clear signs that BP needs to train its personnel better.

Transocean has acknowledged that it does not train its personnel in the conduct or interpretation of negative pressure tests and that its Well Control Handbook does not describe a negative pressure test. Instead, Transocean states that its rig crews learn how to conduct a negative pressure test through general work experience.[110]

Partly because of this, Transocean has been unable to conclude whether its *Deepwater Horizon* rig crew had enough experience to conduct and interpret the negative pressure test on April 20.[111] Transocean is not unique in omitting training for the negative pressure test. Experts have stated that academic training on the negative pressure test may only be included in coursework as time allows.[112]

Transocean has argued that the members of its rig crew were tradesmen, not engineers, and could not have been expected to interpret the complex results of the Macondo negative pressure test. Transocean's training approach certainly supports that view.

However, a negative pressure test essentially consists of underbalancing a well and then watching to see if a hydrocarbon kick enters the well as a result. Transocean expected its rig crew to recognize signs of a kick during complex drilling operations. It appears inconsistent for Transocean to claim that its crew is trained in and skilled in recognizing kick indicators during drilling but is unable to recognize the same kick indicators during controlled testing.

Inadequate Procedures for Macondo

The most conspicuous problem with the negative pressure test procedures at Macondo is that there were almost no written procedures at all. As described in Chapter 4.5, although BP eventually developed temporary abandonment procedures that included a negative pressure test, the procedures stated only when the test would be done in relation to other operations. BP did not explain to the crew or its well site leaders how they should perform or interpret the test. The final M-I SWACO procedure, for instance, said simply, "[c]onduct negative test." After the incident, BP engineering managers opined that the Transocean crew knew how to conduct a negative test, and that these limited instructions should have been adequate.[113] Whether justified or not, the events of April 20 prove that BP's expectation was incorrect.

BP's early plans for abandonment repeatedly failed to mention a negative pressure test at all.[114] On April 12, Morel circulated a draft temporary abandonment plan that did not include a negative pressure test.[115] Morel's omission may have been a mere oversight, but it may also have signaled his unfamiliarity with the test.

Ronnie Sepulvado, one of BP's *Deepwater Horizon* well site leaders who was not on the rig for the negative pressure test, needed to tell Morel that he should include one.[116] Similarly, Kaluza's pre-tour briefing to the rig crew described temporary abandonment procedures that did not include a negative pressure test. This prompted Harrell to state that Kaluza needed to add a negative pressure test.[117] Kaluza's omission, like Morel's, may have signaled unfamiliarity with the test and its importance.

Although Morel and other BP engineers continually refined their temporary abandonment procedures, they never expanded their negative pressure test procedures to explain what pressures or flow volumes the crew should expect to see.[118] Even more importantly, they did not add criteria for determining if the test had passed, nor contingency procedures in case the test failed. Kaluza admitted "[w]e didn't talk about what if the negative test fails."[119] Moreover, several of the BP Macondo team's early descriptions of the negative pressure test (including the one approved by MMS) were written so imprecisely that team members disagree even today about what they mean (as described in Chapter 4.5). Nor were the later descriptions passed along in "Ops Notes" or telephone calls necessarily better. When Hafle called Kaluza to discuss the test on the afternoon of April 20, he "had [the] impression that Kaluza wasn't really clear on neg[ative pressure] test procedure."[120] Unfortunately, neither Hafle nor Kaluza seemed to think this uncertainty was a problem, because they appear to have ended the call without resolving it.

Lindner eventually wrote a displacement procedure for BP that contained the most detailed procedure for running the negative pressure test. Lindner's document spelled out how much spacer and seawater the rig crew should pump into the well before conducting the test. His was the first procedure that reflected BP's decision to use a large combined spacer fluid to help displace mud from the well.[121] But it told rig personnel nothing about expected bleed-off volumes, how to interpret the negative pressure test, or what to do about anomalous pressure readings. It may also have included errors. For example, Lindner's calculations directed rig personnel to pump a volume of seawater that may have been too small to fully displace spacer above the blowout preventer. In retrospect, it is inexcusable that the most detailed written procedures for the negative pressure test were written by a mud engineer in the course of specifying fluid volumes to be displaced prior to the test.

Finally, the men on the rig did not always follow the few clearly written procedures that they had. Beginning April 14, the procedures directed that the negative pressure test would be conducted on the kill line. But rig personnel did not follow this instruction during the first negative pressure test. Instead, they conducted the initial negative pressure test on the drill pipe. This may suggest that in addition to creating better test procedures, BP and Transocean need to ensure that those procedures are followed.

BP Failed to Recognize and Alert Rig Personnel to the Exclusive Reliance on the Negative Pressure Test at Macondo

Both the Macondo well plan and the challenges surrounding the Macondo cement job put a premium on the negative pressure test. BP's temporary abandonment procedures required the crew to severely underbalance the well and to rely solely on the high-risk bottomhole cement as the exclusive barrier in the wellbore to flow while they displaced mud from the riser.

Despite these facts, BP never emphasized to rig personnel the particular importance of the Macondo negative pressure test. BP personnel forgot even to mention the test during relevant communications on at least two occasions. (See "Inadequate Procedures for Macondo" section, above). Had BP properly emphasized the importance of the test and the need for special scrutiny of its results, BP and Transocean personnel on the rig may have reacted more appropriately to the anomalous pressure readings and flows they observed.

Leadership and Communication

Even in the absence of detailed procedures, BP well site leaders should have exercised better judgment and initiative. When they confronted a 1,400 psi pressure reading from the drill pipe and a 0 psi reading from kill line, they should have insisted on probing and fully resolving the issue. Instead, interview notes suggest that they deferred to a toolpusher's explanation without fully understanding, questioning, or testing it.

Kaluza was not on the rig floor during most of the preparations for the test and may have missed the first part of the attempted negative pressure test on the drill pipe. He was in the well site leader's office doing calculations for the planned cement plug.[122] Had he been on the rig floor and participating in the test the entire time, Kaluza would have been in a better position to observe several anomalies, including:

- the excessive pressure (2,325 psi) at the end of the pre-test fluid displacement;
- the pressure changes in the drill pipe and kill line when the rig crew opened the kill line valve at the BOP;
- the rig crew's inability to bleed the drill pipe below 260 psi and the abnormally large volume of fluid flow during that bleed; and
- the drop in the fluid level in the riser.[123]

One BP well site leader who was not on the rig on April 20 stated that his practice during negative pressure tests is to remain on the rig floor from the beginning of preparations until he signs off on the test.[124] Independent experts have stated that well site leaders should certainly be present as seawater is pumped out of the drill pipe during displacement and before the crew begins any bleeds.[125]

Kaluza also apparently never personally analyzed the unusual spacer that the rig crew used during his shift.[126] And notes of his statements to BP investigators suggest that he did not recognize that such a spacer could confound the negative pressure test.[127] One independent expert has stated that it would have been standard industry practice for the well site leader to "personally confirm[] the properties of the final blend."[128]

Most significantly, it appears that neither the BP well site leaders nor the Transocean drilling team ever called shore-based personnel to ask for assistance, to report the anomalous pressure readings, or to check the "bladder effect" explanation. Neither company had specific policies in place that required their personnel to report the results of the test to shore.[129] But both BP and Transocean expected rig personnel to call if they needed help or were uncomfortable.[130] Indeed, BP personnel called to shore on April 19 to discuss the problems the rig crew was experiencing while trying to convert the float collar.[131] Instead, the well site leaders and drilling team relied solely on their own limited experience and training to wrongly interpret the test results as a success. ◗

Chapter 4.7 | Kick Detection

The Chief Counsel's team finds that rig personnel missed signs of a kick during displacement of the riser with seawater. If noticed, those signs would have allowed the rig crew to shut in the well before hydrocarbons entered the riser and thereby prevent the blowout. Management on the rig allowed numerous activities to proceed without ensuring that those operations would not confound well monitoring. Those simultaneous activities did confound well monitoring and masked certain data.

Despite the masking effect, the data that came through still showed clear anomalies.[1] The crew either did not detect those anomalies or did not treat them as kick indicators.

Well Monitoring and Kick Detection

A **kick** is an unwanted influx of fluid or gas into the wellbore. The influx enters the wellbore because a barrier, such as cement or mud, has failed to control fluid pressure in the formation. In order to control the kick, personnel on the rig must first detect it, then stop it from progressing by adding one or more barriers.[2] The crew must then circulate the influx out of the wellbore. If the crew does not react properly, fluids will continue to enter the wellbore. This will eventually escalate into uncontrolled flow from the well—in other words, a blowout.[3]

In order to detect a kick, rig personnel examine various indicators of surface and downhole conditions. These indicators include pit gain, flow-out versus flow-in, drill pipe pressure, and gas content in the mud.[4]

Pit Gain (Volumetric Comparison)

Pit gain is the difference between the volume of fluid pumped into the well and the volume of fluid pumped out of the well. If the well is stable (that is, there are no gains or losses) the two should be equal.

The easiest way to monitor pit gain is to pump fluids into the well from a single pit and route returns from the well into the same pit. This is called **single-pit monitoring**. However, when dealing with several different fluids (mud, spacer, seawater), the crew must use several different pits to prevent the fluids from mixing. In order to monitor multiple pits, the crew can use the active pit system.

Active Pit System. The **active pit system** refers to a computer setting that allows the driller (and others) to select several pits and aggregate their volumes into one "active pit volume" reading. Even though there are several different pits involved, the rig's computer system displays them as a single pit for volume monitoring purposes.[5]

There are several ways to configure the active pit system. In a closed-loop system, the fluids going into the well are taken from the active pit system, and the fluids coming out of the well are returned to the active pit system. Because volume-in should equal volume-out, the active pit volume will stay constant when the well is stable. If the active pit volume increases, that strongly indicates that a kick is under way.[6] A volume increase should be easily detectable by a positive slope in the trend line (seen in Figure 4.7.1) or an uptick in the numerical data.

Figure 4.7.1. Active pit volume in a closed-loop system.

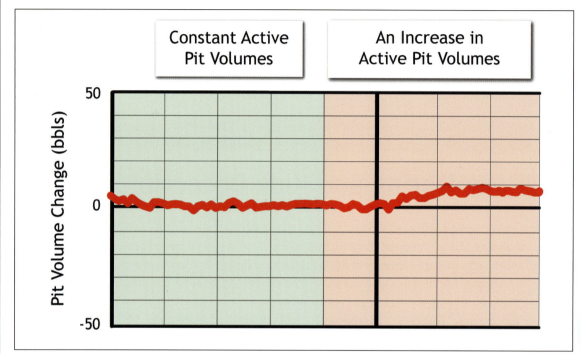

Sperry-Sun data/TrialGraphix

In a closed-loop system, active pit volume will remain constant so long as the well is stable. An increase in active pit volume strongly indicates that a kick is under way.

Monitoring pit gain in a non-closed-loop system is more complex. In a non-closed-loop system, fluids are either taken from or returned to places other than the pits on the rig. For instance, when rig crews use seawater to displace mud from a well, the rig may pump the seawater in from the ocean (and bypass the pits) but still direct mud returns back to the pits.[7] In that case, active pit volume will increase over time because the returns are filling up the pits (seen in Figure 4.7.2).

To monitor pit gain in a non-closed-loop system, rig personnel must manually calculate the volume of seawater pumped into the well (pump strokes × volume per pump stroke) and compare it to the volume of mud returning from the well (measured by changes in pit volume).[8]

Certain kinds of operations can make it impossible to use pit gain as a kick indicator. For example, this happens when return flow from the well goes overboard instead of into a pit. Rig personnel generally cannot measure the volume of flow overboard, so they cannot make a volume-in/volume-out comparison during such operations.

Figure 4.7.2. Active pit volume in a non-closed-loop system.

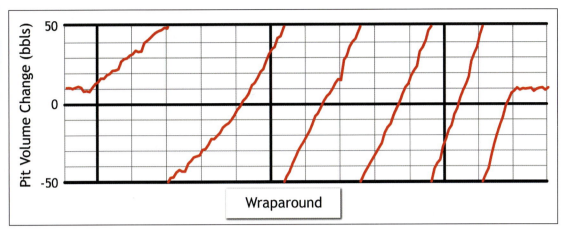

Sperry-Sun data/TrialGraphix

In a non-closed-loop system, active pit volume will increase continuously regardless of the well's stability.

Flow-Out (Rate Comparison)

Flow-in is a calculation of the rate at which fluid is being pumped into the well (pump rate × volume per pump stroke). Because it is calculated from known and reliable values, flow-in has a small margin of error. It is a trusted value.

Flow-out is a measurement of the rate at which fluid returns from the well. It is typically measured by a sensor in the flow line coming out of the well. As a result, the accuracy of the flow-out measurement depends on the quality of the sensor. It is a less reliable value than flow-in.[9]

If the well is stable, flow-in and flow-out should be equal.[10] An unexplained increase in flow-out is a kick indicator. For example, if the pump rate is constant but flow-out increases, the additional flow is likely caused by fluid or gas coming into the wellbore from the formation.[11]

The simplest application of this principle occurs when the rig is not pumping fluids into the well at all. At this point, flow-in is zero, so flow-out should also be zero. Rig personnel can confirm that flow-out is zero in two ways: by reading the data from the flow-out meter and by visually inspecting the return flow line (performing a **flow check**). If rig personnel see flow from the well at a time when the pumps are off, that is an anomalous observation. While such flow can indicate thermal expansion of the drilling fluid, rig heave, or ballooning, it can also indicate that a kick is under way.[12] In any case, further investigation is warranted.

When the rig crew first shuts pumps down, it generally takes some period of time for flow-out to drop to zero. This reflects the time it takes for the pumps to drain and for circulation to come to a stop. During this time period, there continues to be some **residual flow**.[13]

Each rig has its own residual flow-out **signature**—a pattern wherein flow-out dissipates and levels off over the course of several minutes.[14] It is important that rig personnel identify that signature and monitor flow-out for a sustained period of time afterward to confirm that there is indeed no flow after the pumps have been shut down.[15]

Flow checks constitute an important safeguard and "double-check" ensuring that the well is secure. It is therefore a common practice to assign one member of the rig crew to always visually confirm that flow has stopped whenever the pumps have been shut down, and announce it to the rest of the rig's personnel.

Drill Pipe Pressure

Drill pipe pressure is a measurement of the pressure exerted by fluids inside the drill pipe.[16] When the rig pumps are off, drill pipe pressure should remain constant.[17] When the density of fluids in the well outside the drill pipe is higher than the density of fluids inside the drill pipe, drill pipe pressure will be positive. This is because the heavier fluid outside the drill pipe exerts a u-tube pressure on the fluids inside the drill pipe.

When the rig crew turns pumps on, drill pipe pressure will fluctuate depending on the relative densities of fluids inside and outside of the drill pipe and the circulating friction generated by moving those fluids.[18] When the pumps are pushing lighter fluid down the drill pipe to displace heavier fluid outside it, drill pipe pressure should steadily decrease as the lighter fluid displaces the heavier one.

Drill pipe pressure can be a difficult kick indicator to interpret because so many different factors can affect that pressure. For instance, drill pipe pressure might change because of a washout in the drill pipe or wear-out of the pump discharge valves.[19] But such causes should still prompt the driller to stop and check that the rig and well are all right.[20]

In a situation where there are changing fluid densities, changing pump rates, and changing wellbore geometry, close monitoring of drill pipe pressure can be facilitated by advance planning and charts describing what pressures to expect.[21] Unexplained fluctuations in drill pipe pressure can indicate a kick.

Some kicks exhibit an increase in drill pipe pressure,[22] although an increase can also indicate a clog in the pipe or that the crew is pumping the wrong fluids into the well.[23] More commonly, it is a decrease in drill pipe pressure that indicates a kick; lighter oil and gas flow into the annulus around the drill pipe and thereby lower the drill pipe pressure.[24] But a decrease in drill pipe pressure can also indicate a hole in the drill pipe.[25] In any case, unexplained fluctuations in drill pipe pressure are a cause for concern and warrant further investigation.[26]

Gas Content

Gas content refers to the amount of gas dissolved or contained in a fluid. Fluid returns from a well can contain gas for several reasons. Some amount of gas is often present in a well during normal operations, depending on the mud type and the location of the well. And "trip gas" appears when tripping out of the hole and conducting a bottoms up circulation after a trip.

An increase in the gas content of fluid returns over time can indicate an increase in pore pressure,[27] penetration of a hydrocarbon-bearing zone, or a change in wellbore dynamics allowing more effective cuttings removal.[28] But unexplained increases in gas content are always a cause for concern. They can indicate either that a kick is occurring or that wellbore conditions are becoming conducive for a kick.[29]

Sensors and Displays

Rig personnel rely on data that are recorded and displayed by proprietary sensors, hardware, and software. For the *Deepwater Horizon*, Transocean hired National Oilwell Varco (NOV) to provide **Hitec**-brand sensors, driller's chairs, and displays for the rig.[30] BP contracted Sperry Drilling, a Halliburton subsidiary, to conduct additional independent mud logging and well monitoring services.[31]

NOV placed a comprehensive set of sensors on the rig that measured various drilling parameters and surface data, including flow-in, flow-out, pit volume, drill pipe pressure, block position, and hook load.[32] The Hitec system recorded and displayed only the data from the Hitec sensors. Sperry Drilling's **Sperry-Sun** system collected data from many of the Hitec sensors,[33] including the sensors for pit volumes, flow-in, drill pipe pressure, and kill line pressure.[34] It also collected data from separate Sperry-Sun sensors, including Sperry-Sun sensors for flow-out and gas content.[35]

Sperry Drilling and NOV both provided BP and Transocean with proprietary displays consisting of real-time numerical data, historical trend lines, and other features like tables and charts.[36] Each of the systems allowed users to manually set (and constantly adjust) audible and visual alarms for various data parameters, including pit gain, flow-out, and drill pipe pressure.[37] The alarms could be set to trigger whenever incoming data crossed preselected high and low thresholds, and could also be shut off.[38]

While the Hitec and Sperry-Sun data systems displayed similar data, they did so using significantly different visual design (seen in Figures 4.7.3 and 4.7.4).

Figure 4.7.3. Hitec data display.

Figure 4.7.4. Sperry-Sun data display.

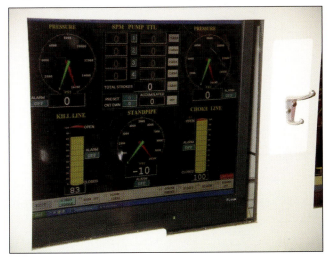

Sambhav N. Sankar

Fred H. Bartlit, Jr.

Photos taken on Transocean's *Deepwater Nautilus*.

Because the two systems in many cases used the same underlying sensors, most of the numerical values should have been close if not identical.[39] Where they displayed data from different sensors, the differences were usually predictable and could generally be dealt with through calibration.[40]

Hitec and Sperry-Sun each had its own flow-out sensor in the return flow line. These flow-out sensors differed in type, location, and format. Hitec had a paddle-type flow-out sensor.[41] As fluid rushed past, it pushed and lifted the paddle. The Hitec system inferred the rate of flow from the degree of paddle elevation. Sperry-Sun, by contrast, used a sonic-type sensor.[42] The sensor emitted a beam to ascertain the height of the fluid. The Sperry-Sun system inferred the rate of flow from the fluid level.[43]

The Hitec flow-out sensor was located in the return flow line before the line forked to either send returns to the pits or send them overboard.[44] The Sperry-Sun sensor was located after the fork, capturing flow-out only when returns from the well were routed to the pits.[45] (Positioning of both sensors is illustrated in Figure 4.7.5.) This means that the Hitec flow-out sensor could register returns going overboard, but the Sperry-Sun sensor could not.[46]

In addition to the data display systems, the rig also had video cameras that monitored key areas and components, including the rig floor and the flow line. The flow line camera (also illustrated in Figure 4.7.5) simply pointed at the flow line. Like the Sperry-Sun flow-out sensor, this camera was located after the fork; rig personnel could use it to observe flow returning to the pits but not flow that had been routed overboard.[47] When returns were sent overboard, rig personnel could still visually inspect for flow but could not do so using the video camera. They had to physically look behind the gumbo box (which was located before the fork).[48]

Figure 4.7.5. Flow-out sensors and flow line camera.

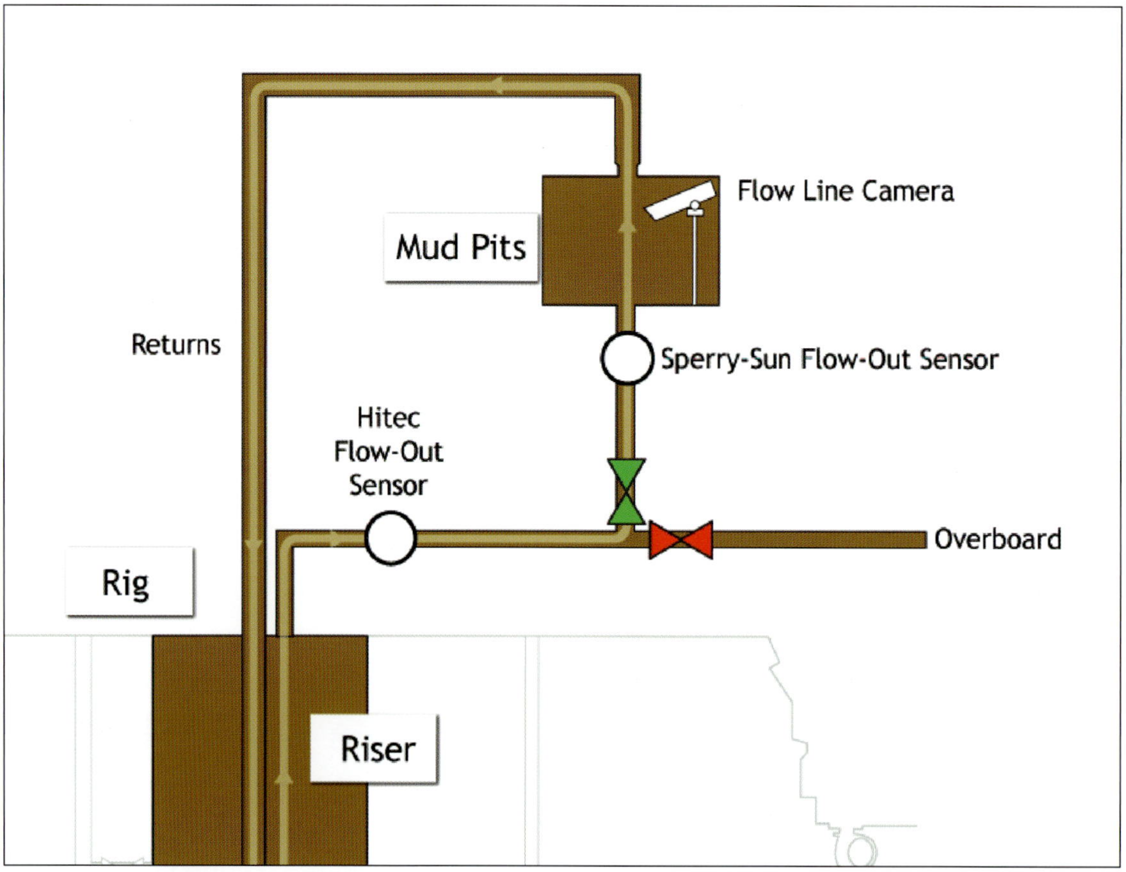

TrialGraphix

The Sperry-Sun flow-out sensor and the rig's flow line camera could not register returns going overboard. The Hitec flow-out sensor could, but data from the Hitec flow-out sensor sank with the rig.

The rig's sensors and display equipment appear to have been working properly at the time of the blowout. There is no evidence that the Sperry-Sun system malfunctioned. It continued recording and transmitting data up until the first explosion. The Hitec system was also "in satisfactory condition," as an April 12 rig condition assessment recorded in some detail.[49]

The crew had expressed some complaints about the driller's and assistant driller's control chairs, known as the "A-chair" and "B-chair" respectively.[50] The computer system powering the chairs' controls and displays had "locked up" or crashed on several occasions.[51] When this happened to the A-chair, the driller's screens would either freeze or revert to a blank blue screen, disabling real-time data display on the screen and requiring the driller to move to the adjacent B-chair.[52] In response, Transocean replaced the chairs' hard drives.[53] This appears to have corrected the problem.[54] The April 12 assessment found that the software on all of the chairs "was stable and had not shown (excessive) crashes."[55] There is no evidence that the chairs malfunctioned on April 20.[56]

Personnel and Places

On the Rig

Rig data are available in various forms to personnel on the rig and onshore. The Hitec data, Sperry-Sun data, and video feeds were all available to personnel on the rig, in real time, anywhere there was a television.[57] Certain individuals had more extensive data displays depending on their level of well monitoring responsibility.

On the *Deepwater Horizon*, the Transocean driller and assistant driller, and the Sperry Drilling mudlogger, were directly responsible for well monitoring.

The **driller** was responsible for monitoring well conditions at all times, interpreting and responding to downhole conditions, and securing the well in a well control situation (see Figure 4.7.6).[58] The driller sat in the A-chair in the drill shack. He normally monitored three screens: two screens in front of him that displayed Hitec data and a screen to the side with

Figure 4.7.6. Transocean's *Deepwater Horizon* Emergency Response Manual.

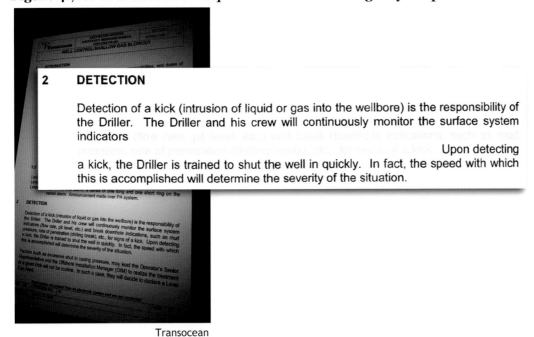

Transocean

Sperry-Sun data.[59] He also had a screen with live video feeds and a window straight ahead with a direct view of the rig floor.[60] The driller was supposed to actively look at his data screens during well operations.[61] He contemporaneously recorded rig activities for each day's daily drilling report.[62]

The driller was the central point of contact for all well control concerns: Anyone with "an understanding of something that may have indicated a well control event, would have called back to the driller, most likely, and informed him." [63] He was the one who had the most information about current operations on the rig and the ability to react to them.[64]

The assistant driller was also responsible for monitoring the well and taking well control actions. He served as a crucial backup and assist to the driller. The assistant driller was expected to have "a comprehensive understanding of well control" and "be able to recognize the signs of a well kick or blowout before it develops into an emergency condition."[65] He assisted the driller in monitoring the drilling instrumentation and recognizing and controlling well conditions.[66] As part of that assistance, he monitored the pit volumes and from time to time would go to the pits and check in with the derrickhand to make sure all was well.[67]

There were two assistant drillers on duty at any one time. One sat in the B-chair, adjacent to the driller in the drill shack.[68] He had access to the same screens as the driller. If there was activity on the deck—like pipe handling—another assistant driller would sit in the "C-chair" in the auxiliary driller's shack.[69] Although the assistant drillers had many responsibilities, at least one should have been monitoring the well at any given time.[70]

The Sperry Drilling mudlogger also monitored the well, serving as a second set of eyes for the Transocean crew.[71] BP specifically contracted the mudloggers for this purpose.[72] It was the mudloggers' duty to continuously monitor operations and provide well and drilling data upon request. They watched the data but did not have any control over rig operations and could not respond directly themselves. If the mudloggers identified problems, they would notify the driller (or drill crew).[73]

The mudlogger sat in the mudlogger's shack, one flight of stairs away from the drill shack.[74] He had 12 monitoring screens arranged in two rows of six. These screens displayed both Hitec and Sperry-Sun data.[75] Among the screens, the mudlogger had a display to the left showing all of the rig's pit volume levels. Below that, the mudlogger had a graphical log and a digital readout of the Hitec numbers.[76] He also had a screen with live video feed from the rig's cameras—he could switch between channels showing the flow line, the rig floor, and other areas.[77] In addition to monitoring the well, the mudlogger performed formation analysis when the rig was drilling and provided data printouts and reports.[78]

Several individuals supervised well monitoring work by the driller, assistant driller, and mudlogger.

The BP **well site leader** had responsibility for overseeing all operations on the well. That responsibility involved delegating duties like minute-by-minute monitoring of data.[79] Some well site leaders did monitor the well during critical operations.[80] To facilitate such monitoring, the well site leaders' office had screens that constantly displayed the Hitec data, Sperry-Sun data, and live video feeds.[81] The Sperry Drilling mudloggers reported to the BP well site leader.

The Transocean **toolpusher** supervised the driller and ensured that all drilling operations were carried out safely, efficiently, and in accordance with the well program.[82] That included confirming that all well control requirements were in place, performing all well control calculations, and assisting in killing the well in emergency situations.[83] The toolpusher was generally on the rig floor at all times, had access to the driller's and assistant driller's monitors, and had a small office inside the drill shack.[84]

The toolpusher reported to the **senior toolpusher**. The senior toolpusher had a similar job description as the toolpusher but was one level higher in the hierarchy.[85] Although he had no continuous role in operations and was not generally on the rig floor, the senior toolpusher was supposed to be consulted when there were anomalies or emergencies. In a well control event, the senior toolpusher organized response actions and acted as a liaison to the well site leader.[86] The senior toolpusher reported to the offshore installation manager.

The **offshore installation manager (OIM)** was the senior-most Transocean drilling manager on the rig and oversaw the entire Transocean crew. He assisted with abnormal or emergency situations.[87] Both the senior toolpusher and OIM had separate offices away from the rig floor, near their living quarters, that included data displays.[88]

Onshore

Onshore, only the Sperry-Sun data were available in real time.[89] The Hitec data and video feeds did not go to shore.[90]

BP personnel could view the Sperry-Sun data in their Houston offices and in an operations room for the *Deepwater Horizon* that had dedicated data displays.[91] They could also view the data over a secure Internet connection.[92] Personnel at Anadarko and MOEX could access the Sperry-Sun data onshore as well.[93] BP, Anadarko, and MOEX appeared to have used real-time data to examine geological and geophysical issues.[94]

Sperry Drilling personnel could access the Sperry-Sun data in their Houston real-time center and Lafayette operations office.[95] They appeared to have used their access to provide customer support and quality control.[96]

None of the entities receiving the Sperry-Sun data onshore appears to have monitored the data for well control purposes.[97] (Transocean did not receive data onshore.[98])

Table 4.7.1. Personnel and places with access to the rig's Sperry-Sun data.

Rig: Responsible for Monitoring Data	Rig: Accountable for Operations	Onshore: Could Access Data in Real Time
• Transocean driller • Transocean assistant driller • Sperry-Sun mudlogger	• BP well site leader • Transocean OIM • Transocean senior toolpusher • Transocean toolpusher	• BP • Anadarko • MOEX • Sperry-Sun

Well Monitoring at Macondo

It is difficult to know exactly what data screens rig personnel were looking at during their final hours on the *Horizon*.[99] There were multiple screens, with multiple data types, and each was highly customizable.[100] This Report relies on the Sperry-Sun historical log for its data analysis because that log is the only surviving dataset and display from the rig.[101]

The Sperry-Sun data log is valuable. This log (or something very close to it) was "the actual log that they were watching on the Horizon"[102]—it was displayed on one of the several screens in front of the driller, assistant driller, mudlogger, and company man. The drill pipe pressure presented on the Sperry-Sun screen was collected from Transocean's Hitec data sensors. Accordingly, the data values shown on the available Sperry-Sun screen formats would also have been shown on the Hitec screens.

Witness accounts suggest that the driller, assistant driller, and mudloggers all watched the Sperry-Sun data log.[103] The numerical values reflected in the data log would have been available on other screens as well.[104] And one can reasonably expect that rig personnel monitoring the well would have had (or should have had) pit volumes, flow-out, flow-in, and drill pipe pressure reflected in the log somewhere on their screens—no matter the format.[105]

At the same time, the Sperry-Sun data have significant limitations. The log is not fully inclusive: It does not contain data from the Hitec flow-out sensor. And scrutinizing the complete log carefully in retrospect is significantly different from monitoring it in real time, while the trend lines are developing.[106]

The First Hour

After cementing the production casing and conducting pressure tests that had been deemed successful, the crew moved on to the remainder of the temporary abandonment procedure. The crew would displace mud and spacer from the riser with seawater. There were several stages in the planned displacement. First, rig personnel would pump seawater down the drill pipe to displace mud from the riser until the spacer fluid behind the mud reached the rig floor. They would then shut down the pumps and conduct a "sheen test." That test would confirm that the crew had displaced all of the oil-based mud from the riser. The crew would then change the lineup of valves to send further returns from the well (spacer) overboard rather than to the mud pits. They would then resume the displacement until all of the spacer was out of the wellbore and the riser was full of nothing but seawater.[107]

At the start of the displacement process, Transocean driller Dewey Revette was in the drill shack's A-chair, monitoring the well. Transocean assistant driller Stephen Curtis was likely in the drill shack's B-chair, also monitoring the well.[108] BP well site leader Don Vidrine was in the drill

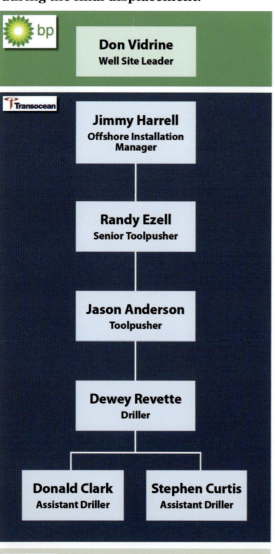

Figure 4.7.7. Rig personnel on duty during the final displacement.

bp

Don Vidrine
Well Site Leader

Transocean

Jimmy Harrell
Offshore Installation Manager

Randy Ezell
Senior Toolpusher

Jason Anderson
Toolpusher

Dewey Revette
Driller

Donald Clark
Assistant Driller

Stephen Curtis
Assistant Driller

HALLIBURTON
Sperry Drilling Services

Joseph Keith
Mudlogger

TrialGraphix

shack to oversee the initiation of the displacement.[109] Donald Clark, the other Transocean assistant driller, was at the bucking unit (a machine for making up pipe) on the port aft deck, working with personnel from Transocean, Weatherford, and Dril-Quip to prepare for setting the lockdown sleeve.[110] Sperry Drilling mudlogger Joseph Keith was in the mudlogger's shack, monitoring the well.[111]

At 8:02 p.m., the crew began displacing the mud and spacer in the riser with seawater.[112] The pumps were not lined up in a closed-loop system. Instead, the crew was pumping seawater from the ocean through the sea chest and into the well. This bypassed the pits. Returns from the well were flowing into the active pits (in this case, pits 9 and 10).[113] As a result, individuals monitoring the well could not rely on the "pit volume change" display.[114] To monitor pit gain, rig personnel would have had to perform volumetric calculations comparing the increase in pit volume (reflecting returns) against the volume of seawater pumped into the well (pump strokes × volume per stroke).[115] There is no evidence, one way or the other, as to whether the crew performed such volumetric calculations.

This setup should not have impaired rig personnel's ability to monitor flow-out versus flow-in. However, the flow-out readings appear to have been more erratic than readings captured the previous day (seen in Figure 4.7.8). This may be because cranes were moving on the rig's deck, causing the rig to sway and thus affecting the level of fluids in the flow line.[116] Otherwise, flow-out appeared normal.[117]

Figure 4.7.8. Erratic vs. normal flow-out.

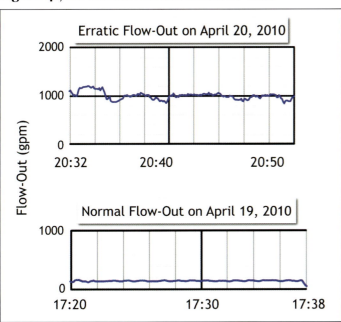

Sperry-Sun data/TrialGraphix

Flow-out readings appear to have been more erratic than normal during the final displacement, perhaps because crane operations were causing the rig to sway.

This setup also should not have impaired rig personnel's ability to monitor drill pipe pressure. The drill pipe pressure appears to have behaved as expected. It rose initially as the pumps turned on and then decreased gradually as lighter seawater replaced the heavier mud and spacer in the riser. At 8:10 p.m., mud engineer Leo Lindner looked at the drilling screen and "thought everything was fine."[118] At 8:16 p.m., the data showed an increase in gas units— not atypical at the start of circulation.[119] The gas readings then tapered off as the last of the mud left the wellbore.[120]

From 8:28 to 8:34 p.m., the crew emptied the trip tank (pit 17), with the fluid going into the flow line and pits with the rest of the returns. This complicated the monitoring of both the pits and flow-out. To accurately monitor either parameter, the crew had to perform calculations to subtract the effect of emptying the trip tank from the pit volume and flow-out readings that appeared on-screen. It is unknown whether the crew did so.

At 8:34 p.m., the crew did three things simultaneously. They (1) directed returns away from the active pits (pits 9 and 10) and into a reserve pit (pit 7); (2) emptied the sand traps into the active pits (pits 9 and 10); and (3) began filling the trip tank (pit 17).[121] Each of these actions further complicated pit monitoring for well control purposes. The active pit system was eliminated as a well monitoring tool. In order to know the volume coming out of the well, the crew had to perform calculations taking into account that returns were going to two different places—the reserve pit (pit 7) and the trip tank (pit 17). In addition, routing returns to the trip tank bypassed the flow-out meter, so the flow-out reading appeared artificially low and had to be added to the rate of entry of fluids into the trip tank to ascertain actual flow-out.[122] Again, it is unknown whether the crew was performing any such calculations. In addition, communication between the rig crew and mudlogger may have broken down at this time: The drill crew did not inform Keith about the switch in pits.[123] Keith did notice a slow gain in the active pits and called M-I SWACO mud engineer Leo Lindner to inquire; Lindner said they were moving the mud out of the sand traps and into the active pits.[124]

Trip Tank. A trip tank is a small tank. Its primary purpose is to hold fluid that the drill crew may need to rapidly send into the well, for example, to compensate for the volume removed when pulling out the drill pipe (known as tripping the pipe). The drill crew also uses the trip tank to monitor the well. The trip tank is situated between the well and the mud pits. When emptied, fluid from the trip tank goes into the return flow line, past the flow-out meters, and into the same pits as the returns from the well. The *Horizon* had two trip tanks.

At 8:49 p.m., the crew again rerouted returns, this time from one reserve pit (pit 7) to another (pit 6). At about this time, the displacement process had underbalanced the well. The combined hydrostatic pressure at the bottom of the well (generated by the mud and spacer still in the riser, the seawater in the riser and the well, and the mud remaining in the well beneath 8,367 feet below sea level) dropped below the reservoir pressure.

Transocean's post-explosion analysis estimates that the well became underbalanced at 8:50 p.m.[125] BP's post-explosion modeling estimates that the time was 8:52 p.m.[126] Given the failed bottomhole cement job, hydrocarbons would have begun flowing into the well at this time.

At 8:52 p.m., Vidrine called BP's shoreside senior drilling engineer Mark Hafle to ask about the procedure for testing the upcoming surface cement plug. Hafle asked Vidrine if everything was OK. Hafle had the Sperry-Sun real-time data up on-screen in front of him. It does not appear that the two discussed the rig crew's handling of the displacement or rig activities complicating well control monitoring.[127]

In retrospect, it does not appear there were (or would have been) any signs of a kick prior to about 9 p.m. Nevertheless, between 8 and 9 p.m., rig personnel did not adequately account for whether and to what extent certain simultaneous operations, such as emptying the trip tanks, may have confounded their ability to monitor the well.

Indications of an Anomaly as Early as 9:01 p.m.

Just before 9 p.m., Keith left the mudlogger's shack to take a short break.[128] He notified the drill crew (by calling Curtis) and then stepped out.[129] He went downstairs, used the restroom, got a cup of coffee, and smoked half a cigarette.[130] He was apparently gone for about 10 minutes before returning to his post.[131]

At 8:59 p.m., the crew simultaneously decreased the pump rate on all three pumps and began emptying the trip tanks.[132] The decrease in the pump rate should have caused a decrease in the flow-out, but because emptying the trip tanks sent additional fluid flowing past the flow-out meter, the flow-out reading actually increased. That increase potentially masked any sign of a kick from the flow-out reading.[133]

At 9:01 p.m., drill pipe pressure changed direction. Instead of continuing to steadily decline, it began to increase. This change in direction was a significant anomaly. If lighter seawater were replacing the heavier mud and spacer in the riser as should have been the case, drill pipe pressure should have continued to drop, as it had done for at least the previous 40 minutes.[134] In retrospect, this change in drill pipe pressure likely indicated that hydrocarbons were pushing heavier mud up from the bottom of the well against and around the drill pipe.

By 9:08 p.m., with the pump rates constant, drill pipe pressure had increased by approximately 100 pounds per square inch (psi). The magnitude of the increase would have appeared subtle on the Sperry-Sun screen showing only trend lines, but it likely would not have been subtle on the numerical displays.[135]

The change in direction was by now clear and clearly anomalous. An individual who saw the drill pipe pressure increase should have been seriously concerned and should have investigated further.[136] But Keith, who would have returned from his break by that time, reviewed the logs for the period he was absent and did not notice any indication of a problem: "I went back over it and looked, and to my recollection, I didn't see nothing wrong."[137]

At 9:08 p.m., after the top of the spacer column reached the rig, the crew shut down the pumps and switched the lineup to route returns overboard.[138] Keith looked at the video feed from the flow line camera and visually confirmed that there was no flow.[139] He likely communicated this to the rig floor.[140] According to Vidrine, who was on the rig floor, everything looked fine.[141]

Everything was not fine. For about a minute after the pumps stopped, flow-out continued beyond the *Horizon*'s typical flow-out signature.[142] This was a kick indicator (Figure 4.7.9 depicts a typical flow-out signature at 4:52 p.m. and the 9:08 p.m. spike). A driller, assistant driller, or mudlogger watching the screen could have seen it.[143] Instead, they thought they had visual confirmation of no flow, based at least on Keith's observations.[144]

Figure 4.7.9. Typical flow-out signature vs. spike at 9:08 p.m.

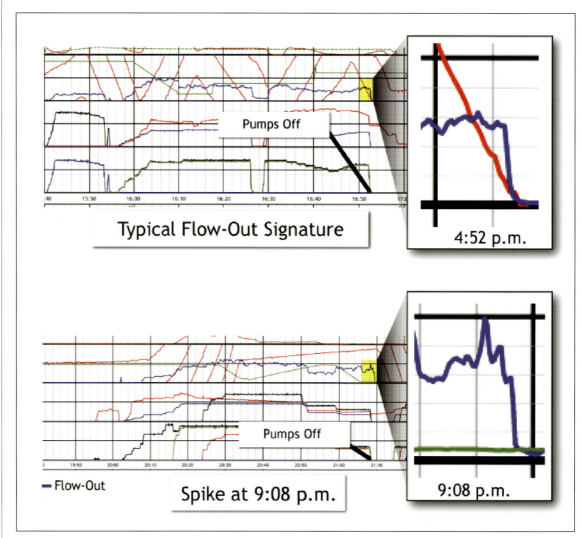

Sperry-Sun data/TrialGraphix

For about a minute after the pumps stopped at 9:08 p.m., flow-out continued beyond the *Horizon*'s typical flow-out signature.

There are several possible explanations for this contradiction: (1) Keith may have seen some flow but attributed it to residual flow; (2) Keith may not have looked at the camera for long enough to realize that it was not residual flow;[145] (3) the flow may have been too modest to detect from the video feed;[146] or (4) the flow may already have been rerouted overboard before Keith performed his flow check.[147] Rig personnel could have performed a secondary flow check by sending someone to physically look behind the gumbo box, but apparently they did not do so. On many rigs (including the *Horizon*[148]), this would have been a common practice, especially if rig personnel had noted anomalies.[149]

By 9:10 p.m., the crew had rerouted returns overboard. Doing so bypassed the pits, the Sperry-Sun flow-out meter, and the gas sensors.[150] That equipment could no longer be used to monitor the well. The flow did not bypass the Hitec flow-out meter, but for some reason— perhaps malfunction, perhaps neglect—data from that meter never alerted the crew to the kick. At about the same time that they rerouted returns overboard, the crew also transferred mud from the active pits (pits 9 and 10) to the reserve pit that had been taking returns from the well (pit 6).

The crew probably made this pit transfer to prepare for cleaning out the active pits (pits 9 and 10).[151] The immediacy of the transfer suggests that the crew did not take the time to compare the volume of fluid pumped into the well with the volume of fluid returned from the well.

Meanwhile, the mud engineers conducted the sheen test and communicated to the drill shack that it passed. Vidrine directed the crew to get in place to start sending returns overboard and ordered the displacement to begin again. He then returned to his office and did paperwork.

During the course of these activities, drill pipe pressure gradually increased. From 9:08 to 9:14 p.m., while the pumps were shut down, drill pipe pressure increased by approximately 250 psi (see Figure 4.7.10). This was a significant anomaly.[152] By 9:14 p.m., the increase would have been noticeable and a cause for concern.[153] The driller apparently missed this increase, perhaps because "having looked and seen 60 seconds of constant pressure…he may have then turned to do the next step in the process which was line up another mud pump to pump down the kill lines."[154] It is unclear why the assistant driller and the mudlogger also missed the increase.[155]

Figure 4.7.10. Drill pipe pressure anomalies from 9:01 to 9:14 p.m.

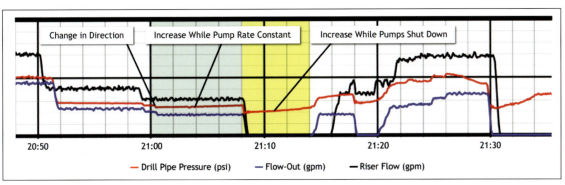

Sperry-Sun data/TrialGraphix

At 9:01 p.m., drill pipe pressure changed direction. By 9:08 p.m., with the pump rates constant, drill pipe pressure had increased by approximately 100 psi. From 9:08 to 9:14 p.m., while the pumps were shut down, drill pipe pressure increased by approximately 250 psi. Each of these changes in drill pipe pressure was an anomaly that should have prompted rig personnel to stop and investigate, but the signs apparently went unnoticed.

At 9:14 p.m., the drill crew turned the pumps back on: first, pumps 3 and 4 at 9:14 p.m., then pump 1 at 9:16 p.m. Keith called Curtis and asked why the drill crew was turning the pumps on gradually and not at full rate. Curtis replied, "That's the way we're going to do it this time."[156] Shortly after 9:17 p.m., the crew also turned on pump 2 to pump down the kill lines. Within seconds of turning on pump 2, the pressure relief valve (PRV) on pump 2 blew.[157] The PRV probably blew because the crew had inadvertently started the pump against a closed kill line valve (a rare but not unheard-of mistake).[158]

After the PRV blew, at 9:18 p.m., the crew shut down the primary pumps (pumps 3 and 4). They left the riser boost pump (pump 1) on. The driller organized a group of individuals including Clark to go to the pump room and fix the PRV on pump 2.[159] In addition, the driller ordered someone to open up the closed kill line valve that had caused the PRV to blow.[160]

At 9:20 p.m., the drill crew restarted the primary pumps (pumps 3 and 4). Transocean senior toolpusher Randy Ezell called the drill shack and spoke with toolpusher Jason Anderson. He asked how the displacement was going. Anderson said, "It's going fine. It won't be much longer...I've got this."[161] From 9:14 to 9:27 p.m., the data did not clearly reflect any anomalies. The return flow bypassed the pits, Sperry-Sun flow-out meter, and gas sensors. Drill pipe pressure appeared to be behaving roughly as expected—increasing as the pumps ramped up and then decreasing as seawater replaced the last of the spacer.[162]

Drill Crew Notices Anomaly but Does Not Treat It as a Kick

By 9:27 p.m., an obvious anomaly appeared. The pressure on the kill line—now discernable because the drill crew had just opened up the previously closed kill line valve—rose to approximately 800 psi.[163] This kill line pressure was anomalous.[164] The crew noticed a "differential pressure" between the kill line (approximately 800 psi) and the drill pipe (approximately 2,500 psi).[165] At 9:30 p.m., they shut down the pumps to investigate.[166]

Around that time, Transocean chief mate David Young went to the drill shack to speak with Anderson and Revette about the timing of the surface plug cement job.[167] Revette, sitting in the driller's A-chair, and Anderson, standing next to him, were speaking to each other.[168] At times, they looked at the driller's screens.[169] Revette noted that they were "seeing a differential."[170] The two men appeared concerned but calm. According to Young, "It was quiet...there was no panic or anything like that."[171]

From 9:30 to 9:35 p.m., drill pipe pressure increased by approximately 550 psi (see Figure 4.7.11). This was another significant anomaly: With the pumps shut off, there should have been no movement in the well.[172] (The increase might have reflected mud continuing to travel up the wellbore with oil and gas below.)[173] Revette and Anderson were intently watching the screens, but they did not shut in the well. Instead, Revette ordered Transocean floorhand Caleb Holloway to bleed off the drill pipe pressure[174]—apparently to eliminate the differential pressure. At 9:36 p.m., Holloway cranked open a valve on the stand pipe manifold to bleed down the pressure.[175] But it was taking longer than usual to bleed off.[176] Revette told Holloway, "Okay, close it back."[177]

Figure 4.7.11. Drill pipe pressure and kill line pressure anomalies from 9:27 to 9:40 p.m.

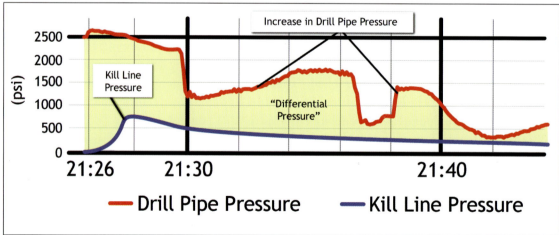

Sperry-Sun data/TrialGraphix

At 9:27 p.m., kill line pressure reached approximately 800 psi. From 9:30 to 9:35 p.m., while the pumps were shut down, drill pipe pressure increased by approximately 550 psi. After the crew attempted to bleed it down, drill pipe pressure again shot up, at 9:38 p.m., by approximately 600 psi. Each of these anomalies was a sign that fluids were moving in the well. Despite observing those signs, the crew did not yet shut in the well.

Once he did, at 9:38 p.m., the drill pipe pressure shot back up. It increased by approximately 600 psi. Again, the increase was a serious anomaly.

By this point, rig personnel had observed several serious anomalies. Each was "a sign that fluids are moving" in the well.[178] Those anomalies should have "caused alarm."[179] But there appears to have been no hint of alarm.

The crew actively investigated the anomalies and performed diagnostic interventions. But it appears that the crew did not perform the most basic kick detection intervention—a flow check. If they had done so, they would have directly seen flow coming out of the well and should have shut in the well.[180] The fact that the crew apparently did not perform a flow check suggests that Revette and Anderson either did not consider or had already ruled out the possibility of a kick.

Anderson thought "it would be a little bit longer" before they figured out the differential pressure and told Young that they probably wouldn't need him for the cement job meeting for another couple of hours.[181] According to Young, Anderson "wasn't sure if they were going to need to circulate."[182] Anderson then left to go to the pump room.[183] Young also left at about the same time.[184] He ran into Holloway, who was coming down from the stand pipe manifold; they spoke for a couple of minutes and joked.[185] There was no sign of concern or hurry.[186]

Not long afterward, Holloway was leaving the rig floor and ran into Curtis. Curtis was on his way to the drill shack. He was in no rush. Curtis and Holloway spoke for a few minutes.[187]

Throughout this period of investigation, the drill crew did not communicate with the mudlogger about the anomaly.[188] Nor did they contact the senior toolpusher, OIM, or well site leader to ask for their help or to notify them that something was amiss.

Mud Overflow and Recognition of the Anomaly as a Kick

Sometime between 9:40 and 9:43 p.m., mud overflowed onto the rig floor, shot up to the top of the derrick, and poured down onto the main deck.[189] By about that time, drill pipe pressure had decreased by approximately 1,000 psi. At 9:41 p.m., the trip tank (pit 18) abruptly gained about 12 barrels in volume. The crew likely routed flow back to the trip tank intentionally to help diagnose whether the riser was static.[190] The gain showed that there was still flow from the well up the riser.

At about the same time, Anderson returned to the drill shack. At 9:41 p.m., he activated the blowout preventer's (BOP's) annular preventer.[191] Drill pipe pressure began to increase (as it should when a well is shut in). By now, gas would already have been in the riser, expanding rapidly on its way to the surface. This may have made it more difficult to successfully activate the blowout preventer. In any case, even if the crew had successfully shut in the well, they should have expected flow from the well to continue at least until all of the gas in the riser had escaped.

Interviews and testimony after the blowout recount what happened next. Anderson called Vidrine to say the crew was getting mud back and had diverted flow to the mud gas separator and closed the annular.[192] Curtis called Ezell and said: "We have a situation. The well is blown out. We have mud going to the crown.... [Anderson] is shutting it in now."[193] Someone, perhaps Revette, called Andrea Fleytas on the bridge, said "We have a well control situation," and hung up.[194] Vidrine started for the rig floor.[195] Ezell did the same.[196] Fleytas turned to Yancy Keplinger and yelled, "We're in a well control situation."[197] Keplinger radioed the *Damon Bankston*, alongside the rig,[198] and told the vessel to disconnect and move off 500 meters: The *Horizon* was in a well control situation.[199]

Although Anderson had activated the annular preventer, that action had not fully shut in the well. Instead of reaching the expected shut-in pressure (approximately 6,000 psi), drill pipe pressure plateaued at about 1,200 psi.[200] In response, the drill crew either tightened the annular to create a seal or activated a variable bore ram.[201] At 9:47 p.m., drill pipe pressure increased dramatically. At this point, the well may have been shut in.[202]

At 9:48 p.m., pit 20 abruptly gained 12 barrels in volume. The data also show an increase in active pit volume (pits 9 and 10) and several upward spikes in flow-out. Flow from gas already in the riser might have been jostling the rig or otherwise overwhelming the rig's systems.[203]

The first explosion happened at 9:49 p.m. At the time, Anderson, Revette, and Curtis were in the drill shack, trying to get the well under control. Vidrine had been on his way to the drill shack but, seeing mud blowing everywhere,[204] turned back toward the bridge.[205] Ezell was at the doorway of his office, on his way to the rig floor. Clark and three others were in the pump room; they had just finished fixing the PRV.[206] Keith was in the mudlogger's shack, apparently surprised that anything went wrong.[207] Transocean OIM Jimmy Harrell was in the shower, with no knowledge that there had been a well control situation.[208]

Technical Findings

The data available to rig personnel showed clear indications of a kick.[209] The change in direction of drill pipe pressure (9:01 p.m.) and its subsequent steady increases (9:01 to 9:08 p.m., 9:08 to 9:14 p.m.) should have been a cause for concern but apparently went unnoticed. Even after the drill crew noticed an anomaly (9:30 p.m.), they do not appear to have seriously considered the possibility that a kick was occurring.

The anomaly the rig crew noticed at 9:30 p.m. and discussed occurred before hydrocarbons had entered the riser and 10 to 13 minutes before mud appeared on the rig floor. If the rig crew had at all considered that a kick might be occurring, they had plenty of time to activate the blowout preventer.

Rig Activities Potentially Confounded Kick Detection

The crew on the *Deepwater Horizon* engaged in a number of concurrent activities during displacement of the riser. Each could have interfered with the data.[210]

First, rig personnel were pumping seawater directly into the well from the sea chest. The crew had to pump water in from the sea chest for the displacement. But pumping it in directly from the sea chest to the rig pumps, thereby bypassing the pits, made it harder for the crew to monitor the pits. It created a non-closed-loop system that made it impossible to detect a kick by visually monitoring pit gain. Instead, pit monitoring required volumetric calculations. The crew could have, and should have, performed those calculations[211]—it was the rig crew's regular practice to do so[212]—but there is no evidence that they did so here. They also could have routed the seawater through the active pit system before sending it down the well.[213] That approach would have preserved visual monitoring of pit gain.

Second, rig personnel sent returns overboard during the latter part of the displacement. Sending returns overboard was an inherent part of the displacement. But pumping it directly from the well overboard—bypassing the pits, Sperry-Sun flow-out meter, and both gas meters—eliminated the crew's ability to monitor the pits and the Sperry-Sun flow-out meter for kick indicators.[214] The crew could still monitor the well by using the Hitec flow-out meter and by physically checking the overboard line whenever the pumps were stopped. But there is no evidence that they did so. The crew could also have lined up the displacement so that it did not confound well monitoring by taking returns to the pits first and then channeling it overboard.

Third, rig personnel were using the cranes. From early in the displacement (about 8:20 p.m.) until the explosion, rig personnel were operating one or both of the cranes.[215] Crane movement can cause the rig to sway,[216] affecting the flow-out levels and pit volumes,[217] and "complicat[ing] kick recognition."[218] Rig personnel can still detect kicks when there is rig sway, but the movement increases the level of background noise in the data and thereby reduces the minimum detectable kick sensitivity with respect to flow-out and pit volumes.[219] The crane movement was not necessary for the displacement. Rig personnel could have waited until the displacement was complete to engage in crane activity.

Fourth, rig personnel appear to have begun emptying the mud pits without first checking for pit gain. During the sheen test, the rig crew began emptying the active pits into reserve pit 6. Until that point, returns from the well had been flowing to pit 6. The problem is, the crew does not appear to have measured the volume in pit 6 before emptying the active pits into it. This suggests that the crew was not mathematically comparing the actual volume of returns to the expected volume of returns to verify that there had been no gain. The apparent reason that rig personnel emptied the active pits was to prepare for cleaning them.[220] It was unnecessary to clean the active pits, or even empty them in preparation for cleaning, during the displacement.

Fifth, rig personnel were emptying the sand traps into the pits.[221] Sand traps separate sand from mud. After a while, they fill up with clean mud. When that happens, the crew empties the mud from the sand traps into the pits. Emptying the sand traps was not problematic by itself. The problem was that the crew emptied them into the active pit system and thereby complicated pit monitoring. The crew could have simplified pit monitoring by using the active pit system to monitor the volume of fluid returning from the well and routing mud from the sand traps to a reserve pit instead.

Sixth, rig personnel were emptying the trip tanks during the displacement. It appears that the crew had to do so at this point in the displacement process.[222] It also appears that the rig's plumbing forced the crew to route flow-out from the trip tank past the flow-out meter.[223] This flow added to pit gain and flow-out, making both figures higher than they would have been otherwise. The crew could nevertheless have preserved pit monitoring and flow-out monitoring if they calculated the effect of emptying the trip tank in this manner, but there is no evidence that they did so. Alternatively, the crew could have stopped displacing the riser while they emptied the trip tanks.

Kick Detection Instrumentation Was Mediocre and Highly Dependent on Human Factors

The data sensors on the rig had several shortcomings. First, the system did not have adequate coverage. For example, there was no camera installed to monitor returns sent overboard and no sensor to indicate whether the valve sending returns overboard was open or closed. Therefore, while video monitoring of flow was possible when returns went to the pits, it was not possible when returns went overboard.

Second, some of the sensors were not particularly accurate. For example, electronic sensors for pit volumes can be unreliable, so much so that the crew would sometimes revert to using a string with a nut to measure pit volume change.[224]

Third, the sensors often lacked precision and responded to movement unrelated to the state of the well. For example, a fluctuation in flow-out might result from crane activity on the rig.[225] These shortcomings can result in rig personnel not receiving quality data and, furthermore, discounting the value of the data they do receive.

The data display systems also had notable limitations. There were no automated alarms built into the displays. Rather, the system depended on the right person being in the right place at the right time looking at the right information and drawing the right conclusions.[226] Although the systems did contain audible and visual alarms, the driller was required to set them manually.[227] He could also shut them off. Manually setting and resetting alarm thresholds is a tedious task and not always done. For example, there is typically no alarm set for flow-in and flow-out because the pumps stop and start so often that the alarms would trigger too frequently.[228]

There was also no automation of simple well monitoring calculations. For example, if the displacement is set up as a non-closed-loop system, and rig personnel want to keep track of volumes, they must perform the calculation by hand (return volume − (pump strokes × volume per pump stroke)). If the rig is emptying its trip tank while taking returns, and rig personnel want to disaggregate the two activities, they must perform the subtraction by hand. Each of those calculations could easily be automated and displayed for enhanced real-time monitoring.

There was also no advance planning or real-time modeling of expected pressures, volumes, and flow rates for the displacement. Although well flow modeling has been employed in post-explosion analysis,[229] there was no comparable modeling technology in place for real-time analysis.[230]

Finally, the displays themselves sometimes made fluctuations in data hard to see.[231] Indeed, in post-explosion reports and presentations, BP has consistently chosen to rotate the vertical Sperry-Sun log and enlarge it so that viewers can understand the data from April 20.

These limitations made well control monitoring unnecessarily dependent on human beings' attentions and abilities.

Management Findings

One of the most important questions about the Macondo blowout is why the rig crew and mudlogger failed to recognize signs of a kick and did not diagnose the kick even when they shut operations down to investigate a well anomaly. The Chief Counsel's team finds that a number of management failures, alone or in combination, may explain those errors.[232]

BP, Transocean, and Sperry Drilling Rig Personnel Exhibited a Lack of Vigilance During the Final Displacement

The evidence suggests that BP, Transocean, and Sperry-Sun personnel on the rig were not sufficiently alert to the possibility that a loss of well control might occur during the final displacement. There are several reasons why this might have been the case. First, kicks are not commonly associated with the temporary abandonment phase of well operations. In a 2001 study of 48 deepwater kicks in the Gulf of Mexico, the vast "majority of kicks occurred during drilling operations."[233] By contrast, only one kick "occurred in association with a well abandon[ment] operation."[234]

Second, confidence in barriers, particularly tested barriers, can make rig personnel overconfident in the well's overall security. A satisfactory negative pressure test generally confirms that the well is secure and that hydrocarbons will not flow into the well during riser displacement operations. Once rig personnel deemed the Macondo negative pressure test a success, they may have believed that a kick was no longer a realistic hazard.[235] Investigations of a 2009 North Sea blowout and a 2009 Timor Sea blowout found that rig personnel were "blinkered" by a successful negative pressure test or drew an "unwarranted level of comfort" from the presence of a barrier.[236] Both attitudes "reflected and influenced a lax approach to well control."[237]

Third, end-of-well activities tend to be marked by a hasty mindset and loss of focus.[238] This can result simply from a desire to finish and move on, particularly when a well has been difficult to drill (like Macondo).[239] Rig personnel have noted in post-blowout interviews that "[a]t the end of the well sometimes they think about speeding up."[240] This may be because "everybody goes to the mindset that we're through, this job is done...everything's going to be okay."[241]

Together, these factors appear to have contributed to reduced well monitoring vigilance, diminished sensitivity to anomalous data, delayed reactions, a failure to undertake routine well monitoring measures (like flow checks and volumetric calculations), and a willingness to perform rig operations in a manner that complicated well monitoring.

Such a lack of vigilance was particularly surprising at this well. Given the risk of a poor bottomhole cement job and the fact that the final displacement would severely underbalance the well, rig management—and the well site leader in particular—should have treated the displacement as a critical operation and personally monitored the data.[242]

Transocean Personnel Lacked Sufficient Training to Recognize That Certain Data Anomalies Indicated a Kick[243]

Several times during the evening of April 20, data anomalies indicated that hydrocarbons were flowing into the well.[244] Despite noticing the anomalies—and taking time to discuss them—the rig crew did not recognize that a kick was under way.

Earlier in the evening, during the negative pressure test, hydrocarbons flowed into the well. Pressure anomalies signaled the kick. But rig personnel did not heed those signals.

During the final displacement, the pressure anomalies reappeared. Although some went unnoticed, the rig crew did recognize an anomaly at 9:30 p.m. and shut the pumps down to investigate. Over the next 10 minutes or so, the crew watched the drill pipe pressure visibly increase—steadily at first (9:30 to 9:35 p.m.) and then, after they attempted to bleed it off, rapidly (9:38 p.m.)—even though the pumps were off. They also saw an anomalous kill line pressure. Each indicator was "a sign that fluids are moving" in the well—in other words, a sign of a kick.[245]

To a skilled observer, those anomalies "would have caused alarm."[246] But there appears to have been no hint of alarm. Instead, the rig crew spent at least 10 minutes "discussing" the "anomaly," "scratching their heads to figure what was happening."[247] Even in retrospect, Transocean's internal investigator asserts that it was "a very strange trend," "a confusing signal," explained only after "months of work."[248]

Transocean leaves open the possibility that its rig crew "did not have the experience" or training to interpret pressure anomalies during the negative pressure test.[249] If true, then the crew likely did not have sufficient training or ability to interpret the recurrence of those anomalies during the final displacement.

Transocean further states that its crew relied on the operator (BP) to make a final assessment of anomalies during the negative pressure test.[250] But when those anomalies reappeared during the displacement, the rig crew did not notify BP rig personnel and ask for their help in interpreting the data.[251]

BP and Transocean Allowed Rig Operations to Proceed in a Way That Inhibited Well Monitoring

BP and Transocean management on the rig allowed simultaneous operations without adequately ensuring that those operations would not complicate or confound well monitoring.[252]

Simultaneous activities can interfere with well monitoring in several ways. First, they can influence data that are used to monitor for kicks (for example, by altering fluid levels) and thereby obscure signals of a kick.[253] Second, they can make it more difficult to interpret data because rig personnel may attribute data anomalies to rig activities instead of a kick. Third, even when simultaneous operations are necessary, such as when changing the lineup of pipes and valves or fixing a mud pump, they can distract rig personnel who would otherwise be monitoring the well.[254] Rig personnel can reduce these difficulties by identifying relevant rig activities, calculating or otherwise predicting their probable effect, and communicating any expected effects to well monitoring personnel. Rig management should ensure that *someone* is watching the screens at all times, despite ongoing activities.

BP, Transocean, and Sperry Drilling Rig Personnel Did Not Properly Communicate Information

Insufficient communication, both prior to and during the final displacement, affected risk awareness and well monitoring on the *Deepwater Horizon*.

BP did not adequately inform Transocean about the risks at the Macondo well, particularly the risks of a poor bottomhole cement job.[255] Transocean argues that if BP had done so, its crew might have demonstrated "heightened awareness."[256] But it is unlikely that this particular communication failure compromised kick detection; the crew would probably have dismissed warnings about cement risks anyhow after the successful negative pressure test.

BP and Transocean did not do enough to ensure that rig personnel were aware of the objectives, procedures, and hazards of the riser displacement operation.[257] The individuals conducting the pre-job meetings should have emphasized that BP's temporary abandonment procedures would leave only a single barrier in the well besides the BOP and would produce an unusually underbalanced well.[258] They should have warned against complacency stemming from the negative pressure test and emphasized that tested barriers can fail.

The pre-job meetings should also have informed well monitoring personnel that certain kick indicators such as pit gain and flow-out would be compromised or unavailable during the planned operations. Well monitoring personnel should have been told that, as a result, they would need to perform volumetric calculations to keep track of pit gain, pay special attention to other parameters (such as drill pipe pressure), and conduct visual flow checks whenever the pumps were stopped.[259] In addition, to facilitate well monitoring, those personnel should have been given a pump schedule for the different phases of the displacement, along with guidance regarding how much deviation from that schedule should be considered anomalous.[260]

Transocean and Sperry Drilling personnel did not communicate effectively about the displacement operation.[261] And the BP well site leader did not play a sufficiently active role in ensuring such communication.[262] Communication broke down between the drill crew and the mudloggers on several occasions. For example, when rig personnel announced early on April 20 that they would be pumping mud to a supply boat, Cathleenia Willis (the mudlogger on shift) told Clark she was concerned that this would limit her ability to monitor pit gain.[263] Clark said he would address the matter but never got back to Willis.[264] Keith reported after the explosion that he was concerned that simultaneous activities would complicate monitoring but never expressed those concerns to others.[265] The drill crew repeatedly failed to inform Keith of various activities that influenced well monitoring data.[266]

Even after the Transocean crew shut down the pumps to investigate an anomaly, they did not inform the Sperry Drilling mudlogger, senior Transocean personnel, or the BP well site leader of the anomaly or ask for their help in resolving it.[267]

The Chief Counsel's team cannot conclude that any one of these problems contributed to the failure to detect the kick. But together they suggest a communication breakdown that made kick detection more difficult. Knowledge of ongoing rig activity "is essential to accurate interpretation of the data."[268] Absent that knowledge, it is difficult to ascertain whether anomalous data are benign or problematic.[269]

While BP and Transocean Management Were Taking Steps to Improve Well Monitoring, These Steps Had Not (Yet) Improved Kick Detection on the *Deepwater Horizon*

BP

BP recognized that well control was critically important to its operations. In a 2009 Major Hazard Risk Assessment, the company identified "Loss of Well Control" as first among the two "major accident risks" in drilling and completions operations.[270]

BP specifically gave the *Deepwater Horizon* a mid-range risk rating for loss of well control[271] and acknowledged the potentially severe consequences of a well control failure: "Catastrophic health/safety incident" with the "potential for 10 or more fatalities," "extensive" and "widespread" damage to sensitive environments, "$1 billion - $5 billion" in financial impact, "severe enforcement action," government intervention, and "[p]ublic and investor outrage."[272]

To address this risk, BP checked to ensure that all drilling and completions workers had well control training and certification, developed tools to further assess the risk ("BowTie diagrams," "Risk Mitigation Plans," "Asset-specific" risk assessments, a "Barrier Assessment Tool"), and emphasized that risk management in this area would be "under continual review."[273] The company also planned to evaluate the effectiveness of barriers with each rig's team and train personnel in the new well control response guide.[274]

BP understood the risks presented by less severe well control events too. An April 14, 2010 presentation to the drilling and completions Extended Leadership Team noted that half of all nonproductive time in the company's offshore drilling sector was the result of "downhole problems (wellbore instability, losses, gains, tight hole) and stuck pipe."[275] The presentation continued: "Post analysis of the...incidents clearly indicates that in most cases[,]...events could have been prevented or decisions influenced if the drilling data that is already generally available had been appropriately presented and analysed." That is, "early warning indicators were usually present albeit invisible in the mountain of data."[276] Therefore, as Macondo senior manager David Sims stated, downhole problems were "low hanging fruit" for decreasing nonproductive time.[277]

Reviews conducted in late 2007 and early 2008 similarly showed that "the quality of monitoring, detection and reaction to downhole hazards during drilling operations" was "variable."[278] In response, BP planned to develop a program to facilitate Efficient Reservoir Access, the "ERA Advisor." This ambitious program would monitor data in real time onshore, generate expert and automated advice in response to that data, and use new software and sensors to track and diagnose the data.[279] The program's goal was to "ensure the *right information is in the right place at the right time*."[280] It would focus, however, on monitoring data during the drilling of the well (not end-of-well activities).[281] BP's Extended Leadership Team developed and endorsed the ERA in 2009; initial pilot testing of the first stage of the system was to begin in the fourth quarter of 2010.[282]

Even before planning its ERA program, BP contracted Sperry-Sun to relay rig data to its Houston offices. But despite recognizing the risks associated with poor well monitoring and the usefulness of onshore assistance, BP did not monitor this data for well control purposes. Even though each of its working rigs had an operations room with dedicated Sperry-Sun data displays,[283] BP typically used these rooms only for meetings and the data were "not ever monitored."[284] Thus, before BP implemented its ERA Advisor system, it failed to take the interim step of ensuring that someone onshore was monitoring the data systems it already had in place.

This is surprising in light of the fact that BP was particularly concerned about well monitoring at Macondo. Less than two months before the blowout, on March 8, 2010, the Macondo well took a kick.[285] The kick occurred while the rig was drilling.[286] The "well kicked for 30 minutes before the trends were obvious enough."[287] The Transocean drill crew and Sperry Drilling mudlogger—indeed, the very same Revette, Curtis, Clark, and Keith—observed a gain in flow-out, a slow gain in the pits, a decrease in equivalent circulating density (ECD), and an increase in gas content.[288] The drill crew stopped the pumps, performed a flow check, and shut in the well.[289] The situation soon went "from bad to worse."[290] There were "[m]ajor problems on the well":[291] The pipe was stuck. BP ultimately had to sever the pipe and sidetrack the well.[292]

After the event, BP involved its in-house Totally Integrated Geological and Engineering Resource team (the "TIGER team"), to conduct an engineering analysis, and (on March 18) distributed a "lessons learned" document to its Gulf of Mexico drilling and completions personnel.[293] BP recommended that its personnel "evaluate the entire suite of drilling parameters that may be indicative of a shift in pore-pressure" (including gas, flow-out, and flow checks), "ensure that we are monitoring all relevant [pore pressure] trend data," "have [pore pressure] conversations as soon as ANY indicator shows a change," "no matter how subtle," and *be prepared to have some false alarms and not be afraid of it.*"[294] The "lessons learned" document also specified that "[b]etter lines of communication, both amongst the rig subsurface and drilling personnel, and with Houston office needs to be reestablished. Preceding each well control event, subtle indicators of pore pressure increase were either not recognized, or not discussed amongst the greater group."[295]

In addition, BP wells team leader John Guide initiated several conversations to address the rig's response to the kick, which he thought was "slow and needed improvement."[296] Guide specifically instructed the BP well site leaders to "up their game."[297] He spoke with Transocean and Sperry Drilling personnel about "tighten[ing] up wellbore monitoring."

The goal of Guide's conversations and of the TIGER team's involvement was to maintain heightened attentiveness "for the remainder of the Macondo well,"[298] up to the point when the *Horizon* unlatched its BOP and left.[299] Evidently, the team fell short of that goal. As Guide conceded after the incident, the Macondo team's heightened attentiveness to well monitoring lasted all the way up until, apparently, the negative pressure test.[300] This is likely because BP's focus, once again, was on monitoring data during the drilling of the well (not end-of-well activities).[301]

Transocean

Transocean also recognized the importance of well control. In a 2004 Major Accident Hazard Risk Assessment, the company gave *Deepwater Horizon* a 5B risk rating for reservoir blowout,[302] meaning that there was a "Low" likelihood of a blowout occurring, but if one did occur, the event would have "Extremely Severe" consequences:[303] "Multiple personnel injuries and/or fatalities," "Major environmental impact," and "Major structural damage and possible loss of vessel."[304] As prevention and mitigation measures, Transocean listed (among other things) well control procedures, training of drill crew, and instrumentation indicating well status.[305]

As discussed in greater detail in Chapter 5, despite those concerns, Transocean did not inform the *Deepwater Horizon*'s crew of lessons learned from an earlier well control event on another rig. On December 23, 2009, Transocean barely averted a blowout during completion activities on a rig in the North Sea. Rig personnel were in the process of displacing the wellbore from mud to seawater.[306] They had just completed a successful negative pressure test, and they had lined up the displacement in a way that inhibited pit monitoring. [307] During the displacement, a critical tested barrier failed, and hydrocarbons came up the wellbore, onto the drill floor, and into the sea.[308]

The incident differed from Macondo in some respects: It occurred during the completion phase of the well, not drilling or temporary abandonment, and the failed barrier was a mechanical valve, not cement.[309] But the incident was identical to Macondo in crucial respects:

- the rig crew underbalanced the well while displacing mud to seawater;

- a successful negative pressure test "blinkered" the crew and produced an improper "change in mindset";

- the crew conducted displacement operations in ways that inhibited pit monitoring; and

- the crew discounted kick indicators by attributing them to other occurrences on the rig.[310]

Transocean nevertheless failed to effectively share and enforce the lessons learned from that event with all relevant personnel. The company held two conference calls and distributed an advisory for its North Sea personnel only. It also posted a shorter advisory about the event on its electronic documents platform—accessible fleetwide—but it did not alert crews of the advisory's existence. Indeed, there is no evidence that anyone on or affiliated with the *Deepwater Horizon* knew of the North Sea incident or read any of its lessons prior to the Macondo blowout. ◆

Figure 4.7.12.   Last two hours of Sperry-Sun data.

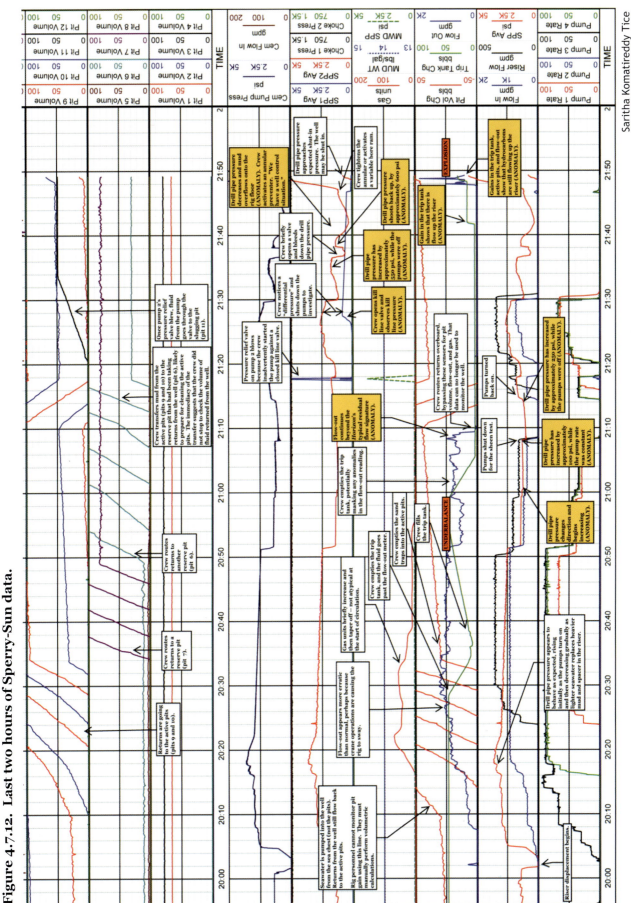

Saritha Komatireddy Tice

Chapter 4.8 | Kick Response

I n the event of an unwanted influx of fluid or gas into the wellbore (a "kick"), the safety of a drilling rig turns on split-second responses by the rig crew.

The *Deepwater Horizon*'s crew did not respond to the April 20 kick before hydrocarbons had entered the riser, and perhaps not until mud began spewing from the rig floor. If the rig crew had recognized the influx earlier, they might have been able to shut in the well. But the crew still had response options even at the point that they eventually did recognize the kick. If the crew had diverted the flow overboard immediately, they might have delayed the ignition and explosion of the gas flowing out of the well. Instead, the crew sent the flow to the mud gas separator.[1] The mud gas separator was not designed to handle this flow volume and was overwhelmed. Sending flow to the mud gas separator, rather than overboard, therefore increased the risk that gas from the well would explode on the rig.

The crew appears to have followed standard Transocean procedures for dealing with hydrocarbon kicks. But those procedures were written to guide the crew's response to routine hydrocarbon kicks. They did not address extreme emergencies like the one the *Deepwater Horizon* crew faced on the evening of April 20. In the future, Transocean and other companies must provide better training and drills to ensure that their crews are prepared to respond quickly to low-frequency, high-risk events like the Macondo blowout.

Well Control Equipment

Blowout Preventer and Emergency Disconnect System

The last piece of equipment that can prevent hydrocarbons from flowing into the riser above the wellhead is the **blowout preventer** (BOP). As Chapter 4.9 explains in more detail, the *Deepwater Horizon*'s BOP had several annular preventers, pipe rams, and shear rams that the rig crew could use to control flow coming from the well from going up the riser.

Most of the barriers in the wellbore, such as drilling mud and cement, block hydrocarbon flow without active supervision by the rig crew. By contrast, BOP elements are typically open during well operations. The BOP does not block flow unless the rig crew spots an influx and closes a BOP element, or an automated backup system activates the blind shear ram. Chapter 4.9 explains the BOP's automated backup systems in detail.

In addition to directly activating the BOP rams, the rig crew can activate the blowout preventer's **blind shear ram** and disconnect the rig from the well using an **emergency disconnect system** (EDS).[2] In accord with Transocean policy, the rig crew had tested the *Deepwater Horizon*'s EDS at surface prior to deploying the blowout preventer at the Macondo well.[3]

Emergency Disconnect System. The crew can activate the emergency disconnect system (EDS) from either the driller's control panel, the toolpusher's control panel, or the bridge.[4] Power and communication signals are sent from the rig to the BOP through multiplex (MUX) cables.[5] The signals initiate a sequence in which pod receptacles de-energize and retract, choke and kill line connectors unlatch, the blind shear ram closes, and the lower marine riser package unlatches from the BOP stack,[6] separating the rig and riser from the well. Once initiated, this sequence typically takes about a minute.[7] Emergency disconnect is not generally considered a well control response. Rather, it is used in emergency dynamic positioning scenarios to separate the rig from the well. The rig may begin to "drift off" from its station if the rig loses power, or the rig may "drive off" if the dynamic positioning system mistakenly directs the rig to move away. The riser would likely be damaged if the rig drifted or drove off, potentially resulting in an uncontrolled release of hydrocarbons into the water.

Once gaseous hydrocarbons move past the blowout preventer, they expand exponentially with decreasing depth[8] and reach the rig within minutes.[9] Timely BOP activation is therefore crucial to drilling safety.[10] If the BOP is activated quickly, little or no gas will enter the riser and travel to the rig. Transocean advises its personnel: "If the volume of gas above the BOP stack is kept small by detection equipment and shut-in, then the gas can be safely handled at [the] surface."[11] If this is not done, the consequences can be severe. On March 14, BP well site leader Jimmy Adams cautioned BP senior drilling engineer Mark Hafle: "Rigs have been burn[ed] down and people killed from gas in the riser."[12]

Diverter and Mud Gas Separator

Transocean's Well Control Handbook warns that "[l]arge amounts of gas above the BOP stack can rise rapidly and carry a large volume of mud out of the riser at high rates."[13] In those situations, the rig's **diverter** becomes the last line of defense. The diverter on the *Deepwater Horizon* sat directly beneath the rig floor.[14] It could prevent gas from flowing uncontrollably onto the drilling rig,[15] in order to "keep combustible gases safely away from sources of ignition."[16]

As Chapter 4.7 explains, mud coming out of the well normally flows up the riser, through the mud cleaning system and into the mud pits. When the rig crew activates the diverter, an annular packer in the diverter closes around the drill pipe (or closes the open hole if no drill pipe is in the hole) and prevents flow up the riser and onto the drill floor. The *Deepwater Horizon's* diverter packer had a 500 pounds per square inch (psi) working pressure rating,[17] meaning that it could safely withstand 500 psi of pressure exerted by fluids flowing up the riser. Although the diverter is designed to handle worst-case scenarios,[18] pressures above the pressure rating could cause it to fail and allow an influx to continue up the riser.

When closed, the packer forced flow to one of two 14-inch diameter **overboard lines**—one going to the port side of the rig, the other to starboard (see Figure 4.8.1).[19] The rig crew could select the direction of overboard flow in order to discharge gas on the downwind side of the rig. The starboard-side overboard line was also connected to another pipe that led to the mud gas separator. The rig crew could close a valve in the starboard line in order to route flow from that line to the mud gas separator.[20]

Figure 4.8.1. Diverter system.

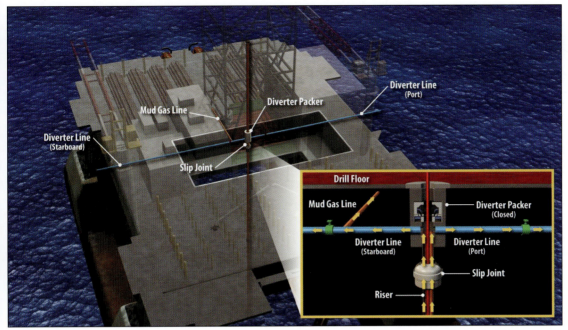

TrialGraphix

On April 20, the rig crew diverted the influx to the mud gas separator rather than sending it overboard. That caused mud and gas to spray onto the rig from the derrick.

A mud gas separator consists of a series of pipes, valves, and a tank. When gas-bearing mud flows into the tank, the mud falls to the bottom of the tank while the gas rises. The mud flows out through a pipe in the tank bottom to the rig's mud pits. The gas flows out through a separate pipe. On the *Deepwater Horizon*, that pipe ran to a vent high atop the derrick where gas could discharge into the open air.

When using the diverter system, the crew's most important decision is whether to send the fluid influx overboard or to send it to the **mud gas separator**.[21] The choice depends on the size of the hydrocarbon influx in the riser.[22] The mud gas separator is the right choice for small quantities of mud and hydrocarbons. By separating mud from gas, it allows the crew to collect and reuse the mud rather than discharge it overboard and pollute the sea. Moreover, it vents gas out of a gooseneck pipe on the derrick at the center of the rig. But sending a large influx to the mud gas separator can create a large flammable cloud of gas over the rig.[23] If a sufficiently large and sustained influx of gas from the riser goes to the mud gas separator, ignition becomes more likely, with the potential for explosion.[24] As a result, it is inappropriate to send large flows through the mud gas separator.[25] In the event of a large hydrocarbon influx into the riser, the crew should send flow overboard through the downwind line.[26]

Kick Response at Macondo

On April 20, gas moved through the *Deepwater Horizon's* open blowout preventer and shot up the riser. As it rose, the gas expanded, pushing the mud and gas faster and faster toward the rig.[27] Sometime between 9:40 and 9:43 p.m.,[28] mud spewed from the rotary table,[29] sprayed onto the rig floor,[30] and shot up and out the crown of the derrick[31] about 200 feet above the rig floor.

A Transocean representative likened the force of the gas to "a 550-ton freight train hitting the rig floor,"[32] followed by a "jet engine's worth of gas coming out of the rotary."[33]

The Rig Crew Sends the Influx to the Mud Gas Separator

After drilling mud began spraying out from the rig floor, the crew activated the diverter system.[34] Transocean toolpusher Jason Anderson was in the drill shack. He called BP well site leader Don Vidrine to say that the crew was taking action in response to mud coming back from the well.[35] It appears that rig personnel had previously set the valves on the diverter system to route diverted flow through the mud gas separator rather than overboard.[36] The crew may have done this to avoid inadvertently discharging oil-based drilling mud or other pollution into the Gulf of Mexico in violation of environmental regulations. Whatever the reason, it appears that the rig crew did not change the valve settings to route the flow overboard in response to the sudden mud influx.

Diverting flow to the mud gas separator stopped the flow of mud onto the rig floor within seconds. Micah Sandell, a Transocean gantry crane operator, testified: "I seen mud shooting all the way up to the derrick...then it just quit...I took a deep breath thinking, 'Oh, they got it under control.'"[37]

Any relief was temporary. Given the size of the influx, routing the influx to the mud gas separator rather than overboard made ignition all but inevitable. The capacity of a mud gas separator depends on the size of the outlet lines,[38] and these lines are generally not large enough to handle very high flow rates.[39] The Macondo blowout therefore quickly overwhelmed the *Deepwater Horizon*'s mud gas separator.[40] Sandell observed: "Then all the sudden the...mud started coming out of the degasser...so strong and so loud that it just filled up the whole back deck with a gassy smoke...loud enough...it's like taking an air hose and sticking it to your ear."[41]

A Weatherford specialist on the rig watched mud come out of the gas vent lines of the mud gas separator.[42] Gas likely entered the line to the mud system, which would have sent gas to the pump room, the mud pit room, and the shaker room.[43] Components of the mud gas separator may have failed at that time as well.[44] There was little wind on April 20,[45] creating "worst-case" conditions for gas dispersion.[46] A flammable gas cloud started accumulating on the rig.

The Rig Crew Activates the Blowout Preventer

In addition to activating the diverter, the crew also attempted to shut in the well with the BOP's annular preventer.[47] (Though there is evidence that the rig crew activated the lower annular preventer at 9:41 p.m., Transocean has recently contended the rig crew activated the upper annular, not the lower annular.)[48] At about the same time, Transocean assistant driller Stephen Curtis called Transocean senior toolpusher Randy Ezell to tell him that the well was blowing out, that mud was shooting through the crown on top of the derrick, and that Anderson was shutting the well in.[49] Pressure data indicate the crew activated a variable bore ram—or tightened the annular preventer—on the BOP at about 9:46 p.m.[50]

Activating the annular preventer and variable bore rams are "normal and appropriate" responses to a typical kick.[51] But this was not a typical kick. By the time the *Deepwater Horizon's* rig crew attempted to activate the BOP, substantial volumes of hydrocarbons probably had already entered the riser, where they would have been rapidly expanding upward toward the rig.[52] The flow rate of mud and hydrocarbons may have been high enough to prevent the annular preventer from sealing.[53]

In addition to activating the annular preventers or pipe rams, the crew could have activated the blind shear ram to cut the drill pipe and shut in the well.[54] The blind shear ram can be activated directly by the rig crew from the control panels, seen in Figures 4.8.2 and 4.8.3.[55] There is no evidence the rig crew attempted to activate the blind shear ram prior to the explosion.[56]

The rig crew's response generally followed the procedures that Transocean's Well Control Handbook specified "upon taking a kick."[57] The "shut-in" procedure in the handbook that applied to the April 20 situation specifies that the rig crew should first close the "annular" and then close "pre-determined rams" later if necessary.[58] The handbook's shut-in procedures do not offer any specific guidance on the use of the blind shear ram. (The handbook elsewhere advises that the blind shear rams may be used "only in exceptional circumstances."[59]) By closing the annular preventer and then a variable bore ram, the rig crew thus appears to have followed Transocean procedures.

Gas Ignites Minutes After Mud Reaches the Rig Floor

The first explosion occurred at about 9:49 p.m.[60] Ezell was on his way to the rig floor when the explosion "threw [him] against the wall in the toolpusher's office."[61] "Debris" covered him.[62] Transocean performance division manager Daun Winslow was smoking in the coffee room when he felt the walls suck in and the roof panels collapse on top of him.[63] The explosion injured several of the rig crew[64] and likely killed the men on the rig floor instantly.

The precise source of ignition may never be known. Most of the equipment on a drilling rig is not classified to protect against ignition.[65] One of the engines likely exploded first—or at least shortly after an initial explosion. Transocean motor operator William Stoner testified that he heard gas hissing and Engine 3 starting to overspeed before the first explosion.[66] The engine revved higher than Mike Williams, Transocean's electronics technician, had ever heard before.[67] Engine 6 was also on and began to rev.[68] Transocean chief mechanic Douglas Brown testified that the first explosion came from the direction of Engine 3.[69] After the explosion, the exhaust stacks, wall, handrail, and walkways around Engine 3 were all missing.[70] Seconds after the first explosion, another explosion occurred.[71] Parts of the rig were in flames.[72] Fewer than 10 minutes, and perhaps as few as six minutes, had elapsed since mud first hit the rig floor.

The Rig Crew Attempts to Activate the Emergency Disconnect System

After the explosions, crew members elsewhere on the rig attempted to activate the emergency disconnect system. Transocean subsea supervisor Chris Pleasant rushed to the bridge and informed Transocean Captain Curt Kuchta that he was activating the emergency disconnect.

Figures 4.8.2 and 4.8.3.
BOP control panels on the rig floor and bridge.

BP

The blind shear ram can be activated from the BOP's control panels on the rig floor and bridge.

Captain Kuchta replied, "[c]alm down, we're not EDSing."[73] Nevertheless, with the backing of Vidrine and Transocean offshore installation manager (OIM) Jimmy Harrell, Pleasant initiated the emergency disconnect at approximately 9:56 p.m.[74] It appears that the panel's electronic signals responded, but there was no indication of hydraulic flow closing the blind shear ram.[75] The low accumulator alarm was sounding, indicating a loss of surface hydraulic power.[76]

The Chief Counsel's team believes that by this time the explosion had already damaged the MUX cables connecting the rig and the blowout preventer, preventing the command from reaching the stack.[77] Pushing the EDS button does not appear to have activated the blind shear ram or the remainder of the emergency disconnect system. This left the rig attached to the riser. Gas continued to flow up the riser, fueling the fires on the rig.[78]

Technical Findings

If the Rig Crew Had Recognized the Kick Earlier, They Could Have Shut in the Well Before Gas Entered the Riser

The crew would have been able to prevent gas from reaching the rig if they had recognized the influx before gas entered the riser and responded by shutting in the well. At that point, closing the annular preventer or the variable bore ram should have controlled the kick and stopped flow. By the time the *Deepwater Horizon* crew actually did recognize the influx and activate the blowout preventer, hydrocarbons had almost certainly entered the riser and begun expanding rapidly upward toward the rig.

The *Deepwater Horizon* crew recognized that there was an anomaly, but they did not identify that anomaly as a kick. If rig personnel suspect a kick, they perform a flow check and shut in the well.[79] The same cannot be said for responses to anomalies. The *Horizon* crew suspected that something was amiss when they shut down the pumps at 9:30 p.m. Over the next 10 minutes or so, they conducted diagnostics and discussed the anomalous pressures they were seeing. Only after hydrocarbons had entered the riser, and about when mud started emerging from the rotary, did the crew act to shut in the well. Apparently, the crew did not suspect a kick until 10 minutes after they detected the anomaly. A more conservative initial approach to the anomaly—of shutting in *first* and investigating *afterward*—would have resulted in rig personnel shutting in the well while hydrocarbons were still confined to the wellbore and thereby preventing the blowout.

By the time the crew activated the annular preventer, mud and hydrocarbons may have been flowing through the BOP at a high enough rate to prevent it from sealing.[80] Data on drill pipe pressure indicate that the annular preventer did not achieve shut-in pressure. Only 1,200 psi registered,[81] well below what would have been required.[82] Later, the drill pipe pressure climbed above 5,500 psi.[83] That appears to have been due either to tightening of the annular or to activation of the variable bore ram.[84] Though the well may have been shut in by 9:49 p.m.,[85] it appears that there was already a substantial volume of gas above the BOP at this time because this is when the first explosion took place.

Previous modifications to the BOP may have compromised the ability of the lower annular preventer to seal the well. (As noted above, Transocean has recently contended the rig crew activated the upper annular and not the lower annular. If true, modifications to the lower annular would not have affected the BOP's performance during the blowout.) As discussed further in Chapter 4.9, BP asked Transocean in 2006 to modify the lower annular to a "stripping" annular.

This change reduced the rated working pressure from 10,000 to 5,000 psi,[86] and allowed the rig crew to raise or lower pipe through the BOP while the annular was closed. The 10,000-psi-rated annular body was not replaced.[87] While the stripping annular would still be able to close in pressures above 5,000 psi, it is not clear whether it would completely seal at these higher pressures.[88]

Diverting Overboard Might Have Delayed the Explosion

The rig crew should have diverted the flow overboard when mud started spewing from the rig floor.[89] The flow of mud at this point was tremendous—it shot 200 feet up to the crown of the derrick. That should have prompted the crew to take immediate emergency measures.

Transocean's Well Control Handbook advises that "at any time, if there is a rapid expansion of gas in the riser, the diverter must be closed (if not already closed) and the flow diverted overboard."[90] The handbook also provides: "[I]f large volumes of gas have entered the riser, it will flow rapidly on its own and there will be no way to control it by adjusting the circulation rate. Then, the surface gas and liquid rates become very high, especially as the gas bubble reaches surface and *the flow* **must** *be diverted overboard.*"[91]

Although mud flow at the rig floor does not always mean that gas is in the riser, the *Deepwater Horizon*'s crew should have assumed that this was the case for two reasons. First, the fact that *mud* was *spewing* from the rig floor after the crew had displaced the well with seawater down to 8,367 feet below sea level should have indicated that hydrocarbon flow had already proceeded a substantial distance up the well. Second, and more significantly, the high mud flow rate and volume should have warned the crew that the kick was severe and prompted them to send the influx overboard.

While the Chief Counsel's team finds that the rig crew should have sent the influx overboard immediately, doing so may not have prevented an explosion. Two factors determine whether diverting flow overboard would have prevented an explosion: (1) the ability of the diverter packer, overboard lines, and other equipment to handle the flow rate and volume, and (2) the way in which gas dispersed away from the rig.[92]

With regard to equipment capabilities, currently available information leads the Chief Counsel's team to conclude that the diverter packer probably would have been able to handle the flow rate and volume during the blowout, though it is not certain. The diverter packer on the *Deepwater Horizon* was rated to withstand 500 psi of pressure. Two post-blowout computer models commissioned by BP for its internal investigation offer perspective on the forces that may have been exerted on the diverter packer during the blowout; the Chief Counsel's team is not aware of any other modeling that has been performed at this time. One model predicts that the maximum pressure exerted on the diverter packer during the blowout was 145 psi,[93] not even close to the packer's limit. Another model indicates that the pressures may have been much higher, peaking at 500 psi.[94] But even under that scenario, the diverter packer probably would not have failed. That model only predicted that the packer would have been subject to 500 psi for an instant,[95] and this type of equipment can generally handle pressures beyond rated capacity for a short period of time. Moreover, if the rig crew had sent the influx overboard, the pressure on the diverter element likely would have been even lower.[96] The Chief Counsel's team therefore believes that the diverter packer probably would not have failed if the rig crew had sent the influx overboard.[97]

Though the diverter packer probably could have withstood the blowout flow rate and pressure, the **slip joint** could have failed. The slip joint sat below the diverter packer, permitting the rig to heave vertically while maintaining the riser connection to the sea floor. It had two modes: a low-pressure mode with a 100 psi working pressure[98] and a high-pressure mode with a 500 psi working pressure.[99] If the slip joint had been in low-pressure mode, it would have been vulnerable to failure.[100] That would have allowed gas to escape into the moon pool area of the rig. Additionally, because the diverter packer does not seal off the riser, there is a possibility that gas could have also traveled up the drill pipe and onto the rig.

With regard to gas dispersion, the calm wind conditions on April 20 would have limited the rate at which gas dispersed away from the rig. The wind speed was low, about 2 to 4 knots.[101] The wind also appears to have been blowing from starboard to port,[102] though the precise direction is difficult to ascertain.[103] Because of this, gas flowing out of the starboard overboard line would have stayed close to the rig and perhaps even blown back onto the rig rather than drifting away.[104] Nevertheless, diverting overboard would have substantially reduced the risk of ignition of the rising gas and given the rig crew more time to respond.[105] An MMS study of offshore blowouts between 1992 and 2006 found that the "success rate for diverter systems was very high...16 of the 20 diverter uses were considered successful because the desired venting of gas was sustained until the well bridged."[106]

The Chief Counsel's team concludes that diverting flow overboard likely would have sent a substantial amount of gas off the rig.[107] This may not ultimately have prevented an explosion but probably would have given the rig crew more time to respond to the blowout. BP has concluded that "diversion of fluids overboard, rather than to the MGS, may have given the rig crew more time to respond and may have reduced the consequences of the accident."[108] Transocean agrees that "diverting overboard might have delayed the explosion...."[109]

Management Findings

Transocean Should Have Trained Its Employees Better on How to Respond to Low-Frequency, High-Risk Events

There are at least three explanations for why the crew did not immediately divert the flow overboard.

- First, the crew may not have recognized the severity of the situation, though that seems unlikely given the amount of mud that spewed from the rig floor.
- Second, they did not have much time to act. At most, the drill crew had six to nine minutes after mud emerged from the rig floor before the first explosion.
- Finally, and perhaps most significantly, the rig crew had not been trained adequately regarding how to respond to such an emergency situation. It appears that the crew followed the procedures for dealing with a kick set forth in Transocean's Well Control Handbook. Those procedures were inadequate given the circumstances.[110]

Transocean has highlighted to the Chief Counsel's team the "extensive curriculum of courses" available to its rig crew, including courses on well control.[111] Transocean contends that the "initial response...was the appropriate first normal response"[112] and that the "crew utilized the proper sequencing."[113] The Chief Counsel's team recognizes that the rig crew may simply have done what it had been trained to do. But that assertion indicates the inadequacy of the crew's training and guidance in the first place.

Though Transocean's protocols provide that a severe influx should be sent overboard, the sequence of "procedures for handling gas in the riser"[114] (Transocean document shown in Figure 4.8.4) specifically recommends the overboard line—instead of the mud gas separator—only in the ninth step after actions such as monitoring for flow and circulating the riser. Here, there was *no time* to get to the ninth step.[115] In the future, well control training should include simulations and drills for low-probability, high-consequence emergency events and well-control protocols should specifically address such emergencies.[116] ◢

Figure 4.8.4. Transocean's "procedures for handling gas in the riser."

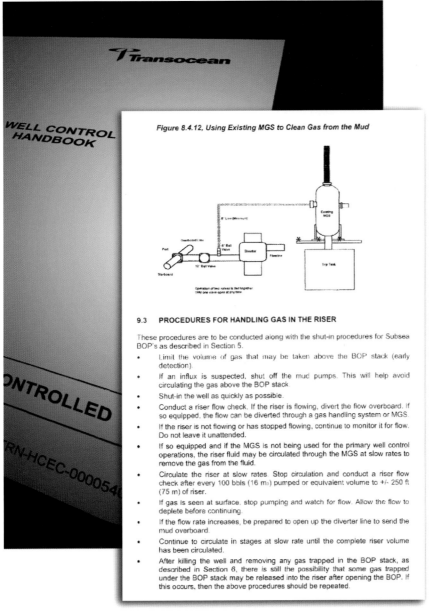

Figure 8.4.12, Using Existing MGS to Clean Gas from the Mud

9.3 PROCEDURES FOR HANDLING GAS IN THE RISER

These procedures are to be conducted along with the shut-in procedures for Subsea BOP's as described in Section 5.

- Limit the volume of gas that may be taken above the BOP stack (early detection).
- If an influx is suspected, shut off the mud pumps. This will help avoid circulating the gas above the BOP stack.
- Shut-in the well as quickly as possible.
- Conduct a riser flow check. If the riser is flowing, divert the flow overboard. If so equipped, the flow can be diverted through a gas handling system or MGS.
- If the riser is not flowing or has stopped flowing, continue to monitor it for flow. Do not leave it unattended.
- If so equipped and if the MGS is not being used for the primary well control operations, the riser fluid may be circulated through the MGS at slow rates to remove the gas from the fluid.
- Circulate the riser at slow rates. Stop circulation and conduct a riser flow check after every 100 bbls (16 m₃) pumped or equivalent volume to +/- 250 ft (75 m) of riser.
- If gas is seen at surface, stop pumping and watch for flow. Allow the flow to deplete before continuing.
- If the flow rate increases, be prepared to open up the diverter line to send the mud overboard.
- Continue to circulate in stages at slow rate until the complete riser volume has been circulated.
- After killing the well and removing any gas trapped in the BOP stack, as described in Section 6, there is still the possibility that some gas trapped under the BOP stack may be released into the riser after opening the BOP. If this occurs, then the above procedures should be repeated.

Transocean

Protocol from Transocean's Well Control Handbook.

Chapter 4.9 | The Blowout Preventer

The **blowout preventer** (BOP) is a routine drilling tool. It is also designed to shut in a well in case of a kick, thereby "preventing" a blowout. As described in Chapter 4.8, the rig crew attempted to close elements of the BOP and to activate the emergency disconnect system (EDS) in response to the Macondo blowout. Automatic and emergency activation systems should have also closed the BOP's blind shear ram and shut in the well. Though preliminary evidence suggests one of these systems may have activated and closed the blind shear ram, the blind shear ram never sealed the well.

The federal government has recovered the BOP from the blowout site, and forensic testing is ongoing. Until that testing is complete, a full examination of blowout preventer failure is impossible. In the meantime, the Chief Counsel's team has made preliminary findings and identified certain technical faults that may have prevented the BOP system from activating and shutting in the well.

Figure 4.9.1. Transporting the *Deepwater Horizon* BOP.

U.S. Coast Guard photo/Petty Officer 3rd Class Stephen Lehmann

The *Deepwater Horizon*'s blowout preventer on the Mississippi River in transit to Michoud, Louisiana, to undergo forensic testing, September 11, 2010.

Figures 4.9.2 and 4.9.3. The *Deepwater Horizon* blowout preventer stack.

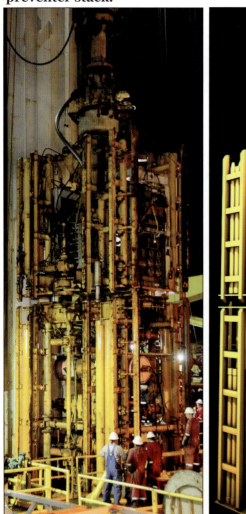

U.S. Coast Guard photo/
Petty Officer 1st Class Thomas M. Blue

TrialGraphix

Left: Photo of the recovered *Deepwater Horizon* BOP.
Right: 3-D model of the *Deepwater Horizon* BOP.

Figure 4.9.4. Blind shear ram.

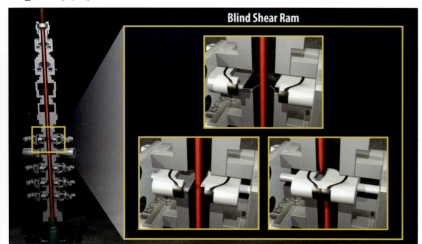

TrialGraphix

Blind Shear Rams

Federal regulations required the *Deepwater Horizon* to have a BOP that included a **blind shear ram** (BSR).[1] The blind shear ram is designed to cut drill pipe in the well (as shown in Figure 4.9.4) and shut in the well in an emergency well control situation.[*] But even if properly activated, the blind shear ram may fail to seal the well because of known mechanical and design limitations. In order for a blind shear ram to shut in a well where drill pipe is across the BOP, it must be capable of shearing the drill pipe.[2] And blind shear rams are not always able to perform this critical function, even in controlled situations.

Blind Shear Rams Cannot Cut Tool Joints or Multiple Pieces of Drill Pipe

Blind shear rams are not designed to cut through multiple pieces of drill pipe or **tool joints** connecting two sections of drill pipe.[3] It is thus critically important to ensure that there is a piece of pipe, and not a joint, across the blind shear ram before it is activated.[4] This fact prompted a 2001 MMS study to recommend every BOP to have two sets of blind shear rams such that if a tool joint prevented one ram from closing, another adjacent ram would close on drill pipe and would be able to shear the pipe and shut in the well.[5] MMS never adopted the recommendation.

The *Horizon*'s blowout preventer had only one blind shear ram. Sections of drill pipe are joined by a tool joint at each interval and are often about 30 feet in length, though some of the drill pipe used on the *Horizon* varied in length.[6] If one of those joints was in the path of the blind shear ram at the time of attempted activation, as portrayed in Figure 4.9.5, the ram would have been unable to shear the pipe and shut in the well.

---

[*] Although not separately depicted in Figures 4.9.3 and 4.9.4, there are hydraulic, power, and communications lines (cables), as well as the choke, kill, and boost lines (pipes) running from the rig to the blowout preventer.

Even if a tool joint did not prevent the blind shear rams from shutting in the Macondo well, the inability to shear tool joints is a recognized and significant limitation. The Chief Counsel's team agrees with the MMS study that installing a second blind shear ram would mitigate this risk and increase the probability of success in shutting in a well.[7]

Figure 4.9.5. Tool joint in the blind shear ram.

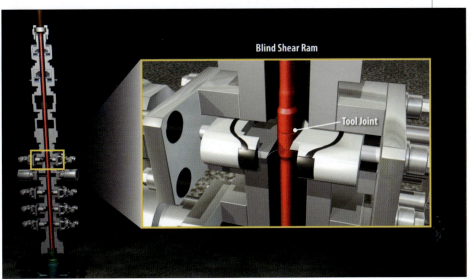

TrialGraphix

Blind shear rams cannot cut tool joints.

Study Finds Deepwater Exacerbates Limitations

A 2002 MMS study conducted by West Engineering Services, a drilling consulting firm, presented "a grim picture of the probability of success when utilizing [shear rams] in securing a well after a well control event."[8] The study found that only three of six tested rams successfully sheared drill pipe under operational conditions.[9] It also found that "operators often do not know how their shear rams would perform in a high pressure environment."[10] These problems worsen in deepwater because, among other things, deepwater operators often use stronger drill pipes that are more difficult to cut.[11] Increased hydrostatic and dynamic pressures in deepwater wells also increase the difficulty of shearing.[12]

Although the study found that these factors were "generally ignored,"[13] it is not certain whether these factors affected the blind shear ram at Macondo.

Deepwater Horizon Blind Shear Ram Testing

Earlier Tests Establish Shearing Ability

The shearing ability of the *Deepwater Horizon*'s blind shear ram was demonstrated on at least two occasions. During the rig's commissioning, the rams sheared a 5.5-inch, 21.9-pound pipe at a shear pressure of 2,900 pounds per square inch (psi).[14] According to pipe inventory records, this was the same thickness and weight of the drill pipe retrieved from the Macondo well.[15] The ram also successfully sheared drill pipe during a 2003 EDS function.[16]

The Rig Crew Regularly Tested the *Deepwater Horizon*'s Blind Shear Ram, but Often at Reduced Pressures

Regulations require frequent monitoring and testing of the BOP blind shear ram both on surface and subsea. This includes testing the blind shear ram on the surface prior to installation[17] and subsea pressure testing after installation.[18] The BOP stack was inspected almost daily by remotely operated vehicle (ROV).[19] Like the positive pressure test, other **pressure tests** of the blind shear ram established that the ram was able to close and seal in pressure.[20] The rig crew also regularly **function tested** the blind shear ram, which tested the ability of the ram to close but did not test its ability to withhold pressure.[21] Subsea pressure and function tests do not demonstrate the ability of the blind shear ram to shear pipe.[22]

MMS regulations include, among other things, requirements regarding the amount of pressure a BOP must be able to contain during testing. MMS regulations normally require rams to be tested to their rated working pressure or maximum anticipated surface pressure, plus 500 psi.[23] However, BP applied and received MMS approval to downgrade test pressures for several of the *Deepwater Horizon*'s BOP elements. The departure that MMS granted allowed BP to test the *Deepwater Horizon*'s blind shear ram at the same pressures at which it tested casing.[24] Though the rig crew tested the blind shear ram to 15,000 psi prior to launch (showing that it would contain 15,000 psi of pressure), subsequent tests were at pressures as low as 914 psi.[25] The rig crew also tested the annular preventers at reduced pressures. MMS regulations require that high-pressure tests for annular preventers equal 70% of the rated working pressure of the equipment or a pressure approved by MMS.[26] BP's internal guidelines similarly call for annular preventers to be tested to a maximum of 70% of rated working pressure "if not otherwise specified."[27] In May 2009, BP filed an application to reduce annular tests to 5,000 psi.[28] In January 2010, BP filed another application to further reduce testing pressures for both annular preventers to 3,500 psi.[29] It is likely BP sought to test equipment at lower pressures in order to reduce equipment wear.[30]

BP's lowered pressure testing regime was both approved by MMS and consistent with industry practice. BOP elements are designed to withstand and should be able to withstand higher pressures even if tested to lower pressures.[31] Nonetheless, low-pressure testing only demonstrates that equipment will contain low pressures. At Macondo, many tests did not prove the blowout preventer's ability to contain pressures in a worst-case blowout scenario.[32]

Blind Shear Ram Activated and Sealed During April 20 Positive Pressure Test

On the day of the blowout, the rig crew used the blind shear ram to conduct a positive pressure test.[33] As discussed in Chapter 4.6, the blind shear rams closed and sealed as expected during the test. This fact suggests that the rams were capable of sealing the well when the blowout occurred. But the evidence on its own is inconclusive that the rams could have functioned in an emergency; during the positive pressure test the crew closed the blind shear rams using a low-pressure hydraulic system, rather than the high-pressure hydraulic system that would have activated the rams in the event of a blowout.

Blind Shear Ram Activation at Macondo

There are five ways the blind shear ram on the *Deepwater Horizon* blowout preventer could have been activated:

- direct activation of the ram by pressing a button on a control panel on the rig;
- activation of the EDS by rig personnel;
- direct subsea activation of the ram by an ROV "hot stab" intervention;[34]

- activation by the automatic mode function (AMF) or "deadman" system due to emergency conditions or initiation by ROV; and

- activation by the "autoshear function" if the rig moves off location without initiating the proper disconnect sequence or if initiated by ROV.

Preliminary information from the recovered blowout preventer suggests the blind shear ram may have been closed and indicates erosion in the BOP on either side of the ram as pictured in Figure 4.9.6.[35] This suggests one of these mechanisms may have successfully activated the blind shear ram but failed to seal the flowing well because high-pressure hydrocarbons may have simply flowed around the closed ram.

Figure 4.9.6. *Deepwater Horizon* blowout preventer's closed blind shear ram (top view).

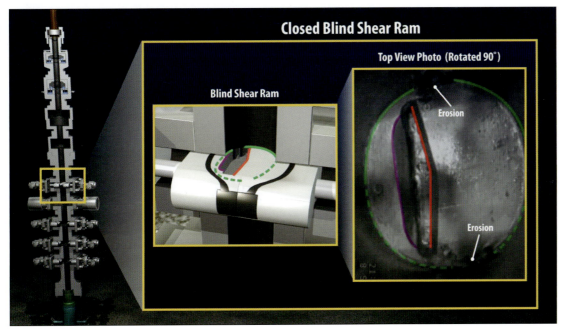

TrialGraphix, BP photo

As discussed in Chapter 4.8, there is no evidence that rig personnel attempted to directly activate the blind shear ram from the rig's control panels. Rig personnel did attempt to activate the EDS system after the explosions, but those attempts did not activate the blind shear ram. Emergency personnel in the days following the blowout were unable to shut in the well by directly activating the blind shear ram using an ROV. At various points in time, the deadman function should have closed the ram. Though Transocean has suggested that this system activated the blind shear ram, faults discovered post-explosion may have prevented the deadman from functioning. BP has suggested that post-explosion ROV initiation of the autoshear system activated the blind shear ram.

It is clear that some of these mechanisms failed to activate; forensic testing will likely confirm which, if any, of these triggering mechanisms successfully activated. Even if activated, none of these mechanisms shut in the flowing well.

ROV Hot Stab Activation at Macondo

Rig personnel can also close the blind shear ram by using an ROV to pump hydraulic fluid into a **hot stab** port on the exterior of the BOP. The hot stab port is connected to the blind shear ram hydraulic system; fluid flowing into the port actuates the ram directly, bypassing the BOP's control systems.

In theory, this function should close the blind shear ram when other methods fail. But an MMS study by West Engineering found ROVs may be unable to close rams during a well control event due to lack of hydraulic power.[36] The study also found that a flowing well may cause rams to erode or become unstable in the time it takes for an ROV to travel from the surface to the BOP on the seafloor.[37]

ROVs deployed at Macondo at about 6 p.m. on April 21.[38] ROV hot stab attempts to shut in the well on April 21 and 22 with the pipe rams and the blind shear ram failed.[39] As discussed below, on April 22 ROVs may have successfully activated the blind shear ram through the AMF/deadman system or autoshear system.[40] But despite these efforts, the blind shear ram did not shut in the well.[41] Efforts to shut in the BOP through an ROV hot stab continued without success until May 5.[42] By May 7, BP had concluded that "[t]he possibility of closing the BOP has now been essentially exhausted."[43]

Efforts to close the BOP stack were frustrated by organizational and engineering problems. In December 2004, Transocean had converted the lower variable bore ram on the BOP into a test ram[44] at BP's request.[45] Because of an oversight that likely occurred during the modification, a hot stab port on the BOP exterior that should have been connected to a pipe ram was actually connected to the test ram, which could not shut in the well.[46] Unaware of this fact, response teams tried to use that hot stab port to shut in the well.[47] For two days, they tried to close a pipe ram but were actually activating the test ram instead.[48] This error frustrated response efforts[49] until crews discovered the mistake on May 3.[50] After discovering the mistake, response crews attempted on May 5 to activate the BOP's pipe rams again, with no success.[51]

None of the attempted hot stab activations prevented the flow of hydrocarbons from the well. The rig crew had tested the hot stab function before installing the *Deepwater Horizon* BOP, in accord with Transocean's Well Control Handbook.[52]

There are a number of possible reasons why ROVs were unable to activate the rams using hot stabs. First, the ram may have activated, but the presence of a tool joint or more than one piece of pipe prevented the ram from shearing the pipe and sealing the well. Second, ROV pumps failed during early intervention efforts.[53] Third, ROVs were incapable of pumping fast enough and as a result were not able to build pressure against a leak in the BOP hydraulic system.[54]

Automatic Blind Shear Ram Activation at Macondo

Transocean and BP both claim an automated backup system activated the blind shear ram. According to Transocean, the automatic mode function activated.[55] According to BP, the autoshear system activated.[56] If activated, neither system sealed the well.

Automatic Mode Function (AMF)/Deadman

The **AMF** or **deadman** system is designed to close the blind shear ram under certain emergency conditions. The system should activate when all three of the following conditions are met:

- loss of electrical power between the rig and BOP;[57]
- loss of communication between the rig and the BOP;[58] and
- loss of hydraulic pressure from the rig to the BOP.[59]

Catastrophic events on a rig can create these conditions, or emergency workers can trigger them by using an ROV to cut power, communication, and hydraulic lines to the BOP (these components are labeled in Figure 4.9.7.).[60] The AMF will not operate unless rig personnel "arm" it at a surface control panel.[61] Notes from response crews and post-explosion analysis of the BOP **control pods** indicate the AMF system on the *Deepwater Horizon* BOP was likely armed.[62]

Figure 4.9.7. AMF system.

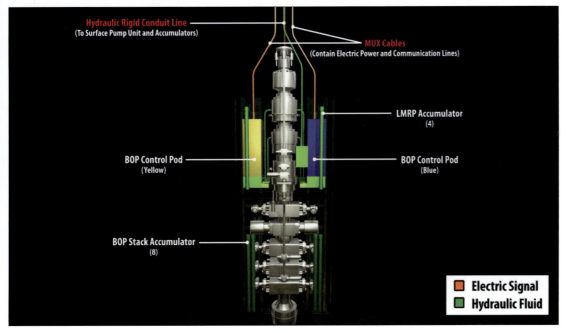

TrialGraphix

The AMF, or deadman, system is activated in emergency conditions.

Based on available information, it appears likely that the explosion on April 20 created the conditions necessary to activate the deadman system. The multiplex (MUX) cables, which carried the power and communication lines, were located near a primary explosion site in the rig's moon pool and would probably have been severed by the explosion.[63] The hydraulic conduit line was made of steel[64] and less vulnerable to explosion damage.[65] However, the BOP would have likely lost hydraulic power at least by April 22 when the rig sank, and the deadman should thus have activated by that date.[66] Response crew personnel also tried to activate the deadman on April 22 by cutting electrical wires using an ROV.[67] According to Transocean, the AMF activated the blind shear ram.[68]

Unclear Whether AMF Activated

It is currently not clear whether the AMF activated the blind shear ram. However, the Chief Counsel's team has identified issues that may have affected the AMF.

First, the universe of available test records may be limited because Transocean destroyed test records at the end of each well.[69] Second, the deadman system was not regularly tested.[70] Although Transocean's Well Control Handbook calls for surface testing the deadman system,[71] based on available evidence the AMF was not tested prior to deployment.[72]

Third, the deadman system relied upon at least one of the BOP's two redundant control pods (yellow or blue) to function. If both pods were inoperable, the system would not have functioned. The rig crew function tested and powered both pods at the surface in February 2010 prior to splashing the BOP.[73] But post-explosion examination revealed low battery charges in one BOP control pod and a faulty solenoid valve in another. If these faults were present at the time of the incident, they would have prevented the deadman and autoshear functions from closing the blind shear ram.

Low Battery Charge in the Blue Pod

In the event that electric power from the rig to the BOP is cut off, the BOP's control systems are powered by a 27-volt and two 9-volt battery packs contained in each pod.[74] These batteries power a series of relays that cause the pod to close the blind shear ram if there is a loss of power, communication, and hydraulic pressure from the rig.[75] BP tests suggest that it takes at least 14 volts of electricity to power the relays,[76] and a Transocean subsea superintendent has stated that the activation sequence may require as many as 20 volts.[77]

Tests on the blue pod conducted by Cameron after the blowout on July 3 to 5, revealed that battery charge levels may have been too low to power the sequence to shut the blind shear ram. The 27-volt battery was found to have only a 7.61-volt charge.[78] One of the 9-volt batteries was found to have 0.142 volts, and the other 9-volt battery had 8.78 volts.[79] If these battery levels existed at the time the deadman signaled the pods to close the blind shear ram, the low battery levels very likely would have prevented the blue pod from responding properly.[80] Transocean disputes whether the batteries were depleted at the time of the explosion. Transocean has suggested battery levels were adequate to power the AMF but, due to a software error, may have been left activated and discharged after the explosion.[81] The Chief Counsel's team has not received evidence in support of this assertion but anticipates ongoing forensic testing of the pods will evaluate expected battery levels at the time of the incident.

Available records suggest that Transocean did not adequately maintain and replace its BOP pod batteries.[82] Cameron recommends replacing pod batteries at least annually, and recommends yearly battery inspection.[83] Transocean itself recommends yearly inspection of batteries.[84]

An April 2010 Transocean ModuSpec rig condition assessment stated that all three pods had new batteries installed.[85] But internal Transocean records suggest that the crew had not replaced the batteries on one pod for two-and-a-half years prior to the Macondo blowout and had not replaced the batteries in another pod for a year.[86] This appears to have been a pattern: Company records show that rig personnel found all of the batteries in one *Deepwater Horizon* BOP pod dead in November 2007.[87]

Table 4.9.1. Control pod battery replacements (based on available records).[88]

Pod	Battery Replacement Dates	Time Between Battery Replacements	Time Between Replacement and Blowout
Pod 1*	January 26, 2006; April 25, 2009	3 years	1 year
Pod 2	May 28, 2004; December 29, 2005; October 13, 2009	1-3 years	6 months
Pod 3	March 26, 2004; November 4, 2007	3 years	2.5 years

*The *Deepwater Horizon* had three pods for its BOP; at any given time, one was the "blue" pod, one was the "yellow" pod, and one remained on the surface.

Solenoid Valve Problems in the Yellow Pod

Control pods also rely on functioning solenoid valves (diagrammed in Figure 4.9.8). The solenoid valves open and close in response to electrical signals and thereby send hydraulic pilot signals from the pods to the BOP elements.[89] The pilot signals in turn open hydraulic valves, which then deliver pressurized hydraulic fluid into BOP rams to close them.[90] Each solenoid activates when electric signals energize one of two redundant coils in the solenoid.[91]

Figure 4.9.8. BOP's electrical schematic.

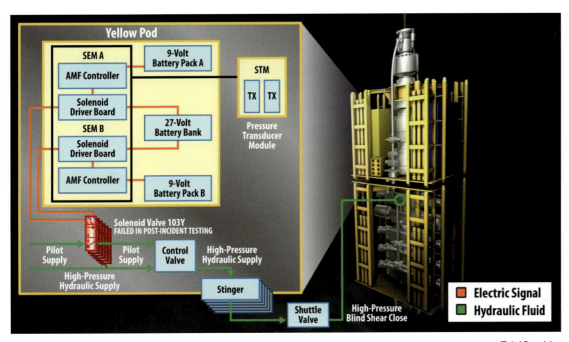

TrialGraphix

Tests on the *Deepwater Horizon*'s yellow pod revealed that the solenoid valve used to close the blind shear ram was inoperable.

According to maintenance records, the yellow pod's solenoids were changed on January 31, 2010.[92] However, tests on the yellow pod conducted by Cameron after the blowout on May 5 to 7[93] revealed that a key solenoid valve used to close the blind shear ram was inoperable.[94]

If this fault existed prior to the blowout, an alarm on the rig's control system should have notified the rig crew and triggered a record entry by the rig's event logger.[95] According to witness testimony, the rig crew believed the solenoid valve in the yellow pod was functioning as of April 20.[96]

Autoshear System May Have Activated but Failed to Shut in Flowing Well

Like the emergency disconnect system (EDS), the **autoshear** function is designed to close the blind shear ram in the event that the rig moves off position. The autoshear is activated when a rod linking the lower marine riser package (LMRP) and BOP stack is severed. The rod can be severed by rig movements; if the rig moves off position, it will pull the LMRP out of place and sever the rod. Rig personnel can also sever the rod directly by cutting it with an ROV.[97] Like the deadman, the rig crew must arm the autoshear system at the driller's or toolpusher's control panel.[98] According to BP's internal investigation, the autoshear function was armed at the time of the incident.[99] Transocean policy required its personnel to surface test the autoshear system before deploying the BOP, and the *Deepwater Horizon* rig crew conducted a test on January 31, 2010.[100]

Response crews used an ROV to activate the autoshear function directly by cutting the rod on April 22 at approximately 7:30 a.m.[101] According to BP, response crews reported movement on the stack, which may have been the accumulators discharging pressure and activating the blind shear ram.[102] Even if the autoshear did activate and close the blind shear ram, the blind shear ram did not stop the flow of oil and gas from the well.

Potential Reasons the Blind Shear Ram Failed to Seal

Figure 4.9.9. Erosion in the BOP.

BP

Erosion above the blind shear ram on the BOP's kill side.

Flow Conditions Inside the Blowout Preventer

Even if the blind shear ram activated, it failed to seal the well. One possible explanation is that the high flow rate of hydrocarbons may have prevented the ram from sealing. Initial photos from the recovered BOP show erosion in the side of the blowout preventer *around* the ram, which was a possible flow path for hydrocarbons, as seen in Figure 4.9.9.[103] Therefore even if the ram closed, the hydrocarbons may have simply flowed around the closed ram.

Presence of Nonshearable Tool Joint or Multiple Pieces of Drill Pipe

As discussed above, the ram may not have closed because of the presence of a tool joint across the blind shear ram. If a tool joint or more than one piece of drill pipe was across the blind shear ram when it was activated, the ram would not have been able to shear and seal the well. Though preliminary evidence suggests these factors may not have impacted the blind shear ram's ability to close, the Chief Counsel's team cannot rule out the possibility of such interference.[104]

Accumulators Must Have Sufficient Hydraulic Power

The *Deepwater Horizon* blowout preventer had subsea **accumulator bottles** that provided pressurized hydraulic fluid used to operate different BOP elements. If the hydraulic line between the rig and BOP is severed, these accumulators must have a sufficient charge to power the blind shear ram.

The lower marine riser package had four 60-gallon accumulator bottles were on.[105] On the BOP stack, eight 80-gallon accumulator bottles capable of delivering 4,000 psi of pressure provided hydraulic fluid for the deadman, autoshear, and EDS systems.[106] These tanks were continuously charged through a hydraulic rigid conduit line running from the rig to the blowout preventer.[107] Should the hydraulic line disconnect, the tanks contained compressed gas that could energize hydraulic fluid to activate the blind shear ram. The rig crew checked the amount of pre-charge pressure in the accumulators prior to deploying the BOP in February.[108] However, the available amount of usable hydraulic fluid in the accumulators at the time of autoshear and AMF activation is unknown. If the charge levels were too low, the accumulators would not have been able to successfully power the blind shear ram.[109]

BP's internal investigation suggests accumulator pressure levels may have been low based on fluid levels discovered post-explosion.[110] Responders discovered 54 gallons of hydraulic fluid were needed to recharge accumulators to 5,000 psi.[111] BP's investigation suggests a leak in the accumulator hydraulic system may have depleted available pressure levels but not to levels that would have prevented activation of the blind shear ram.[112] Response crews observed additional leaks from accumulators during post-explosion ROV intervention.[113]

Leaks

It is relatively common for BOP control systems to develop hydraulic fluid leaks on the many hoses, valves, and other hydraulic conduits in the control system. Not all control system leaks affect the ability of the BOP to function: Because BOP elements are designed to close quickly, a minor leak may slow, but not likely prevent, the closing of the BOP.[114]

Even if a leak is minor, rig personnel must first identify the cause of a leak to ensure that more severe system failures do not occur.[115] Constant maintenance, inspections, and testing are required to prevent and detect such leaks.[116] Leaks discovered during surface testing should be repaired before deployment.[117] If rig personnel discover a leak after deployment, they must decide whether the leak merits immediate repair. Raising and lowering a BOP stack is a complicated operation with risks of its own; taking this action to repair a minor control system leak may actually increase rather than reduce overall risk.[118]

Leaks May Have Been Unidentified Prior to Incident

According to Transocean senior subsea supervisor Mark Hay, the *Deepwater Horizon*'s BOP had no leaks at the time it was deployed at Macondo.[119] Even if no leaks existed when the BOP was deployed, rig personnel identified at least three leaks in the months before the blowout after the BOP was in service.[120] And rig personnel identified several more leaks during response efforts that according to independent experts were not likely created during the explosion.[121] It is possible leaks developed during the response effort. But it is also possible leaks already existed and the rig crew had not identified or analyzed the impact of the leak.

A leak on the ST lock close hydraulic circuit (leak 3 in Table 4.9.2) may have prevented ROVs from pumping enough pressure to fully close the blind shear ram.[122] Both BP and Transocean have suggested that a leak on the ram lock circuit (leak 4 in the table) may be proof that the blind shear ram in fact closed.[123] Ongoing forensic testing will likely determine if leaks on the BOP control system otherwise affected the BOP's functionality, though it is unlikely these leaks prevented the BOP from sealing.

Table 4.9.2. Leaks on the *Deepwater Horizon* blowout preventer (partial list).

	Leak	Time of Identification
1	Test ram, pilot leak on yellow pod open circuit shuttle valve[124]	Pre-explosion (February 23, 2010[125])
2	Upper annular preventer, blue pod leak on the hose fitting connecting the surge bottle to operating piston[126]	Pre-explosion (February 19, 2010[127])
3	ST lock close hydraulic circuit leak (this is in the same hydraulic circuit as the blind shear ram)[128]	Post-explosion (April 25, 2010[129])
4	Blind shear ram ST lock circuit leak[130]	Post-explosion (April 26, 2010[131])
5	Lower annular preventer open circuit[132]	Pre-explosion (date not available[133])

Identified Leaks Not Reported to MMS

Even if forensic testing concludes leaks on the BOP control system did not impact functionality, it is not clear BP and Transocean adequately responded to known leaks. According to Transocean senior subsea supervisor Owen McWhorter, "the only thing I'd swear to is the fact that leaks discovered by me, on my hitch, were brought to my supervisor's attention and the Company man's attention."[134]

Under 30 C.F.R. § 350.466(f), drilling records must contain complete information on "any significant malfunction or problem."[135] This provision may require control system leaks or other anomalies to be recorded in daily drilling reports and thus subject to review by MMS inspectors.[136] At least two of the leaks identified pre-explosion were not listed in daily drilling reports. A pilot leak on the test ram open circuit shuttle valve (leak 1 in the table) was not mentioned in the daily drilling report for February 23.[137] However, the leak was reported in BP's internal daily operations report from February 23 until March 13.[138] BP wells team leader John Guide and BP regulatory advisor Scherie Douglas made the decision not to report the leak to MMS, a failure which Guide admits was "a mistake in hindsight."[139] BP well site leader Ronnie Sepulvado also admits this leak should have been noted in the daily drilling report but stated that it was not reported because the leak did not affect the ability to control the well since it was on a test ram and the test ram was still operable.[140]

The rig crew failed to include at least one other known leak in the daily drilling reports. Although the rig crew discovered a leak on an upper annular preventer hose fitting (leak 2 in the table) on February 19,[141] the leak was not listed on the daily drilling report.[142] Although subsea personnel in the past had been required to produce documentation on the leak so that the leak could be explained to MMS, McWhorter was not asked to produce documentation for this leak.[143] A failure to report these leaks potentially violated MMS reporting regulations.[144]

Inconsistent Response to Identified Leaks

There is little industry guidance as to what constitutes an appropriate response to minor leaks.[145] It appears the rig crew was able to identify the cause and impact of some leaks but not others. Evidence indicates both BP and Transocean personnel assessed the leak on the test ram shuttle valve (leak 1 in the table) and determined the ram would still function properly.[146] Records appear to indicate the rig crew planned to further evaluate this leak when the rig moved from Macondo to the next well.[147]

In response to a leak on an upper annular hose fitting (leak 2 in the table), the rig crew appears to have isolated and monitored hydraulic pressure.[148] The crew eventually measured this leak at 0.1 gallons per minute.[149] Sepulvado noted the leak on his office white board.[150] Although the leak was later erased from the board, Transocean crew questioned whether the leak was resolved and a similar leak was still present during post-explosion ROV intervention.[151] According to witness testimony, the rig crew never determined the source of a leak on the lower annular (leak 5 in the table).[152]

BOP Recertification

Recertification of a blowout preventer involves complete disassembly and inspection of the equipment.[153] This process is important because it allows individual components to be examined for wear and corrosion. Any wear or corrosion identified can then be checked against the manufacturer's wear limits.[154] Because this process requires complete disassembly of the BOP at the surface, it can take 90 days or longer[155] and generally requires time in dry dock.[156] Industry papers suggest that "the best time to perform major maintenance on a complicated BOP control system [is] during a shipyard time of a mobile offshore drilling unit (MODU) during its five-year interval inspection period."[157] The *Deepwater Horizon* had not undergone shipyard time since its commission.[158]

MMS regulations require that BOPs be inspected in accordance with American Petroleum Institute (API) Recommended Practice 53 Section 18.10.[159] This practice requires disassembly and inspection of the BOP stack, choke manifold, and diverter components every three to five years.[160] This periodic inspection is in accord with Cameron's manufacturer guidelines, and Cameron would have certified inspections upon completion.[161]

The *Deepwater Horizon* Blowout Preventer Was Not Recertified

It was well known by the rig crew and BP shore-based leadership that the *Deepwater Horizon* blowout preventer was not in compliance with certification requirements.[162] BP's September 2009 audit of the rig found that the test ram, upper pipe ram, and middle pipe ram bonnets were original and had not been recertified within the past five years.[163] According to an April 2010 assessment, BOP bodies and bonnets were last certified December 13, 2000, almost 10 years earlier.[164]

Although the September 2009 audit recommended expediting the overhaul of the bonnets by the end of 2009 and emails between BP leadership discussed the issue,[165] the rams had not been recertified as of April 2010.[166] A Transocean rig condition assessment also found the BOP's diverter assembly had not been certified since July 5, 2000.[167] Failure to recertify the BOP stack and diverter components within three to five years may have violated the MMS inspection requirements.[168] An April 1, 2010 MMS inspection of the rig found no incidents of noncompliance and did not identify any problems justifying stopping work.[169] The inspection did not identify the fact that the *Deepwater Horizon*'s BOP had not been certified in accordance with MMS regulations.[170]

"Condition-Based Maintenance"

Transocean did not recertify the BOP because it instead applied "condition-based maintenance."[171] According to Transocean's Subsea Maintenance Philosophy, "[t]he condition of the equipment shall define the necessary repair work, if any."[172] Condition-based maintenance does not include disassembling and inspecting the BOP on three- to five-year intervals,[173] a process Transocean subsea superintendent William Stringfellow described as unnecessary.[174] According to Stringfellow, the rig crew instead tracks the condition of the BOP in the Rig Management System and "if we *feel* that the equipment is—is beginning to wear, then we make...the changes that are needed."[175] Transocean uses condition-based monitoring to inspect all of its BOP stacks in the Gulf of Mexico.[176] According to Transocean witnesses, its system of condition-based monitoring is superior to the manufacturer's recommended procedures and can result in identifying problems earlier than would occur under time-based intervals.[177]

The Chief Counsel's team disagrees. Condition-based maintenance was misguided insofar as it second-guessed manufacturer recommendations, API recommendations, and MMS regulations.

Moreover, the decision to forego regular disassembly and inspection may have resulted in necessary maintenance not being performed on critically important equipment. As discussed in Chapter 4.10, the Rig Management System used to monitor the BOP was problematic and may have resulted in the rig crew not being fully aware of the equipment's condition. Given the critical importance of the blowout preventer in maintaining well control, the Chief Counsel's team questions any maintenance regime that could undermine the mechanical integrity of the BOP.

Technical Findings

As discussed above, this report does not make any conclusive findings regarding whether and to what extent the *Deepwater Horizon*'s BOP may have failed to operate properly because forensic testing is still ongoing. At this point, the Chief Counsel's team can only identify possible reasons why the BOP's emergency systems failed to activate.

The possibilities include:

- explosions on the rig may have damaged connections to the BOP and thereby prevented the rig crew from using the emergency disconnect system to successfully activate the blind shear ram;

- ROV hot stab activation may have been ineffective because ROVs could not pump at a fast enough rate to generate the pressure needed to activate the relevant rams; and

- BOP control pods may have been unable to activate the blind shear ram after power, communication, and hydraulic lines were severed; low battery levels in the blue control pod and solenoid faults in the yellow control pod may have prevented pod function.

Even if activated, the blind shear ram did not seal in the well on April 20 or in subsequent response efforts. Possible reasons for failing to seal include:

- the high flow rate of hydrocarbons may have eroded the BOP and created a flow path around the ram;

- the BOP's blind shear ram may have been mechanically unable to shear drill pipe and shut in the well because it was not designed to operate under conditions that existed at the time. For instance, the ram may have been blocked by tool joints or other material that it was not designed to cut;

- subsea accumulators may have had insufficient hydraulic power; and

- leaks in BOP control systems may have delayed closing the BOP, though it is unlikely that they prevented the BOP from sealing. Leaks may have existed on the BOP control system but not been identified. Identified leaks were not reported to MMS and may have been inconsistently monitored.

Management Findings

Whether or not BOP failures contributed to or prolonged the blowout, the Chief Counsel's team has identified several major shortcomings in the overall program for managing proper functioning of the BOP stack.

- MMS regulations require only one blind shear ram on a BOP stack. But blind shear rams cannot cut the joints that connect pieces of drill pipe, which comprise a significant amount of pipe in a well. The Chief Counsel's team agrees with a 2001 MMS study that two blind shear rams would mitigate this risk.

- MMS approved the testing of the *Deepwater Horizon* blowout preventer at lower pressures than required by regulation. Though testing at lower pressures is in accord with industry practice, most tests of the blind shear ram did not establish the ability of the equipment to perform during blowout conditions with large volumes of gas moving at high speed through the BOP into the riser.

- Transocean's practice of destroying test records at the end of each well creates unnecessary information gaps that may undermine BOP maintenance.

- Critical BOP equipment on the *Deepwater Horizon* may have been improperly maintained. The BOP ram bonnets, bodies, and diverter assembly had not been certified since 2000, despite MMS regulations, API recommendations, and manufacturer recommendations requiring comprehensive inspection every three to five years. Transocean and BP's willingness to disregard regulatory obligations on a vital piece of rig machinery is deeply troubling.

Table 4.9.3. Modifications to the *Deepwater Horizon* blowout preventer.

Date	Modification
November 2001	Control pod subsea plate mounted valves changed from 1-inch to 0.75-inch valves.[178]
October 2002	Increased power supply to control pod subsea electronic modules (SEMs) to higher amp. rating.[179]
December 2002	ST locks modified.[180]
January 2003	Three high-shock flow meters were installed in BOP control pods, replacing ultrasonic flow meters.[181]
January 1, 2003	Changed retrievable control pods to nonretrievable control pods.[182] *This required the LMRP to be retrieved to surface in order to perform maintenance on control pods.*[183]
November 2003	New high-interflow shuttle valve replaced on LMRP and BOP stack.[184]
May 2004	Control pod regulators modified.[185]
June 2004	Control pod subsea electronic modules (SEMs) software upgraded by Cameron.[186]
July/August 2004	New rigid conduit manifold installed and riser-mounted junction boxes removed.[187]
August 2004	Cameron conduit valve package replaced with ATAG conduit valve package.[188] *This isolates LMRP accumulators if pod hydraulic power is lost.*[189]
August 2004	Fail-safe panels on choke and kill valves removed from LMRP and BOP stack.[190] *Valves will close only by spring force.*[191]
November 2004	"Add a second pod select solenoid functioned by an existing pod select switch—to add double redundancy to each control pod."[192]
December 2004	AMF/deadman accumulators: "[T]he pre-charge required on the subsea accumulators is 6800 psi while the maximum working gas pressure for subsea bottles is 6000 psi. This will mean different fluid volumes than are normal on the BOP control system."[193] The deadman accumulators "have now become part of the subsea accumulators since the deadman system has been modified.... There will be little appreciable differences in the system operability but it is important to know how the reduced pre-charge and extra accumulators work on the system."[194]
December 2004	Lower variable bore ram converted to test ram.[195] *A test ram holds pressure from above, instead of below.*[196] *Possibly overlooked relabeling ROV hot stab connections, resulting in ROVs activating test ram during post-explosion efforts to close the BOP.*[197]
February 2005	Control pod modified: "[R]eplace all unused functions on pod with blind flanges. Possible failure points resulting in stack pull."[198]
September 2005	Control system pilot regulator: "[R]eplace pilot regulator with a better designed, more reliable regulator leaks. (Gilmore is a larger unit and will require a bracket to be fabricated for mounting.)"[199]
February 2006	Control panel: "Modification to Cameron control software to sound an alarm should be a button stay pushed for more than 15 [seconds]. If a button is stuck and not detected it will lock up panel."[200]

Table 4.9.3 (continued)

Date	Modification
June 26, 2006	Installed new repair kit in autoshear valve. New repair kit came with new rod and the rod was too long, had to use old rod.[201]
July 2006 (proposal for modification approved)	At BP's request, the lower annular preventer was changed to a stripping annular.[202]
January 2007	AMF/deadman—Cameron will remove the SEM from the MUX section to replace the pipe connectors (customer provided) and to install the AMF/deadman modification kit.[203]
September 2008	Riser flex joint replaced.[204]
June 10, 2009	Software changes made to allow all functions that were previously locked out from any of the BOP's control panels to become unlocked whenever the EDS command was issued from any control panel.[205]
August 3, 2009	Autoshear valve replaced with new Cameron autoshear valve.[206]
2010	Combined the following ROV hot stab functions:[207] blind shear ram close; ST lock close; and choke and kill fail-safe valves.

Chapter 4.10 | Maintenance

A deepwater drilling rig like the *Deepwater Horizon* has literally thousands of pieces of equipment that need routine monitoring and repair.[1] The *Deepwater Horizon*'s crew performed more than 550 preventative maintenance jobs each month on the *Deepwater Horizon* and had spent more than 30,000 work hours on maintenance in the 10 months prior to the explosion.[2]

In some respects the *Horizon* appeared to be operating quite well. The rig received several safety awards[3] and a place inside Transocean's "excellence box," which compares rigs based on safety performance and equipment downtime.[4] BP wells team leader John Guide described the rig as BP's most successful in terms of performance,[5] and one reason leaders from BP and Transocean were visiting the rig on the day of the blowout was to recognize the rig's high performance.[6]

It is nevertheless possible that poor maintenance contributed to technical failures. According to pre-explosion BP emails, the rig was "getting old and maintenance has not been good enough."[7] Most notably, Chapter 4.9 of this report explains that certification of blowout preventer (BOP) equipment was overdue and that if blowout preventer maintenance was inadequate, it could have affected the ability to shut in the well. Other issues may have affected maintenance but, based on available information, likely did not contribute to the blowout.

Transocean's Rig Management System

Transocean had in place comprehensive procedures and systems for scheduling, implementing, and monitoring maintenance.[8] Like all Transocean rigs, the *Deepwater Horizon* used the computerized "**Rig Management System II**" (RMS), which Transocean had implemented as a result of its merger with Global Santa Fe.[9] Transocean personnel used RMS to schedule maintenance work based on information including equipment data, maintenance records,[10] information on certification and surveys,[11] and risk assessments.[12] Based on these materials, the automated system generated preventative maintenance[13] items for the rig.[14] The rig crew would perform these tasks and then record their completion in the system.[15] Transocean's goal in using the system was to ensure consistency, consolidate information, and facilitate personnel movement from rig to rig.[16]

While the Chief Counsel's team interviewed *Deepwater Horizon* crew members who found the RMS useful (despite the fact that it "definitely had some bugs in it") and who used it daily,[17] the team also found evidence to suggest that the system had problems. Transocean installed the RMS on the *Horizon* in September 2009,[18] but according one witness it was "still a work in progress" at the time of the blowout.[19] For instance, while the system produced thousands of preventative maintenance orders for Transocean's fleet,[20] many orders were disorganized, erroneous, or irrelevant to individual rig crews. The *Deepwater Horizon*'s rig crew was forced to actively search the system for the *Deepwater Horizon*'s maintenance items and to continually submit requests to remove duplicate maintenance orders or orders meant for another rig.[21] The system also generated work orders for equipment that had already been repaired, leaving the rig crew to determine if work orders generated by the system actually needed to be performed.[22] According to chief engineer Stephen Bertone, the rig crew "went through them as much as [they] could just poking through the system, but...there were still issues with it."[23] According to assistant driller Allen Seraile, the system was chaos at one time.[24] Chief electronics technician Mike Williams described the system as "overwhelming."[25]

The crew expressed confusion regarding the new system and concerns about its implementation. In a March 2010 Lloyd's Register survey, crew members stated that system changes to the RMS and other rig systems were ineffectively implemented.[26] They thought that new systems were introduced too frequently and before the previous system was understood.[27] The rig crew also thought there was insufficient support to implement changes and that system changes required a level of technical capability not typically available throughout the rig.[28] An April 2010 Transocean assessment also found that the maintenance system was not understood by the crew.[29]

Competing Interests Between Drilling and Maintenance

The rig services contract between BP and Transocean specifies that shutting down the rig to perform certain types of maintenance will trigger financial consequences. BP paid Transocean a daily operating rate of $533,495 for the *Deepwater Horizon*, but under the contract BP was not obligated to pay for time in excess of 24 hours each month spent on certain equipment repairs.[30]

The Chief Counsel's team cannot be certain whether these provisions or other financial pressures influenced maintenance decisions. However, some of the rig crew raised concerns that drilling priorities took precedence over planned maintenance.[31] The *Deepwater Horizon* had never been to dry dock for shore-based repairs in the nine years since it had been built.[32] BP and Transocean appear to disagree as to whether financial considerations influenced this decision. While Guide suggested the *Horizon* did not go to dry dock because Transocean insisted on being paid its daily rate during repairs,[33] Transocean operations manager Daun Winslow testified that any necessary repairs would have been made regardless of financial constraints.[34]

Lack of Onshore Maintenance

Some maintenance can only be performed when a rig is moving between well sites or when the rig is brought into shore.[35] But the *Horizon* had never been to dry dock since it was built in 2001. Transocean instead conducted "Underwater Inspection in Lieu of Dry-docking" (UWILD) and other at-sea inspections.[36] In the March 2010 Lloyd's Register survey some of the rig crew expressed concern that the lack of dry dock time could generally undermine equipment reliability.[37] According to the survey, the maintenance department was looking forward to a scheduled dry dock visit in 2011 "to carry out evasive [preventative maintenance] routines that they normally could not do."[38] Lack of time in dry dock may have resulted in a lapse in BOP certification.[39]

Following company policy,[40] Transocean commissioned an inspection in April 2010 to assess equipment and prepare for the rig's scheduled 2011 shipyard maintenance.[41] The inspection found that some problems identified in September 2009 remained unaddressed and identified several new maintenance issues.[42] As of April 2010, Transocean documents listed 35 critical items of equipment that either were in bad condition, had shown excessive downtime, had passed manufacturer wear limits, or that the manufacturer no longer supported.[43] As discussed in Chapter 4.9, the list included BOP elements that had passed their certification date.[44] According to witness testimony, Transocean had decided to extend the *Horizon*'s anticipated time in dry dock because of the number of repairs necessary.[45] The Chief Counsel's team requested but was not able to obtain a list of repairs scheduled for the *Horizon*'s 2011 dry dock visit.

Maintenance Audits and Inspections

The *Horizon* was subject to audits and inspections by various government and private entities, including BP,[46] Transocean,[47] MMS,[48] the Coast Guard,[49] the American Bureau of Shipping,[50] and the Marshall Islands (the ship's flag state in 2010).[51] These audits varied in scope and duration. Both BP and Transocean had a vested interest in keeping the *Horizon* in working order. Witness testimony describing the response to a fall 2009 audit indicates collaboration by both companies to ensure necessary repairs were made.

Transocean Resolved Many Maintenance Issues Identified in the September 2009 BP Audit

In September 2009 BP audited the *Deepwater Horizon*'s drilling equipment and the vessel itself.[52] The audit found 390 maintenance jobs overdue and identified some of those as high-priority items.[53] BP estimated that the work would require 3,545 man-hours of labor.[54] The audit may have overestimated the sheer number of jobs that were overdue because of errors and duplicates in the RMS system, which Transocean had recently installed.[55] BP asked Transocean to undertake certain repairs before allowing the *Horizon* to resume operations.[56] A few days later, BP determined that the rig was operational,[57] and the rig resumed operations on September 22, 2009, five days after the audit ended.[58]

BP and Transocean increased communication and coordination to monitor implementation of outstanding audit recommendations.[59] For example, auditors communicated conditions to the rig crew during the audit itself in order to ensure that certain repairs were made promptly.[60] BP and Transocean held weekly meetings to track progress,[61] and Guide or well site leaders signed off on corrective actions taken in response to the audit.[62] By March 30, 2010, 63 of 70 had been completed, progress BP described as "commendable."[63] Twenty-six other outstanding items were in progress and deemed not safety-critical.[64]

BP and Transocean Believed the Rig Was in Safe Working Order

At the time of the blowout, both BP and Transocean believed the *Deepwater Horizon* was in safe operating condition.[65] Well site leader Ronnie Sepulvado did not believe there were serious outstanding safety issues,[66] and neither he nor the other well site leaders indicated that the vessel was unsafe to operate.[67] Guide recognized that the rig was operating safely and making very good progress on addressing audit items.[68]

An April 1, 2010 MMS inspection of the rig found no incidents of noncompliance and did not identify any problems justifying stopping work.[69] But, as discussed in Chapter 6, the inspection did not identify that the *Deepwater Horizon*'s BOP had not been certified.[70]

Maintenance Findings

Inspections, audit programs, and statements by rig- and shore-based leadership indicate that BP, Transocean, and government regulators believed the *Deepwater Horizon* was in safe operating order at the time of the blowout. With the exception of potential BOP maintenance issues, the Chief Counsel's team found no reason to believe that maintenance problems may have contributed to the blowout. However, the Chief Counsel's team believes the following issues may have compromised the rig's maintenance regime:

- Transocean's RMS system may have complicated routine maintenance and monitoring. The rig crew appears to have been confused about the system, and the system issued duplicate and erroneous maintenance instructions; and

- the fact that the *Deepwater Horizon* had never been in dry dock may have delayed or prevented certain repairs that could only have been done onshore. ◆

Chapter 5 | Overarching Failures of Management

The Macondo disaster was not, as some have suggested, the result of a coincidental alignment of disparate technical failures.[1] While many technical failures contributed to the blowout, the Chief Counsel's team traces each of them back to an overarching failure of management.

Better management would have identified the risks at Macondo and prevented the technical failures that led to the blowout. In Chapter 4, the Chief Counsel's team identified particular management failures associated with each technical failure. This chapter synthesizes those findings into higher-level observations about the management system in place at Macondo.

The management breakdown at Macondo affected many of the operational aspects of designing and drilling the well. The Chief Counsel's team observed at least the following management failures: (1) ineffective leadership at critical times; (2) ineffective communication and siloing of information; (3) failure to provide timely procedures; (4) poor training and supervision of employees; (5) ineffective management and oversight of contractors; (6) inadequate use of technology; and (7) failure to appropriately analyze and appreciate risk. Ultimately, the companies placed undue reliance on timely intervention and human judgment in light of their failure to provide individuals with the information, tools, and training necessary to be effective.

BP's and Transocean's corporate guidance documents, in place before the blowout, show that they recognize how important each of these management areas is to safe and effective oil and gas exploration.[2] (Halliburton declined to provide management documents to the team.) The fact that failures in these areas led to the Macondo blowout reinforces the companies' conclusions about their importance. It also underscores the importance of management follow-through to ensure that policies affect cultures and day-to-day routines.

This chapter discusses each of these various failures in turn. The management observations in this chapter are limited to the Macondo well, which has been the focus of the Chief Counsel's investigation. The failures at Macondo were not inevitable, and the Chief Counsel's team sets them out here in the hope that they will not be repeated.

Leadership

The first principle of BP's operating management system (OMS) is leadership. OMS calls for "operating leaders [who] are competent, exhibit visible, purposeful and systematic leadership and are respected by the organizations they lead."[3] BP further expects that "operating leaders create and support clear delegation and accountability."[4] Often this did not happen at Macondo. The Chief Counsel's team observed conflict between managers and confusion about who was accountable for critical decisions.[5] The team responsible for key decisions at Macondo did not always appear to be acting with a consistent and shared purpose.

In March, for example, operations to control the well after a kick led to disagreements between BP's managers on the Macondo team. BP engineering team leader David Sims wrote BP wells team leader John Guide: "We cannot fight about every decision.... I will hand this well over to you in the morning and then you will be able to do whatever you want."[6] Sims later explained this and other comments as "coaching" and stated that Guide's performance was atypical during this time period.[7] Guide himself appears to have acknowledged the concern and responded that he would "consult the team and make well thought out decisions."[8] Nonetheless, the comments suggest management friction during a critical operation, and leadership problems on the Macondo team did not end in March.

At the beginning of April, BP conducted a major reorganization of its exploration business unit, including the BP Macondo team, creating separate reporting structures for engineering and operations. Prior to the reorganization, the unit had been organized by project—all of the engineers and operations personnel for a given well reported to the same manager. Thus, Guide (representing operations) and Sims (representing engineering) both reported to the same person, BP wells manager Ian Little. BP senior drilling engineer Mark Hafle and drilling engineer Brian Morel reported to Sims; the well site leaders reported to Guide.[9]

The reorganization separated engineering and operations into distinct functional groups within the business unit. As of April, the wells team leader reported to a wells operation manager, and the engineering team leader reported to a separate engineering manager. BP also moved key personnel. BP promoted Sims from engineering team leader to wells operation manager. Instead of being Guide's peer, he was now Guide's supervisor. Gregg Walz, who had no prior experience with the Macondo well before March 2010, took over for Sims as engineering team leader. Walz now reported to new engineering manager John Sprague.[10]

The reorganization caused delays and distractions. Shortly before the reorganization, BP vice president of drilling and completions Pat O'Bryan questioned Gulf of Mexico managers about recent subpar performance, asking, "What's getting in the way...reorg uncertainty?"[11] Sims later shared that there were "challenges" associated with the reorganization and that "it may have taken a little more time to ensure that there was alignment between Ops and Engineering teams."[12] In an interview with Commission staff, Sims acknowledged that Walz may have been taking longer than usual to make engineering decisions as he came up to speed in his new role.[13]

Figure 5.1. BP internal presentation slide.

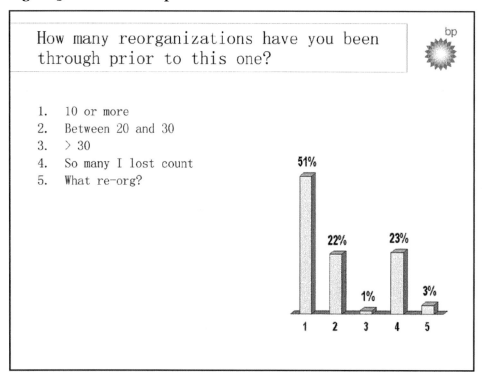

BP

Guide agreed. He told BP investigators that it was "easier" and "faster" to make decisions under the old structure.[14]

The reorganization also led to questions about authority and accountability, and apparent friction between team leaders Guide and Walz. Hafle noted that "no one argues with John Guide," but after the reorganization, Guide expressed confusion about his own authority to Sims and to Sprague.[15] In an April 17 email to Sims, Guide asked, "Everybody wants to do the right thing, but, this huge level of paranoia from engineering leadership [i.e. Walz] is driving chaos.... What is my authority? With the separation of engineering and operations I do not know what I can and can't do."[16]

Sims responded, "I don't think anything has changed with respect to engineering and operations," but went on to note, "If you don't agree with something engineering related, and you and [Walz] can't come to any agreement, [Sprague] or me gets involved." Guide later observed that the resolution of an issue by Sprague *or* Sims was precisely his concern.[17] While Little had previously been responsible for engineering and operations on his own, now there were two separate leaders for each team, each of whom had a different supervisor of their own. To find an individual who had responsibility for both engineering and operations, the Macondo team had to go all the way up to O'Bryan, the head of drilling and completions for the Gulf of Mexico.

The Chief Counsel's team does not presume to know whether the reorganization improved BP's previous management structure, but it is clear that the way BP handled authority and accountability created confusion during the Macondo project. For example, the BP team did not know who was accountable for important practices associated with safety. After the blowout, Hafle told BP investigators that he had no idea who was accountable for ensuring compliance with BP's standards on drilling safety.[18] Sims told BP investigators, "this accountability is not well documented" and "it is more like 'we are all accountable.'"[19]

Saying that everyone is accountable can be beneficial in certain instances, such as with respect to personal safety and "stop-job" authority, but can lead to a diffusion of personal responsibility for process safety. For example, BP has admitted that its internal engineering standards required the Macondo team to conduct a formal risk assessment of the annulus cement barriers in the well, and that such an assessment might have led the team to run a cement evaluation log.[20] Yet nobody on the team appears to have brought up the relevant Engineering Technical Practice (ETP) on zonal isolation.[21] There also appears to have been confusion about who was accountable for ensuring the adequacy of the cement slurry design, determining the risks attendant to changes in operations, and assessing the competence of personnel assigned to perform the negative pressure test.[22]

Though it is understandable that no one would wish to take ownership of the well after the blowout, the Chief Counsel's team found many instances in which nobody was taking ownership before the blowout.

Communication

Inadequate communication and excessive compartmentalization of information contributed to the Macondo blowout. Individuals making decisions regarding one aspect of the well, such as onshore engineers, did not always communicate critical information to others, such as the well site leaders, who were making related decisions on other aspects of the well. When faced with anomalous data, decision makers often failed to seek counsel from others with expertise and instead made decisions based on incomplete information. BP and Transocean also failed to communicate lessons learned from other wells that could have assisted the decision makers at Macondo. Below are a few examples.

Information Compartmentalization

Information about drilling at Macondo was compartmentalized both within and between companies. In several instances, the BP onshore engineering team was aware of risks with the Macondo job but failed to communicate those risks to its own employees on the rig or the contractor personnel who might have helped mitigate those risks. The cementing and temporary abandonment processes provide key examples.

Cement jobs inevitably involve some uncertainty, but this job was particularly tricky. Due to equivalent circulating density (ECD) concerns, BP did not perform a full bottoms up prior to the cement job, it used foam cement, it pumped a smaller volume of cement than normal, it circulated the cement at lower flow rates than normal, and it used an overall slurry having a density approaching the density of the drilling mud in the annulus. BP pumped the job knowing that it had had difficulty converting the float equipment and that post-conversion circulating pressures had been unexpectedly low. And BP used fewer centralizers than called for by Halliburton's model. BP also decided to rely heavily on its difficult cement job soon after pumping it, by using temporary abandonment procedures that forced rig personnel to rely on the cement as the only constant barrier during riser displacement.

Despite knowing all of these cementing-related risks, BP's onshore team did not emphasize them to the individuals conducting the negative pressure test (including its own well site leaders). It also did not emphasize these risks to the individuals who were monitoring the well for kicks during riser displacement (Transocean and Sperry Drilling personnel), much less involve those individuals in discussions about how to mitigate the risks of cement failure.[23]

While rig personnel should always assume for well monitoring purposes that the bottomhole cement (or any other barrier) might fail, BP's onshore team should have, and easily could have, alerted the well site leaders and rig crew that cement failure at Macondo might be more likely than normal and instructed them to be extra vigilant regarding any odd pressure readings.

Chapter 4 is replete with similar examples.

Experts

BP did not always use its internal technical experts effectively.

For example, BP asked an in-house cement expert to help redesign the cement job to address ECD worries and thereby allow BP to use the long string production casing rather than a liner. During that process, the Macondo team asked the expert only for his general opinions about the suitability of foamed cement. Though Guide believed that the expert had "vetted" the cement program,[24] nobody on the Macondo team consulted the expert after April 14, and he never saw any laboratory testing data for the cement until after the blowout.

The Macondo team similarly did not consult completion engineers before reaching a decision on whether to run a long string or a liner. On April 15, one of the completion engineers wrote Morel:

"Yeah, well no one told us what the actual decision was, so we thought y'all were going with the liner...."[25] BP is now developing standards on how to consult internal experts and hiring more cementing experts.

Calling Shore

BP did not provide adequate guidance on when staff on the rig should consult onshore personnel. BP had a communications plan that described instances when rig personnel should call shore.[26] The plan did not address the negative pressure test specifically, and its general criteria for calls to shore did not apply clearly (if at all) to the negative pressure test on April 20.[27] After the incident, Hafle said that the communications plan "was not well written."[28] Another BP employee poked fun at its "weird drawings with boxes & arrows."[29]

It does not appear that the well site leaders ever contacted BP onshore personnel to discuss their inability to bleed off drill pipe pressure during the negative pressure test. They did not seek a second opinion from Sims or O'Bryan, both of whom are engineers and *were on the rig* during the negative pressure test as part of the VIP visit. Instead, according to their own accounts, the well site leaders accepted an explanation from a Transocean toolpusher who had no more training on test procedures than they had.

Less than one week after the blowout, BP well site leader Bob Kaluza wrote the following email to Guide and a colleague explaining how the "bladder effect" could account for the 1,400 pounds per square inch (psi) on the drill pipe:

> I believe there is a bladder effect on the mud below an annular preventer as we discussed. As we know the pressure differential was approximately 1400-1500 psi across an 18 ¾" rubber annular preventer, 14.0 SOBM plus 16.0 ppg [pounds per gallon] Spacer in the riser, seawater and SOBM below the annular bladder. Due to a bladder effect, pressure can and will build below the annular bladder due to the differential pressure but can not flow – the bladder prevents flow, but we see differential pressure on the other side of the bladder.
>
> Now consider this. The bladder effect is pushing 1400-1500 psi against all of the mud below, we have displaced to seawater from 8,367' to just below the annular bladder where we expect to have a 2,350 psi negative pressure differential pressure due to a bladder effect we may only have a 850-950 psi negative pressure until we lighten the load in the riser.
>
> When we displaced the riser to seawater, then we truly had a 2,350 psi differential and negative pressure.[30]

O'Bryan responded to the forwarded email as follows:

> Mike,
>
> ??
> ??
> ??
> ??
> ???

Regards,

Pat

It thus appears that, had Kaluza brought the "bladder effect" explanation to O'Bryan's attention on April 20, events likely would have turned out differently. Guide, Walz, Sims, and BP operations engineer Brett Cocales have each told the Chief Counsel's team that the "bladder effect" does not exist, that it would not account for the pressure readings seen that night in any event, and that they would have insisted on further testing before declaring the test a success had the well site leaders called to shore.[31] While these statements are self-serving, they are believable in this instance—everyone who has testified before the Joint Investigation Panel or spoken with the Chief Counsel's team has agreed the "bladder effect" does not exist and would not explain the pressure readings observed that night.

While O'Bryan appears to have been incredulous at Kaluza's explanation of the "bladder effect," BP management itself is to blame for failing to make clear to its well site leaders that they must call back to shore when confronted with unexpected results on a critical test.[32] After the fact, BP and Macondo team members have said the well site leaders on the *Deepwater Horizon* should have called back to shore on April 20. But they have been unable to point to any specific company policy, written or otherwise, that would have required the well site leaders to seek that second opinion.[33] When asked whether BP had any relevant policy at the Commission's November 8, 2010 hearing, BP's Mark Bly answered, "It's an expectation that if people feel they don't understand what is going on or they need help, that they will escalate and call back. So absolutely...I don't know if it's the policy. It's sort of the behavior that we expect from people."[34]

Given the importance of the negative pressure test, calls back to shore should be required as a matter of course regardless of the whether results appear anomalous. BP has apparently now instituted just such a policy.[35]

Sharing Lessons Learned

Transocean failed to communicate to BP and its rig crew lessons learned from a similar near miss on one of its rigs in the North Sea four months prior to the Macondo blowout. On December 23, 2009, gas entered the riser while the North Sea rig was displacing a well with seawater during a completion operation. As at Macondo, the crew had already run a negative pressure test on the lone static barrier between the pay zone and the rig and deemed it successful.[36] The tested barrier failed during displacement. Hydrocarbons flowed into the well, and mud spewed from the rig floor. Unlike at Macondo, the crew was able to shut in the well before a blowout occurred but not until nearly one metric ton of oil-based mud had spilled into the ocean.[37] The incident cost Transocean 11.2 days of additional work and more than 5 million British pounds.[38]

Transocean subsequently created an internal presentation for a March conference call reviewing the near miss. It warned that "[t]ested barriers can fail" and that "risk perception of barrier failure was blinkered by the positive inflow test [negative pressure test]."[39] It pointed out that "[f]luid displacements for inflow test [negative pressure test] and well clean up operations are not adequately covered in our well control manual or adequately cover displacements in under balanced operations."[40] The presentation concluded with a slide titled "Are we ready?" and "What if?" which contained the following bullet points: "[h]igh vigilance when reduced to one barrier underbalanced," "[r]ecogni[z]e when going underbalanced—heightened vigilance," and "[h]ighlight what the kick indicators are when not drilling."[41] However, the call only involved toolpushers operating in the North Sea.

On April 5, 2010, Transocean issued an advisory setting forth anticipated amendments to its Well Control Handbook in light of the North Sea incident.[42] The advisory sought "to clarify the requirements for monitoring and maintaining at least two barriers when displacing to an underbalanced fluid during completion operations."[43] It noted that a Transocean rig recently experienced a well control event "due to a failure of a tested mechanical barrier."[44] To prevent a recurrence, the advisory required the drill crew to identify:

> (1) the volumes to be pumped, (2) the planned displacement rate(s), (3) the position of the fluid interface(s) at all times, (4) the resultant U-tube pressures in the well at all times and, (5) most importantly the point at which the completion fluid will become under-balanced with respect to formation pressure.[45]

The advisory ended with an apt warning: *"Do not be complacent because the reservoir has been isolated and inflow tested. Remain focused on well control and maintain good well control procedures."*[46]

There are two problems with the advisory. First, it unduly limits the amendment to the "Completions" section of the handbook despite the fact that it should apply equally to temporary abandonment procedures such as those at Macondo. Second, it does not appear that anyone associated with the *Deepwater Horizon* ever received the advisory prior to the blowout.[47]

Transocean points out that it posted the advisory to an online, e-document platform accessible to the *Deepwater Horizon* crew.[48] But Transocean never alerted Macondo personnel to the posting, and there is no indication anyone actually saw it.

Transocean issued a more extensive advisory on April 14, less than one week before the Macondo blowout.[49] The new advisory described the North Sea incident and listed error-inducing conditions, missed opportunities, root causes, and contributing factors. Among the error-inducing conditions, it noted that the "drill crew did not consider well control as a realistic event during the...displacement operation as the [downhole barrier] had been successfully [negative pressure] tested," and the displacement was set up as "an open circulating system" nullifying pit monitoring.[50] The advisory admonished rig management that "[t]ested barriers can fail and risk awareness and control measures need to be implemented," "[s]tandard well control practices must be maintained through the life span of the well," and that well programs must "specify operations that induce underbalance conditions in the well bore."[51] As one Transocean executive noted after the incident, reading the advisory would "increase the awareness of anybody in the drilling industry."[52]

But Transocean circulated its April 14 advisory only to North Sea personnel, even though the lessons applied globally.[53] The company labeled the advisory in a narrow way, describing the North Sea event as a "Loss of Well Control During Upper Completion."[54] Transocean's operations manager for the Gulf of Mexico admitted that personnel involved in drilling operations might not read an advisory labeled this way.[55] Again, there is no evidence that anyone involved with Macondo or the *Deepwater Horizon* ever saw the April 14 advisory.

Transocean argues that alerting the crew to the advisory was unnecessary because the advisory simply restates good well control practice already known to the crew.[56] The Chief Counsel's team does not agree. There is no evidence the rig crew on the night of April 20 followed any of the five steps mandated by the advisory. Asked whether he knew on April 20 that "monitoring the displaced volume alone is inadequate and does not satisfy the requirement for a known monitored column of fluid," Transocean's rig manager for the *Deepwater* Horizon, Paul Johnson, answered, "No. I'm thinking hard and clear about this, no."[57]

Transocean has stated that the North Sea incident and advisory were irrelevant to what happened in the Gulf of Mexico. The December incident occurred during the completion phase, in the North Sea, and involved the failure of a different tested barrier.[58]

Transocean's post-blowout reliance on these cosmetic differences is not an answer; to the contrary, these arguments only further reinforce the Chief Counsel's team's conclusions about the compartmentalization of information. The relevant facts of the Macondo and North Sea incidents are the same. Indeed, the North Sea incident may have had greater implications and relevance in deepwater. There is no reason why the lessons learned in the North Sea would not apply to the Gulf of Mexico or non-completion operations. Had Transocean adequately communicated the lessons from the North Sea to the crew of the *Deepwater Horizon* prior to April 20, events at Macondo may have unfolded differently.

Procedures

BP failed to provide its well site leaders and the rig crew with clear, detailed, and timely procedures. Instead, the evidence shows that BP's onshore Macondo team was rushing to design and provide procedures in order to keep up with operations on the rig. As a consequence, BP employees on the rig were not always sufficiently informed about upcoming operations.

The most obvious example is BP's temporary abandonment procedure. On April 12, for example, BP well site leader Murry Sepulvado wrote Morel: "Brian we need procedures for running casing, cementing and T&A work, we are in the dark and nearing the end of logging operations."[59] As set forth in detail in Chapter 4.5, the procedure changed repeatedly in the eight days between that email and the day of the blowout. It is not clear to the Chief Counsel's team why BP had not finalized and vetted the procedure much earlier in the process. The BP Macondo team instead waited until the last minute.[60]

BP ultimately did not send out the final "Ops Note" to the rig crew until the morning of April 20, meaning that once the well site leaders and rig crew did receive the temporary abandonment procedures, they had precious little time to digest and understand them (see Table 5.1 for breakdown of changes to temporary abandonment procedure).[61] BP could have at least ameliorated that problem by providing detailed guidance in the Ops Note to its well site leaders explaining how to, among other things, conduct the negative pressure test.

Contrary to the apparent views of BP's shore-based team, negative pressure test procedures are not self-evident to rig personnel, particularly in a case like Macondo in which the crew would have to displace and monitor a variety of different types of fluids. Sprague testified that, in order to interpret a negative pressure test, a well site leader would need to know the following: the hydrostatic pressure of fluids in the drill pipe, choke, and kill lines; bottomhole pressure; volumes and densities of fluids in the well, drill pipe, choke, and kill lines; and wellbore and drill string geometry.[62] Sprague acknowledged that "If you have more time to write detailed procedures, there is a greater chance that the result...might be more successful."[63]

Table 5.1. Timeline of changes to the temporary abandonment procedure.

Date	Event
April 12	BP well site leader Murry Sepulvado emails BP drilling engineer Brian Morel (copying BP wells team leader John Guide) stating, "Brian we need procedures for running casing, cementing and T&A work, we are in the dark and nearing the end of logging operations."[64]
April 12	Morel emails BP's subsea well engineers and asks for details on setting the lockdown sleeve: "I need a procedure this morning, do you have one available?"[65] He also emails the subsea wells team leader separately and explores the possibility of not setting the lockdown sleeve at all during temporary abandonment.[66]
April 12	Morel sends Murry Sepulvado and BP well site leader Ronnie Sepulvado (copying Guide) a first draft of the well plan for the final casing string, cement job, and temporary abandonment procedure.[67] The plan does not include a negative pressure test.[68] It calls for setting the lockdown sleeve in mud before setting the surface cement plug and setting the surface cement plug at ~ 6,000 feet below sea level rather than the eventual 8,367 feet. Morel says in his email, "This isn't perfect yet, but I wanted to get everyone a copy so you can ensure all the equipment required for our upcoming operations is offshore in time. Please let me know if you have any questions or suggestions how to improve the procedure."[69]
April 13	Ronnie Sepulvado emails Morel (copying Guide) saying "We need to do a negative test before displacing 14# mud to seawater."[70]
April 13	Morel emails back Murry Sepulvado and Ronnie Sepulvado (copying Guide) saying, "I will add details to the program. Currently my thoughts are negative testing with base oil to the mud line, you both ok with that?" Murry Sepulvado replies, "Base oil sounds good to me."[71]
April 14	Morel emails Ronnie Sepulvado and the rig clerk a different procedure.[72] This procedure sets the cement plug before the lockdown sleeve and in mud instead of seawater. The cement plug is moved from ~ 6,000 feet to 8,367 feet below sea level. And Morel adds a negative pressure test.
April 15	Morel emails the onshore team from the rig, saying that "Recommendation out here is to displace to seawater at 8300' then set the cement plug. Does anyone have issues with this? If we do a negative test prior to this with base oil to the wellhead the shoe will see about 360 psi less after the hole is displaced. Thoughts?"[73] Hafle later replies, "Seems ok to me. I really don't think [MMS} will approve deep surface plug. We'll see. Did permit look ok?"[74]
April 15	Morel changes the temporary abandonment procedures in the well plan.[75] It now calls for running a negative pressure test with base oil to the wellhead after the cement job, then running the drill pipe to 8,367 feet and displacing with seawater, then setting the cement plug, and then finally setting the lockdown sleeve.[76] It contains a contingency, however, in case the MMS does not approve the deeper cement plug, calling for setting the lockdown sleeve first before setting the cement plug at 5,800 feet below sea level.[77]
April 16	Hafle emails the temporary abandonment procedure permit request to Heather Powell of regulatory affairs, asking her to submit it to the MMS. The submission includes BP's request to set the surface cement plug 3,000 feet below the mudline, which is 2,000 feet lower than otherwise allowed by MMS regulations.[78] At 10:54 a.m., Powell sends back the approved permit, meaning that MMS approved the request in less than 80 minutes.[79]
April 17	Morel emails the onshore Macondo team asking, "Anyone know if there are any requirements in the MMS regs for a negative test, can't find any specifics?"[80]
April 20	Morel sends an email titled "Ops Note" to the well site leaders and onshore team. Unlike the earlier application submitted to MMS, the Ops Note calls for first running the drill pipe to 8,367 feet and displacing with seawater to above the blowout preventer (BOP) before running the negative pressure test "with seawater in the kill...[at] ~2350 psi differential."[81]

Planning problems extended beyond the temporary abandonment procedures. In April, Walz emailed Guide: "I know the planning has been lagging behind the operations and I have to turn that around."[82] Weeks earlier, on March 2, Cocales reassured a well site leader after the rig crew had problems interpreting procedures sent by Morel: "We will work on getting you guys any changes in the future sooner so you will have time to review."[83] And the difficulty appears to have extended beyond Macondo. In a meeting of the leadership team for drilling in the Gulf of Mexico, O'Bryan worried that "just in time delivery of well plans" had contributed to problems on other rigs.[84]

As detailed in Chapter 4, the pace and number of last-minute changes at Macondo apparently prompted Guide to write the following email to Sims on the morning of April 17, just three days before the blowout:

> David, over the past four days there has been so many last minute changes to the operation that the WSL's have finally come to their wits end. The quote is "flying by the seat of our pants." Moreover, we have made a special boat or helicopter run every day. Everybody wants to do the right thing, but, this huge level of paranoia from engineering leadership is driving chaos. This operation is not Thunderhorse. Brian has called me numerous times to make sense of all the insanity. Last night's emergency evolved around 30 bbls [barrels] of cement spacer behind the top plug and how it would affect any bond logging (I do not agree with putting the spacer above the plug to begin with). This morning Brian called me and asked my advice about exploring other opportunities both inside and outside of the company.
>
> What is my authority? With the separation of engineering and operations I do not know what I can and can't do. The operation is not going to succeed if we continue in this manner.[85]

Rather than react with alarm or stop work on the rig, Sims wrote back:

> John, I've got to go to dance practice in a few minutes. Let's talk this afternoon.
>
> For now, and until this well is over, we have to try to remain positive and remember what you said below – everybody wants to do the right thing. The WSLs will take their cue from you. If you tell them to hang in there and we appreciate them working through this with us (12 hours a day for 14 days) – they will. It should be obvious to all that we could not plan ahead for the well conditions we're seeing, so we have to accept some level of last minute changes.
>
> We've both [been] in Brian's position before. The same goes for him. We need to remind him that this is a great learning opportunity, it will be over soon, and that the same issues – or worse – exist anywhere else.
>
> I don't think anything has changed with respect to engineering and operations. Mark and Brian write the program based on discussion/direction from you and our best engineering practices. If we had more time to plan this casing job, I think all this would have been worked out before it got to the rig. If you don't agree with something engineering related, and you and Gregg can't come to an agreement, Jon or me gets involved. If it's purely operational, it's your call.

I'll be back soon and we can talk,

We're dancing to the Village People!

Sims has subsequently explained that he believed Guide was expressing temporary frustration and that he saw no cause for alarm. Emails from Guide later the same day support this view.[86] But once the well site leaders reported that last-minute changes were causing chaos and confusion on the rig, there was simply no reason why BP could not have stopped operations temporarily in order to allow planning to catch up.

Employees

Drilling is as much about people as it is about hydrocarbons and equipment. About 30 people designed the Macondo well. Roughly 130 others worked on the drilling rig at any given time. Success in oil and gas exploration depends on effective management of employees, yet the Chief Counsel's team observed poor management of staffing and inadequate training at Macondo.

People especially mattered at Macondo because BP, Transocean, and Halliburton placed heavy reliance on human judgment. For instance, during displacement of the riser with seawater, BP relied on the bottomhole cement as the only barrier in the wellbore. But awareness of whether that barrier was in place—because of the negative pressure test—depended on human judgment. Another barrier, the blowout preventer (BOP), also relied on human judgment because of the importance of kick detection and kick response. Yet, the companies failed to provide the rig crew and well site leaders exercising that judgment with adequate training, information, procedures, and support to do their jobs effectively.

Staffing

BP did a poor job of managing staffing and work assignments at Macondo. BP provided little support to a junior drilling engineer charged with critical design decisions and did not effectively seek input from technical experts. BP also sent a well site leader from another rig out to the *Deepwater Horizon* without properly determining if he was capable of substituting for one of the rig's veterans. BP did not supervise and support its employees as necessary to ensure safe operations.

Oversight

There were significant gaps in supervision and oversight at Macondo. In some cases, a single person made critical decisions and performed critical activities without checks—either by supervisors or other companies.

For example, BP relied very heavily on Morel to design not only the well itself, but also the cement program and temporary abandonment procedures at Macondo. Morel received his engineering degree in 2005, after which he started full time with BP. His first deepwater well was Mad Dog in 2007. BP assigned him to the exploration group in 2008, where he helped to plan two wells before being transferred to Macondo to work alongside Hafle—a much more senior drilling engineer who had been working on deepwater drilling since 1993.[87] The Chief Counsel's team found little evidence that Hafle closely reviewed Morel's work in the last few weeks before the blowout. Indeed, none of BP's shore-based engineers appear to have reviewed Morel's temporary abandonment procedures carefully.

While Morel appears to have been talented and capable, it is not apparent why the team would put so much on his plate without additional supervision and mentoring.[88]

Temporary Substitutions

BP mishandled the substitution of Kaluza for regular well site leader Ronnie Sepulvado. Sepulvado needed a temporary replacement in order to attend well control training school onshore (per MMS regulations and BP policy). BP could have sought dispensation to allow Sepulvado to remain on the rig throughout the critical temporary abandonment phase but did not. BP instead substituted Kaluza, who was serving as well site leader on the *Pride*, a moored rig in BP's Thunder Horse field.[89]

It does not appear that BP undertook any significant effort to assure that Kaluza was qualified for the tasks he would be overseeing at Macondo. Whenever there is transfer or loss of personnel with specific knowledge or experience from a project, BP's internal guidelines require management to submit the change through a management of change (MOC) process, which requires sign-offs from multiple managers.[90] BP did not do so for Kaluza,[91] even though he had not been a well site leader on the *Deepwater Horizon* previously, did not know the history of the Macondo well, and his relief (BP well site leader Don Vidrine) had himself only been on the *Deepwater Horizon* for a few months.[92]

Training

BP and Transocean inadequately trained their personnel. BP did not train its well site leaders how to properly conduct and interpret a negative pressure test. Transocean did not adequately train its rig personnel regarding kick monitoring during end-of-well, nondrilling activities, such as temporary abandonment. It also did not adequately train its crews how to respond to emergency situations such as those that occurred on the night of April 20. Inadequate training set employees up for failure in the face of events outside their expertise and experience.

Nondrilling Situations

BP and Transocean failed to provide its personnel any formal training in how to perform or interpret a negative pressure test. This failure is symptomatic of a broader inattention to end-of-well, nondrilling activities generally. For instance, Transocean's Well Control Manual does not contain a section on monitoring or controlling the well during temporary abandonment procedures, focusing instead on drilling activities (and to a lesser extent, completion operations).[93]

The phenomenon is not limited to Transocean or BP. Like Macondo, the Montara blowout off the northern coast of Australia occurred after the production casing cement job had been pumped.[94] The Montara blowout lasted 10 weeks beginning on August 21, 2009, and spewed between 400 and 1,500 barrels per day of oil and gas into the Timor Sea.[95]

At least one independent expert has testified that in his experience it is not unusual for crew members to let down their guard or lose focus during end-of-well activities.[96] BP subsea wells supervisor Ross Skidmore, who has more than 30 years' experience in the industry, admitted that once the final cement job has been poured, there is a tendency to think "everything is going to be okay" and to begin thinking about the next job.[97]

Emergency Situations

As discussed in Chapter 4.9, Transocean did not adequately train its rig crew how to respond to emergency well control situations, such as a severe blowout. Transocean required regular well control drills, but none focused specifically on emergency situations—how to recognize an emergency and what steps to take immediately upon recognizing it.[98] Transocean's Well Control Handbook provides little guidance on emergency situations, focusing instead on how to handle and circulate out more routine kicks. For instance, the handbook contains a section on "procedures for handling gas in the riser," which provides for the *possibility* of diverting a severe influx of hydrocarbons overboard as the ninth step in a lengthy diagnostic process.[99]

Transocean likewise did not adequately train or drill its dynamic positioning officers (DPOs) on how to respond to emergency situations. DPOs monitor a panel on the bridge that visually and audibly indicates whenever area-specific combustible gas, toxic gas, or fire alarms go off on the rig. The DPO acknowledges the alarms, contacts the affected area, and determines whether to initiate the general alarm to alert the entire rig (such as when more than one gas or fire alarm in contiguous areas goes off).[100]

Andrea Fleytas was the Transocean DPO on duty in charge of the alarm panel at the time of the blowout. After feeling a first jolt and noticing multiple combustible gas alarms sounding throughout the rig, she did not immediately hit the general alarm.[101] At the time, she received a call from the engine control room asking what was going on but did not instruct them to shut down the engines despite the multiple combustible gas alarms sounding throughout the rig.[102]

Asked why she hesitated, Fleytas said, "It was a lot to take in. There was a lot going on."[103] Fleytas said that Transocean provided no formal training or simulations on how to respond to combustible gas alarms.[104] She testified further that Transocean had not trained her to instruct the engine room to shut off the engines when combustible gas alarms were sounding.[105]

It is imperative that companies train and drill for emergency situations precisely because they occur so rarely.[106] There is no on-the-job training, as with more common events. Transocean senior toolpusher Randy Ezell told the Chief Counsel's team that he has worked on 60 to 75 wells during his career and has never seen anyone close the blind shear rams or use the emergency disconnect system (EDS) for well control purposes. He only had to engage the EDS twice in nine years on the *Deepwater Horizon*, both times when the rig had drifted off-site. He has never witnessed anyone divert flow overboard. He only saw the diverter used twice in his nine years on the *Deepwater Horizon*—both times to send returning flow to the mud gas separator.[107]

Contractors

At one point in time, operators owned their own oil rigs and directly employed the people who worked on them. But economic pressure and the complexity of offshore technology have pushed the industry away from that system. Modern offshore oil drilling now involves a team effort between an "operator" (which may have other oil company partners) and many specialized contractors and subcontractors. As Chapter 2 explains, Macondo involved just such a team effort. When the well blew out on April 20, only a handful of the 126 people on the rig worked for BP.[108]

The rest worked for one of the dozens of contractors and subcontractors associated with the project.

It is not necessarily problematic to use contractors to drill wells. Nor is it necessarily problematic to rely on specialized contractor expertise; drilling operations cannot be performed safely without their help, and Transocean and Halliburton are among the largest and best-regarded contractors in the oil and gas industry. But while the operator-contractor-subcontractor relationship can be beneficial in many ways, it also creates the potential for miscommunication and misunderstanding.

BP and its various subcontractors appear to have lost sight of that danger, compartmentalizing information that would have been useful to other companies carrying out their respective tasks. The onus fell on BP to ensure that its contractors were providing all of relevant information to the respective decision makers. As the party responsible for designing the well and well plan, the operator is best positioned to understand the big picture and how decisions and issues regarding one aspect of the well might affect decisions and issues regarding another.

BP's Oversight of Contractors

BP, like most offshore operators, relied heavily on its contractors to advise its engineers regarding important decisions. But BP did not adequately supervise its Macondo contractors in several instances.

The most egregious instances of inadequate supervision concern cementing. After the blowout, BP representatives and officials described Halliburton as "one of the, if not the leading cementing contractor in the world"[109] and contended that it relied on Halliburton's expertise to highlight cementing concerns.[110] But documents from before the incident show that BP's own employees were well aware that Halliburton's cementing services could be problematic. For instance, Chapter 4.4 discusses a 2007 auditing report prepared for BP, which concluded that Halliburton's "chemists and senior lab technicians do a very good job of testing cement slurries, but they do not have a lot of experience evaluating data or assisting the engineer on ways to improve the cementing program."[111] One of BP's top cement experts also described "the typical Halliburton profile" as "operationally competent and just good enough technically to get by."[112]

More importantly, BP engineers had specific concerns about Halliburton cementing engineer Jesse Gagliano, the Halliburton employee working on the Macondo well. Documents show that before the blowout, BP engineers thought Gagliano was not providing "quality work"[113] and was not "cutting it."[114] They highlighted that Gagliano had a habit of waiting too long to conduct crucial cement slurry tests. Three days before the blowout, Morel complained that he had "asked for these lab tests to be completed multiple times early last week and Jesse still waited until the last minute as he has done throughout this well."[115] Morel found "no excuse" for the tardiness.[116]

BP had known of problems with Gagliano for years[117] and "tried to work around" his shortcomings.[118] By the time of the Macondo blowout, BP had even asked Halliburton to reassign Gagliano.[119] Given this history, while waiting for his replacement, BP should have done more to supervise Gagliano's work, especially his work on the difficult production casing cement job at Macondo. At the very least, BP's management should have ensured that their own internal experts or senior Halliburton personnel double-checked Gagliano's cementing plan and foamed cement slurry design. Instead, BP's engineers admitted that they did not review his work "line by line"[120] and never fully utilized their in-house cementing expertise. They did not insist that he report the final April 18 lab results in a timely manner, let alone review those results before allowing Halliburton to pump the final Macondo job.

BP did not even review the February 10 slurry test results that it *did* have. If BP had properly examined those results, it would have seen that the slurry had failed the foam stability test. The Macondo team had consulted BP cementing expert Erick Cunningham on other issues at Macondo. But it appears that nobody at BP ever showed him the foam stability slurry design or lab testing data. Instead, the Macondo engineering team focused exclusively on reducing ECD in order to mitigate the risk of lost returns without ever considering whether the slurry design was itself adequate to achieve zonal isolation.

Contractors' Deference to BP

If BP did not adequately review the work of some of its contractors, the converse problem was that many of BP's contractors were unduly deferential toward BP's design decisions. A Weatherford centralizer technician described the prevailing view as "Third party, we do what the company man requests."[121] In several instances, BP's contractors expressed private reservations about the plans and procedures at Macondo but did not more forcefully communicate to BP that there were better ways to do things.

Again, the failures of communication surrounding the cement job are a good case study. As self-described cementing experts, Halliburton had primary responsibility for designing and pumping the bottomhole cement. It should have alerted BP to any potential problems with that job. Yet, Halliburton often buried its analyses in highly technical reports (including laboratory tests and computer modeling) and never drew BP's attention to the importance of certain data.

Despite touting its cementing expertise in promotional materials, Halliburton adopted a posture of extreme deference throughout the Macondo project. Prior to the incident, Halliburton mentioned two concerns to BP. First, Gagliano mentioned that using a small number of centralizers could lead to cement channeling while admitting that he "did not think there would be a well control issue."[122] Second, a Halliburton cementing technician on the rig briefly suggested that a full bottoms up would be advisable.[123] But Halliburton never raised a host of other concerns to BP. It never pointed out that BP's plan called for a low total cement volume, noted that BP was using a relatively low flow rate, or argued that BP should perform a cement bond log. When asked why, Gagliano explained that this was not Halliburton's role. He said "we do not recommend running a [cement] bond log"[124] and, anyway, he "was never asked."[125] With full knowledge of all of these problems, Halliburton instead pumped the cement job and reported that the job had been "pumped as planned."[126]

Halliburton failed to highlight the importance of foam stability testing to the Macondo team and to communicate test data. In other contexts, Halliburton has argued that its job is merely to do what the operator says and pump the job as directed. But that posture is inconsistent with Halliburton's decision to selectively report stability testing data to BP, as discussed in Chapter 4.4. It is also inconsistent with Halliburton's failure to provide *any* April foam stability testing information to BP before pumping the job. If Halliburton's position is that the operator directs all aspects of the job, then Halliburton should provide the operator with *all* of the information needed to exercise that authority responsibly.

Chapter 4.4 also discusses the numerous concerns with Halliburton's internal management of its slurry design process. Halliburton does not appear to have: (1) ensured that internal experts reviewed the Macondo slurry design; (2) ensured that Gagliano conducted timely lab tests; or (3) ensured that it otherwise adequately addressed BP's concerns about Gagliano's performance. Halliburton's refusal to provide documents that illuminate its internal policies and procedures cannot conceal these defects.

Lack of Clarity About Contractor Expertise and Responsibility

BP and Transocean have sparred since the blowout regarding the relative competence of Transocean rig workers to interpret negative pressure test data. But whatever the formal allocation of responsibility was or should have been, BP personnel certainly *believed* that Transocean personnel were not only competent to interpret those test results, but experienced and worthy of consultation. Based on the accounts of BP's well site leaders, the Transocean rig crew that participated in the test also believed they were competent to interpret it.

Chapter 4.6 explains that BP's well site leaders appear to have accepted a facially implausible explanation of the negative pressure test results from Transocean personnel. This was due in part to BP's inadequate well site leader training. But it was also due to the fact that Transocean personnel were experienced and the BP well site leaders thus believed they could rely on Transocean personnel. Kaluza and Vidrine both appear to have deferred to Transocean toolpusher Jason Anderson's experience. And Guide told the Chief Counsel's team emphatically that the Transocean personnel were in fact capable and competent to recognize the problems with the well during the negative pressure test.[127] Again, even if BP's expectations were justifiable, they were mistaken.

Transocean has argued after the fact to the Chief Counsel's team that its driller and toolpusher were merely "tradesmen" and not competent to interpret a negative pressure test. If that is the case, it is unclear why they would have advocated the "bladder effect" explanation. The Chief Counsel's team also finds it difficult to believe that the driller and toolpusher would be any less competent than the well site leaders to interpret a negative pressure test. During a negative pressure test, the crew underbalances the well to see if it leaks—in other words, whether the well kicks. Transocean agrees that its crew is expert in monitoring for and identifying kicks, even in underbalanced situations. Hence the rig crew did not call the BP well site leaders for advice when they noticed anomalous pressure readings during the displacement of the riser but instead relied on their own expertise to determine whether there was a kick.

Regardless of whether Transocean personnel were competent to interpret the negative pressure test, BP failed to adequately ensure that its well site leaders exercised independent judgment regarding the test results, or to resolve uncertainties before proceeding. In the absence of a clearly defined decision process and success criteria, BP's well site leaders appear to have tried to create consensus by accepting the explanation of the rig crew rather than independently verifying the explanation the rig crew had provided.

Technology

Deepwater operators employ exceedingly sophisticated technology to drill wells. But BP and its contractors had neither developed nor installed similarly sophisticated technology to guard against a blowout.

Displays, Sensors, and Instrumentation

The well monitoring equipment on the *Deepwater Horizon* was inadequate. For example, the data displays depended not only on the right person looking at the right data at the right time, but also that the person understood and interpreted the data correctly.[128] During the displacement, many signs of the kick could have been missed if monitoring personnel were distracted or not paying full attention.

As discussed in Chapter 4.7, the Chief Counsel's team believes that rig workers could benefit from systems that employ *automated* alarms, similar to those in airline cockpits, to call attention to potential kick indicators.[129] Such systems should also inform mudloggers of crucial events—such as a change to the active pit system or a change in fluid routing. On the *Deepwater Horizon*, the mud logger depended on direct communication or guesswork to learn what was happening elsewhere on the rig.[130]

As further discussed in Chapter 4.7, the Chief Counsel's team was surprised to find that rig personnel had to perform basic well monitoring calculations by hand, instead of having automated systems to help monitor, for instance, net flow from the well.[131] The Chief Counsel's team was also surprised by inadequacies in the sensors and instrumentation for detecting kicks on the *Deepwater Horizon*.[132] For instance, there was no camera installed on the rig to monitor flow on the overboard line—a person had to look behind the gumbo box to perform a visual confirmation of flow.[133] Flow sensors could be thrown off by listing seas, crane movement or other activity on the rig.[134] Where data are unreliable, the crew is more likely to discount kick indicators.

Finally, there was no equipment dedicated to identifying the presence of hydrocarbons in the wellbore during nondrilling activities. The oil and gas industry has developed sophisticated sensors that can be installed in drilling tools to detect kicks while actively drilling. But the Chief Counsel's team found no evidence that BP or anyone else in the industry has tried to adapt such sensors for routine well monitoring purposes. For instance, such sensors could be developed, and installed in the BOP or the wellhead to detect gas and other hydrocarbons before they enter the riser.

Utilizing Data and Equipment

BP and the other companies did not adequately use the data displays and monitoring equipment they did have. For instance, BP paid Sperry Drilling to gather and send real-time drilling and other data from the rig back to shore. Prior to the blowout, BP maintained large conference rooms in its Houston headquarters dedicated to each of its Gulf of Mexico wells. The room for the Macondo well had numerous monitors displaying the Sperry-Sun real-time data. The onshore team also could access the data remotely over the internet. But BP had no policy requiring full-time, or even part-time, monitoring from shore.[135]

As discussed in more detail in Chapter 4.7, BP itself apparently recognized the value of having engineers monitor data from onshore. As of the time of the blowout, BP had planned over the next four years to implement the Efficient Reservoir Access (ERA) advisory system.[136] The goal was to create a system that integrated real time drilling and mud logging data, displayed it to the driller in a more user-friendly and useful manner and simultaneously sent it to a drilling engineer or specialist on shore who could provide real time support.[137] "The primary objective of the ERA

Advisor" was "to facilitate the management of real time drilling data and its integration with drilling recommended practices and expertise to ensure the *right information is in the right place at the right time*."[138] Among the goals of the program were "[t]o maximize the use of available real time data and expertise to inform while-drilling decisions" and "[t]o minimize flying blind by improving the quality of real time data...."[139] Among other things, the system would "integrat[e]...expertise across multiple sites and multiple disciplines."[140]

While BP did not plan to have the system up and running until November 2013,[141] it clearly recognized the value of having a second set of eyes onshore—with engineering skills—monitoring well data and supporting rig personnel. Yet, the Macondo team did not use the real-time monitoring equipment it already had in place, relying instead on its well site leaders to alert onshore team members when and if there were issues.[142]

BP explained the disconnect by noting that it is difficult for onshore monitoring personnel to understand the significance of data without knowing what is happening on the rig. But these challenges can be overcome. Redundant shoreside monitoring would clearly have helped in several instances at Macondo—for instance, during the negative pressure test.[143] BP's explanation is also inconsistent with the entire premise for developing and deploying the ERA advisory system.

Risk

Deepwater drilling is a challenging and risky endeavor. It is also a competitive and potentially lucrative business that demands constant attention to economic considerations. Balancing the need to address risk with the need to manage costs is a constant struggle for operators.

The Chief Counsel's team finds that BP and Transocean did not have adequate procedures in place to properly account for risk or to assess the overall impact of decisions that appeared to relate only to one part of the well project. *As a result, understandable cost pressures drove decision making* and allowed some operational redundancies to be purged as inefficiencies. (Again, Halliburton declined to provide documents that would have allowed further insight into its operations at Macondo.)

Risk Assessment

The companies involved at Macondo failed to rigorously analyze the risks created by key decisions or to develop plans for mitigating those risks. *This appears to have biased decisions in the last month at Macondo in favor of cost and time savings while increasing the risk of a blowout.*

BP

Despite making multiple changes over the last nine days before the blowout, the Macondo team did not formally analyze the risks that its temporary abandonment procedures created. The Macondo team never asked BP experts such as subsea wells team leader Merrick Kelley about the wisdom of setting a surface cement plug 3,000 feet below the mudline to accommodate setting the lockdown sleeve or displacing 8,300 feet of mud with seawater without first installing additional physical barriers. It never provided rig personnel a list of potential risks associated with the plan or instructions for mitigating those risks.

BP's management system did not prevent such ad hoc decision making. It required relatively robust risk analysis and mitigation during the planning phase of the well but not during the execution phase.

Almost every decision the Chief Counsel's team identified as having potentially contributed to the blowout occurred during the execution phase.[144]

BP's Beyond the Best Common Process sets forth BP's procedures for selecting, designing, and drilling wells in the Gulf of Mexico.[145] It lays out a five-stage process: (1) Appraise, (2) Select, (3) Define, (4) Execute, and (5) Review. The first two stages consist of identifying and selecting a well site. BP plans and permits the well during the Define stage. During the Execute stage, BP and its contractors actually drill and complete the well. Finally, once drilling and completion is done, there is a Review stage to evaluate the project and to identify areas for improvement.[146] The engineering team is primarily accountable during the Define stage, although the wells operation team is involved. The wells operation team takes over primary accountability during the Execute stage, with engineering continuing to support planning and design decisions.

Before proceeding from one stage to the next, a well must satisfy certain "gate" requirements. For instance, before moving from the Select to Define and from the Define to Execute stages, the well concept, design, and plan must undergo a rigorous peer review process, which consists of "a multi-discipline assessment by an external team of how the balance between risk and value is being managed" and is led by a member of the functional drilling and completion excellence team.[147]

There is not, however, any such peer review process during the Execute stage.[148] The decision whether and to what extent to perform any formal risk analysis is left largely up to the team's discretion, in particular the wells team leader.[149] For instance, BP's MOC process—which imposes risk analysis, mitigation plan and approval requirements—continues to govern decision making during the Execute stage.[150] But the MOC process only applies to decisions to deviate from the well plan approved during the Define stage, not to drilling procedures (such as temporary abandonment procedures).[151]

As a result, after spudding the Macondo well, BP invoked the MOC process only a handful of times. It invoked the process for only three decisions after the *Deepwater Horizon* took over drilling in February.[152] Those three decisions were: (1) the change from a 16-inch to 13⅝-inch casing string; (2) the early total depth decision; and (3) the decision to employ the long string instead of a liner.[153] And some members of the team thought an MOC was unnecessary for the long string decision because the original approved well plan had a long string production casing.[154]

After the blowout, Walz observed that the MOC process was "not in place" and "not clear" for the Macondo team.[155] BP investigators summarized Walz's view of the team's culture as follows: "Performance – not require[d] procedures – do what we have been doing."[156] None of the other key decisions identified in this Report, such as those regarding centralizers, cement slurry design, temporary abandonment procedures, or simultaneous operations went through the MOC process.

BP was aware that its risk assessment process had flaws, but it acted too late to remedy the gap. In 2008, BP's own internal review found that risk assessment required improvement in the Gulf of Mexico. The review noted the "need for stronger major hazard awareness" and stated that "[r]isk assessment processes/results are not integrated."[157] The review went on to state: "As we have started to more deeply investigate process safety incidents, it's become apparent that process safety major hazards and risks are not fully understood by engineering or line operating personnel. Insufficient awareness is leading to missed signals that precede incidents and response after incidents; both of which increases the potential for, and severity of, process safety related incidents."[158] Though BP later rolled out more robust risk assessment procedures in 2010,[159] the procedures were not in place for Macondo. In an interview after the incident, Sprague discussed a new requirement to evaluate the effectiveness of each barrier in a well but noted that it was ready only by the time of the incident.[160]

Problems with risk assessment practices appear to have affected decision making at Macondo in a number of ways. First, they allowed decision makers to avoid systematically identifying the risks their procedures created and the steps necessary to mitigate those risks. Second, the absence of formal risk assessment enabled late and rushed decision making. Third, the lack of rigorous risk assessments led decision makers to solve problems in isolation instead of considering the cumulative impact their solutions might have on the rest of the project. As discussed above, following the lost circulation event at the pay zone, BP's shoreside Macondo team focused almost exclusively on avoiding further lost returns and no longer considered the more general goal: effective zonal isolation. The team designed a cement job that decreased the risk of lost returns but increased the risk of cementing failure. The primary criterion the team used to determine the success of the cement job was whether there had been lost returns. Seeing none, they sent the Schlumberger crew home. With one problem solved, they moved to the next.

Transocean

Transocean's crew appears never to have undertaken any risk analysis nor to have established mitigation plans regarding their performance of simultaneous operations during displacement after the negative pressure test.[161] It is not clear what, if any, steps the crew took to ensure that they could continuously and reliably monitor return volumes during the displacement prior to sending the spacer overboard, or flow-out after they began sending the spacer overboard. There is no indication the crew calculated expected pressures during the displacement.[162] Internal Transocean reviews show that it did not believe that the rig crews could identify and mitigate all risks on their own. A Lloyd's Register audit of Transocean in 2010 found: "[Rig crews] don't always know what they don't know. Front line crews are potentially working with a mindset that they believe they are fully aware of all the hazards when it is highly likely that they are not."[163]

Transocean's crew seems to have concluded prematurely that risks had receded after the negative pressure test. Once the test had been declared a success, the driller and toolpusher appear to have put any concerns about the test behind them rather than increasing their vigilance. They did not immediately shut in the well upon observing unexpected pressure readings; they did not keep the mudlogger apprised of all pit changes and fluid movements and do not appear to have monitored data more closely in his absence.

After the March 8 kick on the *Deepwater Horizon*, Guide asked Transocean rig manager Paul Johnson to consider how to improve the rig crew's hazard awareness. Johnson wrote back: "I thought about this a lot yesterday and asked for input from the rig and none of us could come up with anything we are not already doing.... You can tell them what the hazards are, but until they get used to identifying them their selves, they are only following your lead.... Maybe what we need is a new perspective on Hazard recognition from someone outside the industry."[164]

Bias in Favor of Time and Cost Savings

On any drilling rig—no matter who is the operator—"time is money."[165] BP leased the *Deepwater Horizon* at a rate of about $533,000 per day.[166] The high daily cost made the rig the single greatest expense for drilling the Macondo well.[167] It also gave BP a strong incentive to improve drilling efficiency.

The Chief Counsel's team observed that the Macondo team understandably made individual decisions consistent with an orientation toward efficiency but did not step back to consider what the safety implications of those decisions were when taken together. In the absence of a stronger emphasis on risk assessment and process safety during the Execute stage, engineering and operations decisions tilted toward cost and time savings. The risk register for the Macondo well exemplifies the problem. Though the register was intended to help the team identify potential problems with the well and the consequences of those hazards, it did not include safety as an element.[168] The risk register focused exclusively on the impact risks might have on time and cost. (And there is no indication the Macondo team even used it once the well entered the Execute stage.)[169]

Examples of Decisions That Increased Risk and Saved Time

BP's employees made a number of important decisions that increased risk at Macondo. BP did not run a cement evaluation log, nor did it perform further well integrity tests after the unexpected results of the negative pressure test. BP did not install additional barriers during temporary abandonment, nor did it elect to install the surface cement plug closer to the wellhead. The list goes on. Chapter 4 of the Chief Counsel's Report provides background and detail on these decisions.

Many of the decisions that increased risk also saved time. Take BP's decision-making process about how many centralizers to use. When Gagliano recommended obtaining additional centralizers, Morel responded that it was "too late" to get more centralizers to the rig.[170] It is never "too late" if one is willing to stop operations and wait for the right equipment. Guide informed the Chief Counsel's team that he himself had suggested waiting at one point, but in emails before the incident he argued against using additional centralizers not only because they might hang up, but "also it will take ten hrs to install them."[171] (Guide explained to the Chief Counsel's team that further delaying casing installation would have raised risks of its own. The Chief Counsel's team notes that BP had left the wellbore open for several days at this point in order to log the wellbore, and while that entails some risk, there was no systematic discussion of this risk, or the pros and cons of waiting for additional centralizers.)

As shown in Table 5.2, the decision about centralizers is not an isolated example of time pressure apparently influencing well design or operations at Macondo.

Table 5.2. Examples of decisions that increased risk at Macondo while potentially saving time.

Decision	Was There a Less Risky Alternative Available?	Less Time Than Alternative?	Decision Maker
Not Waiting for More Centralizers of Preferred Design	Yes	Saved time	BP onshore
Not Waiting for Foam Stability Test Results and/or Redesigning Slurry	Yes	Saved time	Halliburton (and perhaps BP) onshore
Not Running Cement Evaluation Log	Yes	Saved time	BP onshore
Using Spacer Made From Combined Lost Circulation Materials to Avoid Disposal Issues	Yes	Saved time	BP onshore
Displacing Mud From Riser Before Setting Surface Cement Plug	Yes	Unclear	BP onshore
Setting Surface Cement Plug 3,000 Feet Below Mudline in Seawater	Yes	Unclear	BP onshore (approved by MMS)
Not Installing Additional Physical Barriers During Temporary Abandonment Procedure	Yes	Saved time	BP onshore
Not Performing Further Well Integrity Diagnostics in Light of Troubling and Unexplained Negative Pressure Test Results	Yes	Saved time	BP (and perhaps Transocean) on rig
Bypassing Pits and Conducting Other Simultaneous Operations During Displacement	Yes	Saved time	Transocean (and perhaps BP) on rig

Meticulous Tracking of Time and Cost

Each day of drilling counted to BP, and BP counted the cost of each day. BP's common process for well design and operations required engineers to set out "detailed time and cost estimates" for "the operational procedures for drilling" the well.[172] The estimates were based on prior drilling performance on other wells.[173] During drilling, BP had its team share with the rig crew every day how long each task should take.[174] The actual time to complete a task would then be recorded and performance shared with the rig crew.[175]

The *Deepwater Horizon* followed this process meticulously.[176] The rig had a database of the "fastest times" to complete "each task the rig carries out."[177] The engineers used the database to "construct a time estimate for the well being planned."[178] Every day, "the actual times for each operational task" were "checked against the Best of the Best data."[179] A spreadsheet accounted for all of the rig's time, from servicing the rig to running the drill pipe.[180] If an activity was "non productive time," then it was marked as such with a brief description of the cause.[181] BP may have linked this information into a worldwide database.[182]

BP tracked not only the time to complete each task, but also the cost of every item to drill the Macondo well.[183] The list runs from $15 for one cargo box to $533,000 for one day of the rig's time.[184] About 10,000 items were accounted for, tallied, and listed by day.[185] Daily costs varied from over $4 million on March 17, 2010, to as low as $6,300 during the planning of the well in 2009.[186]

By the time of the blowout, the Macondo well had taken longer to drill and cost much more than BP had anticipated. BP had spent more than $142 million on the well.[187] The original plans for Macondo set out a price tag for the well of $96 million.[188] Because the well kept going over budget, BP had to return to its partners three times to authorize supplemental expenditures.[189] The final authorization anticipated that the well would cost as much as $58 million more than planned.[190] The Macondo well had also fallen at least 38 days behind schedule.[191]

Comparable wells had taken less time and had cost considerably less to drill. Sims testified that days per 10,000 feet (a common industry metric) was the most important metric for drilling performance.[192] Hafle estimated that the well had taken about 70 days for each 10,000 feet drilled.[193] That performance put Macondo in the bottom 10% of wells drilled (more than 10 days per 10,000 feet slower than the threshold for that category).[194] The well's total cost also placed it in the bottom 10% of comparable wells.[195] So did the amount of what BP classified as "non-productive time."[196] (Nonproductive time is another common industry metric).

It is unclear as to the full extent to which these cost and time overruns impacted personnel and decisions onshore or on the rig. One well site leader remarked that the cost of the Macondo well was a concern and that he was aware the rig was running behind.[197] However, he and others have almost uniformly stated that cost and time pressure was not an issue and that they did not feel more pressure to hurry to get things done than would otherwise be the case.[198]

Cost accounting is a necessary and reasonable part of running a business. Nonetheless, given the many decisions that increased risk but saved time and money, it is a reasonable inference that cost and time overruns had an effect, conscious or unconscious, on decision making.

Well Design and Operations Guidance

At the Commission's November hearing, Steve Lewis testified: "[T]he pressure to make progress is actually inherent in the business. And it takes a stated, conscious management presence to counter that...drillers drill against each other. We want to be the fastest, best driller there is."[199]

Like many other operators, BP's guidance on well design and operations placed a premium on drilling quickly. The Beyond the Best Common Process[200] emphasized the achievement of the "technical limit" for drilling a well.[201] The term technical limit means "what drilling times might be possible if everything works perfectly."[202] Achievement of the technical limit depends on the elimination of "non productive time" and "invisible lost time."[203] Though BP did not expect its engineers to achieve the "technical limit" (at least not yet),[204] they were told that the company's aspiration was to achieve the "Technical Limit as quickly as possible."[205] BP asked its engineers to accomplish times faster than what had been done before—"the 'best of the best.'"[206]

For well design, the emphasis on drilling performance and technical limit meant that BP engineers were expected to carefully account for how long it would take to drill each well. Engineers were asked to consider, "Could the well be constructed more efficiently?"[207] That question appears to have been important to the team that designed the Macondo well.

In an interview with the Chief Counsel's team, Sims shared that he was always thinking about how to drill wells faster.[208] He replied "yeah, that's safe to say" when asked whether Morel, the engineer who designed the temporary abandonment procedures at Macondo, was "always thinking about cost and efficiency."[209] Guide's supervisor flagged in his 2009 mid-year evaluation: "John needs to...take safety performance to the same level as drilling performance."[210]

BP's focus on driving down the time to drill wells could result in a tendency to treat redundancies as inefficiencies. Tasks that took additional time would have counted against the rig's time and cost performance.[211] In the absence of sufficient checks and balances, adding cost that did not immediately appear *necessary* to the safety of the well might not be judged fairly. A cement evaluation log may have been perceived as unnecessary when a negative pressure test was planned not long after. A mechanical plug or additional cement plug may have seemed inefficient when there was cement already at the bottom of the well. The problem is exacerbated for very low-frequency events, which might allow poor decisions to go unnoticed for many years where a particular type of failure (especially one that requires multiple things to go wrong) happens only rarely.

Personnel Evaluations and Incentives

BP provided incentives to its drilling personnel. For more senior personnel, the annual bonuses exceeded $100,000 on top of salaries over $200,000.[212] BP based the annual bonuses and promotions in part on performance evaluations.

The performance evaluations for the Macondo team emphasized, among other things, drilling performance. The Gulf of Mexico's metrics for drilling targeted days per 10,000 feet drilled and performance against AFE as priorities.[213] The AFE is the Approval for Expenditure, a metric for how much BP planned to spend on a well. Early in 2010, Sims listed delivering the wells "at or below" the targeted times as the "#1" priority for him and for Guide in the coming year.[214] O'Bryan also had drilling efficiency in his performance contract for 2010.[215]

The BP team that drilled Macondo had a history of focusing on cost and performance in their performance evaluations. Guide's list of key indicators for 2008 specified "performance," measured by days per 10,000 feet of drilling.[216] After that, Guide had "All Well Objectives delivered at a cost less than AFE."[217] Guide highlighted that "[o]perational performance has been top quartile," meaning that the rigs had outperformed most other BP rigs in how long it took to drill a well,[218] and observed that one well "set numerous industry and B[P] drilling records and finished 32day's / 10K."[219] In 2009, Guide's supervisor noted that Guide had "championed the every dollar counts culture."[220] "Every dollar counts" became a priority at BP during diminished demand for oil in 2008.[221] Guide noted in his self-evaluation that "[d]aily operational decisions now include the cost component."[222]

Sims provided the same level of detail for drilling performance. In 2007, he noted in his interim review that "we have done a good job of delivering fully evaluated wells under time and cost targets."[223] In 2008, Sims observed that the "Kodiak well finished under AFE cost and with top quartile performance."[224] He also highlighted that the "Freedom well finished the original scope under AFE time and budget.[225] In 2009, Sims highlighted when the time to complete a well was "top quartile" and when wells finished "under AFE."[226]

Importantly, BP's performance evaluations and internal standards also emphasized the importance of safety. BP's code of conduct provided: "BP is committed to providing a safe place of work for everyone—that includes stopping work if we ever have concerns about HSSE [health, safety, security, and the environment]. BP will not tolerate retaliation against anyone who in good faith stops work for HSSE issues—it's better to be safe than sorry."[227] BP also had in place "Golden Rules of Safety."[228] The Golden Rules emphasized that "Safety is a legitimate personal expectation and a constant individual responsibility."[229]

Though safety was important at Macondo, BP's approach was strongest with respect to easily measured personal safety metrics, such as injuries, rather than process safety risks of low-frequency, high-consequence events such as a blowout. BP put safety first on individual employees' performance evaluation forms,[230] but the metrics for safety encompassed only a subset of the risks of drilling. Guide's evaluation in 2009, for example, put safety at the top of the list of key performance indicators, measured by recordable injuries.[231] The well site leaders had similar standards, which emphasized recordable injuries and safety meetings.[232]

It is not apparent whether and to what extent BP has or assesses safety metrics regarding drilling procedure or well design. BP expected full compliance with its mandatory engineering policies.[233] But BP lacked a systematic way to assess whether engineers complied with those policies, especially after the peer review process was complete and the well entered the Execute stage.[234] BP did not track how employee decisions impacted process safety or risk.

It is perhaps not surprising that BP's performance evaluations relied on easy-to-track metrics such as injuries and safety meetings to account for an employee's commitment to safety. It would be difficult after the fact to analyze whether an employee's decisions actually increased the risk profile of a project unnecessarily. That is all the more reason why it was critically important for BP to have in place at all stages of the well a formal risk assessment system for evaluating drilling decisions that could increase the overall risk profile of the project.

Closing

As this review of management practices at Macondo demonstrates, the blowout occurred in large part because the companies diffused knowledge, responsibility for, and ownership of safety among themselves and among groups of people. The people onshore and on the rig had a false sense of security. They did not recognize the need for individual leadership in addressing the multiple anomalies and uncertainties that they observed. Instead, they relied on many ambiguous "dotted line" relationships within and between the companies and personnel involved.

To prevent an incident at Macondo from ever happening again, it will not be enough merely to add regulatory personnel. Just putting more inspectors on the *Deepwater Horizon* would not have prevented this blowout.

Nor will it be enough to issue new prescriptive regulations or write more voluminous safety manuals. Adding a new "don't do this either" rule after every accident ensures staying behind the curve.

What the men and women who worked on Macondo lacked—and what every drilling operation requires—was a culture of leadership responsibility. In hostile offshore environments, individuals must take personal ownership of safety issues with a single-minded determination to ask questions and pursue advice until they are certain they get it right. ◗

Chapter 6 | Regulatory Observations

The Commission's full report examines in depth the history and current status of Minerals Management Service (MMS) regulatory programs, and makes specific recommendations for regulatory reform.* In Chapter 3 of that report (displayed in Figure 6.1), the Commission finds that:

- MMS had a built-in financial incentive to promote offshore drilling that was in tension with its mandate to ensure safe drilling and environmental protection;

- revenue increases dependent on deepwater drilling came with increased safety and environmental risks, but those risks were not matched by greater, more sophisticated regulatory oversight;

- MMS was unable to maintain up-to-date technical drilling-safety requirements to keep up with industry's rapidly evolving deepwater technology. As drilling technology evolved, many aspects of drilling lacked corresponding safety regulations; and

- at the time of the blowout, MMS systematically lacked the resources, technical training, or experience in petroleum engineering that is critical to ensuring that offshore drilling is being conducted in a safe and responsible manner.

Figure 6.1. Chapter 3 of the Commission's full report.

OSC

Full report available at www.oilspillcommission.gov.

This portion of the Chief Counsel's Report is more modest. It focuses solely on the role that MMS regulations in force at the time of the blowout played in guiding design and process decisions at Macondo.

MMS Background

MMS, now the Bureau of Ocean Energy Management, Regulation, and Enforcement, employs approximately 600 individuals to run operations in the Gulf of Mexico region.[1] About one-fifth of that staff is distributed among five district operations offices. Each district office has a small

---

* The Minerals Management Service (MMS) was renamed the Bureau of Ocean Energy Management, Regulation, and Enforcement (BOEMRE) on June 18, 2010. For ease of reference, this Report uses the former name, MMS.

cadre of engineers, including drilling engineers. Drilling engineers review drilling permit applications.

The MMS office that supervised drilling in Mississippi Canyon Block 252 was the New Orleans District office. The New Orleans District office (in the bottom row of Figure 6.2) reviewed 25% to 30% of all permits submitted for the Gulf of Mexico.[2] The office had one designated drilling engineer for the review of permits. That individual thus reviewed several hundred permits each year[3] and approved BP's initial application for permit to drill (APD) a well at Macondo, as well as subsequent applications modifying that permit.

Figure 6.2. BOEMRE organizational chart for the Gulf of Mexico region in July 2010.

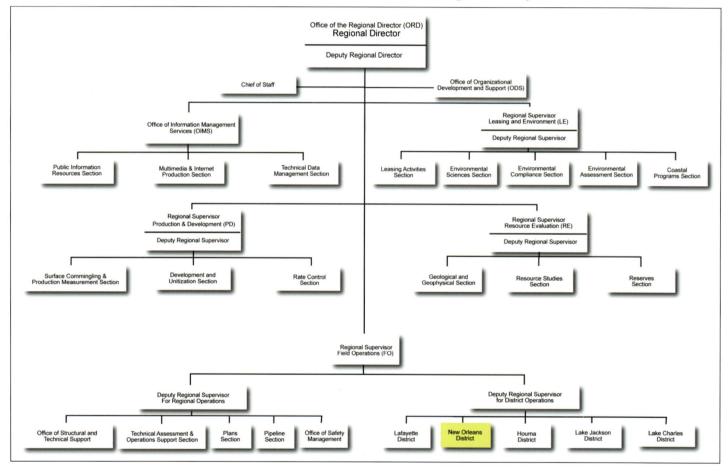

TrialGraphix

The New Orleans District office supervised drilling in the region of the Gulf of Mexico that contained the Macondo well.

APD and APM. An operator's first submission to the MMS for permission to begin drilling is an **application for permit to drill,** or **APD.** Subsequent changes to the initial permit are requested through an **application for permit to modify,** or **APM.**

MMS regulations are located in title 30 of the Code of Federal Regulations, parts 201 to 299. The regulations governing review and approval of drilling operations are primarily located in part 250.

MMS Regulations Did Not Address Many Key Risk Factors for the Blowout

MMS regulations in force at the time of the Macondo blowout did not address many of the key issues that the Chief Counsel's team identified as risk factors for the blowout.

Deepwater Drilling Conditions

At the time of the blowout, most MMS prescriptive and performance-based regulations applied uniformly to all offshore wells regardless of their depth. The regulations did not impose additional or different performance requirements for deepwater wells. Indeed, MMS personnel stated that it had become routine for them to grant certain specific exemptions from regulatory requirements, mostly related to blowout preventer (BOP) testing, in order to accommodate the needs of deepwater operations.[4]

While MMS regulators routinely reviewed an operator's predictions about shallow drilling hazards, they did not review an operator's predictions of drilling conditions in deeper areas.[5] For instance, while MMS regulations required operators to submit predicted pore pressure and fracture gradient charts along with well permit applications,[6] MMS personnel did not review the data in those charts, let alone verify, for example, whether the predictions aligned with offset data from other wells in the area. MMS personnel were not aware of any instances in which the agency had rejected a permit application because of questionable predictions regarding subsurface conditions. (Indeed, MMS personnel rarely questioned any statements or predictions contained in permit applications.[7])

MMS regulations did require BP to submit for approval all of the well design changes that it made in response to drilling conditions or external events.[8] As a result, BP submitted more than 10 separate drilling permit applications for Macondo. For instance, BP submitted a revised permit application after Hurricane Ida forced BP to replace the *Marianas* with the *Deepwater Horizon*.[9] It submitted additional revisions after it was forced to stop drilling when the March 8, 2010 kick caused a stuck drill pipe and forced BP to continue drilling with a sidetrack.[10] BP also submitted revised casing schedules each time drilling conditions required it to alter its overall casing plan. BP did not always explain the need for well design changes. For instance, it did not specifically explain to MMS that it decided to stop drilling earlier than planned and declare a shallower total depth because of the early April lost returns event.

Well Design

At the time of the Macondo blowout, MMS regulations covered only very basic elements of well design. The regulations required operators to submit information on the pore pressure and fracture gradient they expected to encounter, and the maximum pressures to which they expected casing strings and well components to be exposed.[11] The regulations also required operators to specify the weight, grade, and pressure ratings of casing they planned to install, and generally to ensure that casing would "[p]roperly control formation pressures and fluids."[12]

MMS regulations did not authorize, prohibit, or restrict the use of long string production casings. They did not specify any minimum number of annular barriers to flow. They did not address any issues related to annular pressure buildup (APB), nor authorize or prohibit any particular APB mitigation approaches. Regulations did not specify design measures that would facilitate containment or capping measures in the event of a blowout; for instance, the regulations did not address the use of burst or collapse disks in casing design, nor require the use of a protective casing. (Chapter 4.2 discusses these issues in greater detail.)

In several instances, MMS personnel involved at Macondo recognized risks that might be posed by certain Macondo design features or advantages of certain design features not used at Macondo. In each instance, however, the individuals refrained from suggesting or requiring changes. They explained that their role was to check compliance with specific regulatory requirements and not to provide more generalized design advice to operators. One explained that if he were to recommend for or against a particular well design or design feature, he might be held responsible if that approach caused problems.[13]

Cementing Design

MMS regulations contained several provisions that address the use of cement in offshore oil wells, but they were quite general. MMS personnel identified four regulations that address cement and cementing: 30 C.F.R. § 250.415, 420, 421, and 428. Section 250.415 states only that an operator must discuss in its cementing program the type and amount of cement it plans to use. Section 250.420 adds little of relevance: It provides that cement "must properly control formation pressure and fluids." Section 250.421 is the most prescriptive: It specifies minimum cementing volume requirements for each type of casing (conductor, surface, intermediate, and production) and states that for a production casing, cement must extend at least 500 feet above all hydrocarbon-bearing zones.

MMS personnel stated that the only cementing requirement they routinely policed was the linear coverage requirement in 30 C.F.R. § 250.421.[14] The Chief Counsel's team noted, however, that BP's permit applications did not contain information that would allow meaningful review of this issue. While BP's permit applications did include the height of the top of the cement column, they did not include the height of the top hydrocarbon-bearing zone. Without this information, it would be difficult to determine whether BP planned to pump enough cement to cover the annular space 500 feet above that zone.

MMS cementing regulations did not address several issues that proved important at Macondo.

- The regulations did not require the use of casing centralizers, nor specify minimum standoff percentages or other centralization criteria.

- The regulations did not address the possibility of cement contamination, nor specify any measures to reduce the likelihood of contamination (such as the use of wiper plugs or spacer fluids).

- While at least one regulation recommended the use of float valves,[15] the regulations did not specify whether or how to evaluate float valve conversion or performance.

- The regulations did not require BP to conduct or report cement slurry tests, nor specify any criteria for test results.

- The regulations did not address the use of foamed cement (or any other specialized cementing technology) at all. The regulations did not require BP to inform MMS that it would be using foamed cement, nor specify any technical criteria for foamed cement or foamed cement testing.

Cement Evaluation

Section 250.428 of the MMS regulations (displayed in Figure 6.3) is the only one that addresses directly the possibility of a failed cementing job. That section states that if an operator has "indications of an inadequate cement job (such as lost returns, cement channeling, or failure of equipment)" the operator should:

1. Pressure test the casing shoe;

2. Run a temperature survey;

3. Run a cement bond log; or

4. Use a combination of these techniques.

Figure 6.3. MMS regulation 30 C.F.R. § 250.428.

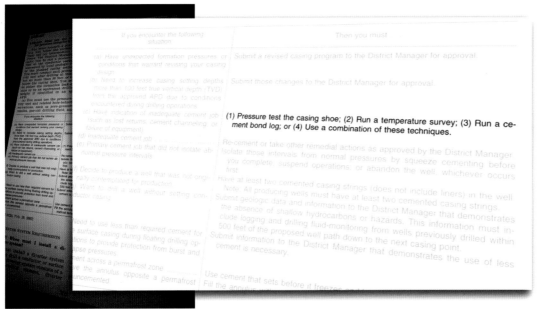

MMS

This regulation applied at Macondo but had little practical effect. The pressure and volume indicators that rig personnel examined did not provide any "indications of an inadequate cement job" during the cementing process at Macondo. But as discussed in Chapter 4.3, these indicators provide little direct information about cementing success. And while the regulation states that indications of "cement channeling" should trigger remedial efforts, it is extremely difficult to determine if cement has channeled based on surface indicators. Finally, the regulation's remedial measure requirement is quite modest; MMS personnel admitted that an operator could satisfy it by conducting a *positive* pressure test on the casing shoe,[16] even though such a test would not examine cementing success. (As Chapter 4.6 explains, BP did conduct a positive pressure test after cementing.)

Negative Pressure Test Procedures

The most notable gap in MMS regulatory structure at the time of the incident was the lack of any regulation requiring negative pressure tests before temporary abandonment.

Representatives of the companies involved and other industry experts uniformly agreed that it is crucial to negative pressure test wells like Macondo before temporary abandonment. But while MMS regulations contain numerous requirements for pressure tests, such as 30 C.F.R. § 250.423 and 250.426, they do not require negative pressure tests on any well, let alone specify how such tests should be done.[17] (In its April 16, 2010 application for permit to modify, or APM,[18] BP told MMS that it would conduct a negative pressure test.)

Since the blowout, MMS has promulgated interim regulations that would require negative pressure tests. MMS regulation 30 C.F.R. § 250.423(c) not only requires negative pressure tests on intermediate and production casing strings, but it also requires operators to submit test procedures and criteria with their APD to MMS. It also requires all test results to be recorded and available for inspection.[19] While this regulation does not specify how the test is to be conducted or interpreted, the requirement to file procedures and criteria may prompt operators to establish best practices.[20]

The Chief Counsel's team notes that negative pressure tests are not necessary at every well. Some operators make a practice of ensuring that wells are "overbalanced" even after they remove the riser and, with it, the balancing pressure generated by the mud column in the riser. (This pressure is called the "riser margin.") If a well is overbalanced, hydrocarbons cannot flow out of the formation into the well even if bottomhole cement and any casing plugs were to fail. Negative pressure tests are therefore not necessary for overbalanced wells.

Well Control

MMS regulations do address the importance of well control. Notably, 30 C.F.R. § 250.401 (displayed in Figure 6.4) required operators to:

- "Use the best available and safest drilling technology to monitor and evaluate well conditions and to minimize the potential" for a kick. Section 250.105 defines the term "best available and safest technology" to mean technologies that the MMS director "determines to be economically feasible wherever failure of equipment would have a significant effect on safety, health, or the environment."

- Ensure that the drilling crew "maintains continuous surveillance on the rig floor" from the start of drilling operations until the well is temporarily or permanently abandoned, unless the well is shut in with BOPs.

- Use personnel who have received well control training that meets regulatory requirements.

- "Use and maintain equipment and materials necessary to ensure the safety and protection of personnel, equipment, natural resources, and the environment."

Figure 6.4. MMS regulation 30 C.F.R. § 250.401.

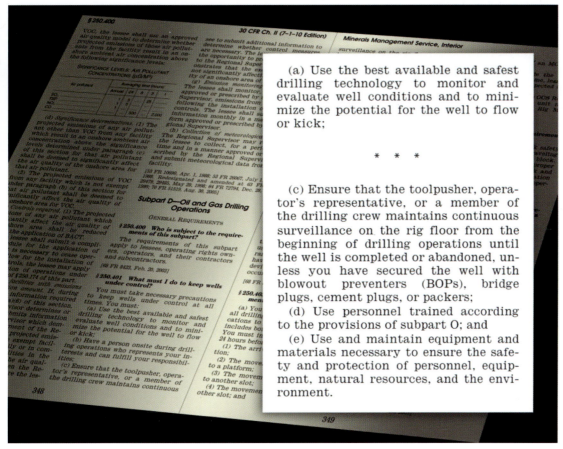

MMS

The Chief Counsel's team observed that all key personnel involved in the blowout had received MMS-approved well control training. However, the training they received appears to have focused primarily on initial kick response during drilling operations—that is, on the process of shutting in a well and circulating the kick out. Numerous individuals stated that well control training typically does not involve extensive instruction in the subtleties of kick detection and kick indicators. Additionally, as discussed further in Chapter 4.8, the training does not appear to have covered the proper emergency response to full-scale blowouts.

Despite the aspirational language of the "best available and safest technology" requirement, the *Deepwater Horizon* rig did not include any devices designed specifically to help rig personnel detect the presence of hydrocarbons in the wellbore during nondrilling procedures such as temporary abandonment. The rig's drilling equipment did include sophisticated instruments that could detect kicks during the course of actual drilling. But these instruments were part of the rig's bottomhole assembly (the lower part of the drill string). They were not present in the wellbore during cementing and temporary abandonment procedures.

Temporary Abandonment Procedures

MMS regulation 30 C.F.R. § 250.1721 specifies the requirements that an operator must satisfy before temporarily abandoning a well. The section contains several provisions, but the only one that appears to have guided decision making at Macondo was subpart (d), which states that an operator must: "[s]et a retrievable or a permanent-type bridge plug or a cement plug at least 100 feet long in the inner-most casing" and that the top of the plug must be "no more than 1,000 feet below the mud line."

BP's last regulatory submission for Macondo was an April 16 APM (seen in Figure 6.5) in which it requested permission to set an unusually deep cement plug 3,000 feet below the mudline—2,000 feet deeper than section 250.1721 would otherwise require.[21] BP made its request by submitting the short numbered list of temporary abandonment procedures discussed in Chapter 4.5. The company stated that it would "[s]et a 300' cement plug (125 cu. ft. of Class H cement) from 8367' to 8067'," explaining its rationale in just 40 words:

> The requested surface plug depth deviation is for minimizing the chance for damaging the LDS sealing area, for future completion operations.

> This is a Temporary Abandonment only.

> The cement plug length has been extended to compensate for added setting depth.

Figure 6.5. BP's April 16 application for permit to modify.

Temporary Abandonment Procedure
Macondo – MC 252 #1
Deepwater Horizon

Current Status:

Making wiper trip prior to running a long string of 9-7/8" x 7" production casing

Forward Plan:

Run casing to 18,300' +/- per approved APD. Test casing to 2500 psi per approved AP

Temporary Abandonment Procedure: *(estimated start time Sunday, April 18, 2010)*

1. Negative test casing to seawater gradient equivalent for 30 min. with kill line.
2. TIH with a 3-1/2" stinger to 8367'.
3. Displace to seawater. Monitor well for 30 min.
4. Set a 300' cement plug (125 cu.ft. of Class H cement) from 8367' to 8067'.

   The requested surface plug depth deviation is for minimizing the chance for damaging the LDS sealing area, for future completion operations.

   This is a Temporary Abandonment only.

   The cement plug length has been extended to compensate for added setting depth.

5. POOH.
6. Set 9-7/8" LDS (Lock Down Sleeve)
7. Clean and pull riser.
8. Install TA cap on wellhead and inject wellhead preservation fluid (corrosion inhibitor) below TA cap.

MMS granted BP's request less than 90 minutes after BP submitted it.[22] MMS regulation 30 C.F.R. § 250.141 is the regulation that gives MMS officials authority to grant departures from otherwise-applicable regulatory requirements. That regulation states, in relevant part:

> You may use alternate procedures or equipment after receiving approval as described in this section…[a]ny alternate procedures or equipment that you propose to use must provide a level of safety and environmental protection that equals or surpasses current MMS requirements.

The Chief Counsel's team asked MMS personnel why they believed that BP's request was appropriate. The individual involved in the decision explained that he granted the request after speaking with BP and learning that BP needed to set the surface plug deep in order to accommodate the setting requirements for its lockdown sleeve. He explained that he viewed it as beneficial for BP set its lockdown sleeve during temporary abandonment procedures but admitted that he had no training or expertise in lockdown sleeve procedures or best practices.[23] Neither MMS individuals nor MMS regulations addressed:

- the fact that BP relied on a single wellbore barrier during temporary abandonment;
- the extent to which BP had underbalanced the well during temporary abandonment activities;
- whether BP could or should set its surface cement plug in drilling mud, or whether BP should satisfy additional requirements before displacing drilling mud from the wellbore in order to set its surface cement plug in seawater;
- whether the BOP could be open during riser displacement operations or plug cementing; or
- whether alternatives besides a deep surface plug could accommodate lockdown sleeve setting requirements.

BOP Testing

BP applied for and received MMS approval to test several elements of the *Deepwater Horizon* BOP at lower pressures than regulations would normally require. One departure MMS granted allowed BP to test the *Deepwater Horizon*'s blind shear ram at the same pressures at which it tested casing.[24] Other departures permitted the rig crew to test the annular preventers at reduced pressures.

MMS regulations required that high-pressure tests for annular preventers equal 70% of the rated working pressure of the equipment or a pressure approved in an APD.[25] BP filed an APD in which it asked permission to reduce testing pressures for the *Deepwater Horizon*'s annular preventers to 5,000 pounds per square inch (psi) in October 2009.[26] In January 2010, BP filed an APD asking permission to reduce testing pressures for both annular preventers to 3,500 psi.[27] MMS granted these requests, which were consistent with industry practice.

MMS personnel did occasionally refuse BP's requests for testing modifications. For instance, after dealing with a February lost circulation event, BP asked MMS to allow rig personnel to delay scheduled BOP testing. This would have allowed the crew to immediately run the next casing string, in order to prevent further losses if the lost circulation treatment broke down during testing. On March 10, however, MMS personnel rejected that request.[28]

General Observations on Macondo Permitting Process

In reviewing BP's permit applications, the Chief Counsel's team noted that several of them included fairly obvious clerical or calculation errors. In some instances, it appears that MMS personnel did not recognize these errors. While none of the errors proved consequential—BP in most cases corrected them before proceeding—the fact that they escaped MMS attention raises concerns about the thoroughness with which the agency conducted even the relatively modest review it did undertake.

The most notable series of errors comes in three APDs that BP submitted to permit its long string production casing: APDs 9511, 9513, and 9515, filed on April 14 and 15. Among the errors in those APDs are the following:

- **APD 9511:** BP's written well design information states on one page that the company plans to use a single 7-inch long string production casing while another page states that it will use a 9-inch casing. Meanwhile, its well schematic states that it plans to use a tapered long string casing that includes a 9-inch section and a 7-inch section.

- **APD 9513:** BP corrected its mistaken statements about the long string diameter and explained that it would be using a tapered long string casing. However, this submission omitted the 9-inch liner that it had already installed from the well design information. Again, this liner appears in the well schematic.

- **APD 9515:** BP corrected its mistake again, this time noting that the well would include both the prior 9-inch liner and a long string casing tapering from 9 inches to 7 inches.

- **APDs 9511, 9513, 9515:** In all three of these permit applications, BP's written well design information states that it will pump 150 cubic feet of cement to cement the long string production casing in place. That volume equates to just 26 barrels—less than the 60 barrels of cement BP actually planned to pump, and far less than would have been necessary to meet MMS linear coverage requirements.

MMS personnel approved all three of these permit applications despite these errors.

BOP Recertification

MMS regulation 30 C.F.R. § 250.446(a) requires that BOPs be inspected in accordance with API Recommended Practice 53 § 18.10.3.[29] This practice requires disassembly and inspection of the BOP stack, choke manifold, and diverter components every three to five years.[30] This periodic inspection is in accord with Cameron's manufacturer guidelines, and Cameron would certify when the inspections were completed.[31]

The rig crew and BP shore-based leadership recognized that the *Deepwater Horizon*'s blowout preventer was not in compliance with certification requirements.[32] BP's September 2009 audit of the rig found that the test ram, upper pipe ram, and middle pipe ram bonnets were original and had not been recertified within the past five years.[33] According to an April 2010 assessment, BOP bodies and bonnets were last certified on December 13, 2000, almost 10 years earlier.[34] An April 2010 Transocean rig condition assessment also found the BOP's diverter assembly had not been certified since July 5, 2000.[35] Failure to recertify the BOP stack and diverter components within three to five years would appear to have violated MMS inspection requirements.[36]

An MMS inspection of the *Deepwater Horizon* on April 1, 2010 did not mention overdue BOP equipment certification.[37] When visiting a rig, inspectors use a "potential incidents and noncompliance" (PINC) list as an inspection checklist.[38] Although the PINC list contains guidelines not intended to supersede regulations,[39] inspectors consider the PINC list a comprehensive list of inspection items.[40] Because the list does not include verifying compliance with 30 C.F.R. § 250.446(a),[41] inspectors may simply have not checked whether the *Deepwater Horizon*'s BOP had been disassembled and inspected in accord with regulations.

Ethical Considerations

In recent years various bodies have concluded that certain MMS offices and programs have violated ethical rules or guidelines. In the wake of the *Deepwater Horizon* disaster, some questioned whether ethical lapses played any role in causing the blowout. The Chief Counsel's team found no evidence of any such lapses. ♦

Endnotes

Chapter 1

[1] Executive Order No. 13543, National Commission on the BP Deepwater Horizon Oil Spill and Offshore Drilling, 75 Fed. Reg. 29,397, May 21, 2010.

[2] *Ibid.*

[3] Press Release, National Commission, Fred Bartlit Named Chief Counsel, July 22, 2010.

Chapter 3

[1] Internal Department of Interior document (OSC-DWH BOEM-WDC-B01-00002-00004).

[2] Internal BP document (BP-HZN-MBI 13494).

[3] David Sims (BP), interview with Commission staff, December 14, 2010.

[4] Internal BP document (BP-HZN-MBI 13494).

[5] U.S. Geological Survey official, interview with Commission staff, October 21, 2010.

[6] Internal BP document (BP-HZN-OSC 4259).

[7] Internal BP document (BP-HZN-OSC 4663).

[8] Testimony of Paul Johnson (Transocean), Hearing before the Deepwater Horizon Joint Investigation Team, August 23, 2010, 352. However, some forms of maintenance, including BOP maintenance, were exempt from this provision. Internal BP document (BP-HZN-MBI 4254).

[9] Testimony of David Sims (BP), Testimony of before the Deepwater Horizon Joint Investigation Team, August 26, 2010, 191; Internal BP document (BP-HZN-MBI 126763).

[10] Nonpublic BP document (presentation to the Commission, August 9, 2010), 6.

[11] CNN Money, "Global 500," *Fortune,* July 11, 2010.

[12] *Ibid.*

[13] Eduard Gismatullin, "BP Makes 'Giant' Oil Discovery in Gulf of Mexico." *Bloomberg,* September 2, 2009.

[14] Internal BP document (BP-HZN-OSC 7093).

[15] Internal BP document (BP-HZN-BLY 47279).

[16] Internal BP document (BP-HZN-MBI 100315).

[17] Testimony of David Sims, 123.

[18] Internal BP document (BP-HZN-MBI 190507).

[19] Testimony of David Sims, 123.

[20] Internal BP document (BP-HZN-MBI 190507).

[21] Sims, interview.

[22] Internal BP document (BP-HZN-MBI 190507).

[23] Internal Transocean document (TRN-HCJ 93526).

[24] Testimony of Brett Cocales (BP), Hearing before the Deepwater Horizon Joint Investigation Team, August 27, 2010, 11.

[25] Internal BP document (BP-HZN-MBI 177777).

[26] *Ibid.*

[27] Transocean, "Our Company," accessed November 17, 2010, http://www.deepwater.com/fw/main/Our-Company-2.html.

[28] Internal Transocean document (TRN-USCG-MMS 30428).

Chapter 4.1

[1] Such evidence includes, but is not limited to, the following: pressure and fluid data from BP's September 16, 2010 *Development Driller III* relief well intersection of the Macondo well, ongoing analysis of cement "rocks" found on the *Damon Bankston* after the blowout, and information on the internal and external pressure testing of the connections used in the Macondo well.

[2] BP and Halliburton agree that cement in the annulus did not isolate the hydrocarbon-bearing zones. Transocean has so far deferred to BP and Halliburton on this issue. Testimony of Mark Bly (BP), Hearing before the National Commission, November 8, 2010, 196-97, 326, 328-29; Testimony of Richard Vargo (Halliburton), Hearing before the National Commission, November 8, 2010, 196-97, 326, 328-29; Testimony of Bill Ambrose (Transocean), Hearing before the National Commission, November 8, 2010, 196-97, 326, 328-29; BP, Deepwater Horizon *Accident Investigation Report* (September 8, 2010), 54.

[3] Internal BP document (BP-HZN-MBI 138868).

[4] Steve Lewis (Expert witness), interview with Commission staff, September 15, 2010.

[5] John Smith, *Review of Operational Data Preceding Explosion on Deepwater Horizon in MC252* (July 1, 2010), 20; Testimony of Richard Vargo, 271; BP, Deepwater Horizon *Accident Investigation Report*, app. M, 3, 15-19.

[6] BP, Deepwater Horizon *Accident Investigation Report*, 38.

[7] *Ibid.*, 74; Confidential industry expert, interview with Commission staff.

[8] Confidential industry expert, interview.

[9] Testimony of Mark Hafle (BP), Hearing before the Deepwater Horizon Joint Investigation Team, May 28, 2010, 77; Confidential source, interview with Commission staff.

[10] John Smith (Expert witness), interview with Commission staff, September 17, 2010. Of 34 loss of well control incidents in offshore wells in U.S. federal waters from 1992 to 2002, 19 (56%) were caused by annular flows associated with the cementing process. American Petroleum Institute, *Recommended Practice 65, Part 2: Isolating Potential Flow Zones During Well Construction* (May 2010), 57 ("API RP 65 part 2"). The API has identified annular flow as a "common problem" with "grave consequences." *Ibid.*, 48. Indeed, on August 16, 2000, MMS presented safety concerns on uncontrolled annular flows to a new API Work Group on Annular Flow Prevention and Remediation. In response, the group developed two recommended practices to document industry best practices to improve zonal isolation, and help prevent annular flow incidents prior to, during, and after cementing operations, to mitigate and prevent annular flows. *Ibid.*

[11] John Martinez, William McDonald and C. Ray Williams, *Study of Cementing Practices Applied to the Shallow Casing in Offshore Wells* (October 1980), ii.

[12] For example, in the August 21, 2009 blowout at the Montara wellhead platform, hydrocarbons entered through the cemented casing shoe. David Borthwick, *Report of the Montara Commission of Inquiry* (The Montara Commission of Inquiry, Australia, June 2010), 6-7. The May 19, 2010 loss of well control incident at Gullfaks C involved a hole in the 13⅜-inch casing. Statoil, *Gullfaks C Report* (April 11, 2010), 6. Similarly, a BP injection well in Azerbaijan sustained a leak in its tubing as a result of debris in the seal of a connection; the leaking connection had been made up improperly. Steve Morey (BP), interview with Commission staff, December 23, 2010.

[13] John Guide (BP), interview with Commission staff, September 17, 2010; David McWhorter (Cameron), interview with Commission staff, August 10, 2010; Doug Blankenship (DOE), interview with Commission staff, October 26, 2010; Testimony of Richard Vargo, 274-75; Internal BP document (BP-HZN-CEC 18892). Some of the containment operations were designed with annular flow as a starting assumption. Stephen Wilson, *Macondo Radius of Failed Zone at Intercept Depth* (June 22, 2010). Indeed, BP used OLGA well flow modeling to model the flow rates for both scenarios (flow up the production casing and flow in the annulus around the production casing). Internal BP document (BP-HZN-BLY 48212).

[14] Confidential industry expert, interview; Confidential source, interview.

[15] BP, Deepwater Horizon *Accident Investigation Report*, 68-69.

[16] Testimony of Bill Ambrose, 185-86.

[17] Testimony of Richard Vargo, 187-94, 266, 270. Halliburton has declined to provide any documentation or written explanation of its theory of annular flow.

[18] Tommy Roth (Halliburton), interview with Commission staff, August 17, 2010; Testimony of Richard Vargo, 187-94, 266, 270.

[19] The Chief Counsel's team takes the three companies' conclusions with a healthy grain of salt. BP and Halliburton in particular each advance flow path explanations that happen to coincide with their own financial and litigation interests. Nevertheless, the companies' explanations and evidence in support of their theories have been helpful to the Chief Counsel's team's analysis.

[20] BP conducted several well interventions after the April 20 blowout and before the October forensic operations. These included the August static kill and the September bottom kill, where BP pumped mud and cement into the Macondo well. These operations could have affected the condition of the Macondo well annulus.

[21] Erosion is expected in light of the force and speed of hydrocarbon flow during the blowout. Testimony of Bill Ambrose, 244.

[22] Trace amounts of hydrocarbons in the annulus might be expected even where flow went through the shoe track and up the production casing, if the annular mud had been exposed to the formation and stray hydrocarbons displaced some of the mud in the annulus during flow. Blankenship, interview; David Trocquet (MMS), interview with Commission staff, October 1, 2010. Alternatively, it is possible that the annular mud was never exposed to the formation or hydrocarbon flow if, for example, the cap cement remained intact as a barrier. BP legal team, interview with Commission staff, January 12, 2011.

[23] Internal BP document (BP-HZN-NAE 2412).

[24] The BP report estimates the density of hydrocarbons as 5.18 ppg at 239 degrees Fahrenheit at 12,000 psi. BP, Deepwater Horizon *Accident Investigation Report*, app. G, 215.

[25] Internal BP document (BP-HZN-NAE 2412).

[26] This flow-out occurs because the gas compresses until the pressures balance. If the annulus is full of liquid and able to withstand the increased hydrostatic pressure of the fluid inside the production casing, there would be no flow. Steve Lewis (Expert witness), email to Commission staff, December 28, 2010.

[27] Internal BP document (BP-HZN-NAE 2412).

[28] Blankenship, interview.

[29] BP, Deepwater Horizon *Accident Investigation Report*, 56.

[30] Internal BP document (BP-HZN-NAE 2412).

[31] Blankenship, interview.

[32] Internal BP document (BP-HZN-NAE 2419).

[33] Internal BP document (BP-HZN-NAE 2426).

[34] *Ibid.*

[35] *Ibid.*

[36] Internal BP document (BP-HZN-NAE 2436).

[37] Doug Blankenship, interview. For the sake of completeness, it is worth noting that on September 22, 2010, BP had Schlumberger perform an acoustic log of the fluid in the annulus of the Macondo well from the wellhead to 9,318 feet using an isolation scanner. The results suggested there was lighter density fluid in the annulus outside of the 9⅞-inch production casing. Internal BP document (BP-HZN-NAE 2406). Given that the results of the perforation test and the sampling do not support the presence of hydrocarbons, it is likely that the log (which is an indirect measurement of annular fluid density) was erroneous.

[38] Testimony of John Sprague (BP), Hearing before the Deepwater Horizon Joint Investigation Team, December 8, 2010, part 2, 91.

[39] Testimony of Bill Ambrose, 244 "[The flow in t]his particular case, just to put it in perspective, was a 550-ton freight train hitting the rig floor. Things happened very quickly, and then it was followed by what we estimate to be a jet engine's worth of gas coming out of the rotary." *Ibid.*

[40] Internal BP document (BP-HZN-NAE 2438).

[41] Dril-Quip representatives, interview with Commission staff, October 27, 2010.

[42] *Ibid.*

[43] *Ibid.* The flow passages are approximately 1 inch in diameter. Given that diameter, the total area for flow coming through the 18 flow passages absent erosion would have been only 14 square inches. *Ibid.*

[44] Dril-Quip legal team, email to Commission staff, November 2, 2010.

[45] Blankenship, interview; Dril-Quip representatives, interview.

[46] Nonpublic BP document (presentation to Commission staff on flow rate estimates, October 22, 2010), 16; Nonpublic BP document (presentation to Commission staff on BOP, September 8, 2010), 7; Blankenship, interview; Internal BP document (BP-HZN-NAE 2432-35).

[47] Dril-Quip representatives, interview.

[48] Blankenship, interview; Dril-Quip representatives, interview. Dril-Quip has made representations to the Commission staff that the evidence is "conclusive." Dril-Quip representatives, interview.

[49] Dril-Quip representatives, interview. "It looks pretty good." Testimony of Richard Vargo, 273.

[50] Dril-Quip representatives, interview.

[51] *Ibid.*

[52] Internal BP document (BP-HZN-NAE 2388); Testimony of John Sprague, 89.

[53] Internal BP document (BP-HZN-NAE 2388); Testimony of John Sprague, 89. "The hanger was properly landed and the seal assembly was in the place it was supposed to be in." *Ibid.*

[54] Dril-Quip representatives, interview; Merrick Kelley (BP), interview with Commission staff, October 22, 2010.

[55] Kelley, interview.

[56] Internal BP document (BP-HZN-NAE 2392). "TEST 9 7/8″ PRODUCTION CASING, WELHEAD CONNECTOR & UPPER BSR. WITH 13.2 PPG SOBM. TEST TO 250 PSI LOW FOR 5 MINUTES STRAIGHT LINE ON CHART & 4,100 PSI FOR 30 MINUTES STRAIGHT LINE ON CHART. PRESSURE UP DOWN KILL LINE @ ½ BPM. TOTAL BBLS PUMPED 6, ISIP 4,270 PSI, FINAL SIP 4,158 AFTER 30 MINUTES. BLED OFF PRESSURE. RECOVERED 5.25 BBLS. TEST LOWER BSR TO 250 PSI LOW & 4,100 PSI HIGH FOR 5 MINUTES EACH STRAIGHT LINE ON CHART." *Ibid.*

[57] Benjamin Powell (BP legal team), letter to Commission staff, November 1, 2010, 2; Testimony of John Sprague, 90. "Prior to running the lockdown sleeve, we actually tested the seal assembly and hanger to 4100 PSI and got a good test, which indicated to us that the seals were intact." Testimony of John Sprague, 90.

[58] Internal BP document (BP-HZN-NAE 2398).

[59] Testimony of John Sprague, 90. "We ran the lockdown sleeve, set it, pressure tested it, I think, to 5400 PSI, which meant both the lockdown sleeve and the seal assembly and the hanger had integrity." *Ibid.*

[60] Powell, letter.

[61] Donald Godwin (Halliburton), letter to Commission staff, December 9, 2010, 1.

[62] *Ibid.* Halliburton references the relevant time frame as 00:22 to 00:36. *Ibid.* These timestamps are mere reference points on the chart contained in Halliburton's post-cement-job report, not the actual times of the relevant events during the cement job. Lewis, email, December 28, 2010; Internal Halliburton document (HAL_28543). "Graph created from Sperry's data, time of events are not correct." Internal Halliburton document (HAL_28543).

[63] Steve Lewis (Expert witness), email to Commission staff, December 12, 2010. "Well bore geometry, when related to a finite element hydrostatic balance calculation reveals a sequence of changing pressure balance over the last eighty one minutes of the displacement which parallels the trend and approximates the magnitude pump pressure response during this period.... A mathematic analysis of the hydrostatic balance of

the MC-252#1 cement displacement produces results which move in concert with the observed surface pressure." *Ibid.*

[64] *Ibid.*

[65] Internal Halliburton document (HAL_11005). The model's predicted numerical pressures do not appear to match precisely with the observed data. This may be due to the predictive nature of the model or imprecise inputs. Testimony of John Gisclair (Halliburton), Hearing before the National Academy of Engineering, September 26, 2010, 40. But the model does predict the general downward trend observed.

[66] Internal BP document (BP-HZN-MBI 129053)("Just wanted to let everyone know the cement job went well. Pressures stayed low, but we had full returns the entire job, saw 80 psi lift pressure and landed out right on the calculated volume."); Internal Halliburton document (HAL_28538)("Cement job pumped as planned."); Mark Bly (BP), interview with Commission staff, September 8, 2010; Kris Ravi (Halliburton), interview with Commission staff, September 19, 2010 (circulating pressures appear to be in the correct ranges).

[67] Testimony of Richard Vargo, 276. "All of that information has to be explained as to why the pressure drop occurred during the displacement of the cementing operation and potentially why there isn't a lot of damage here." *Ibid.*

[68] Dril-Quip representatives, interview.

[69] *Ibid.*

[70] *Ibid.*

[71] Internal BP document (BP-HZN-OSC 8844); Internal BP document (BP-HZN-OSC 8845).

[72] Internal BP document (BP-HZN-OSC 8842); Internal BP document (BP-HZN-OSC 8844).

[73] Internal BP document (BP-HZN-OSC 8842); Internal BP document (BP-HZN-OSC 8840)("The data set supports a hypothesis of a casing and drill pipe flow path."); Internal BP document (BP-HZN-OSC 8843); Internal BP document (BP-HZN-OSC 113099)("Flowpath tracker indicate that flow path was inside the casing only.").

[74] New evidence about the geometry of the wellbore continues to emerge after the static kill. For example, in October, BP obtained new information about the presence of drill pipe above the crossover joint. BP legal team, interview.

[75] The observed data showed a divergence and then a spike in surface pressure after several hundred barrels were pumped. Internal BP document (BP-HZN-OSC 8842); Internal BP document (BP-HZN-OSC 113132). Some have suggested that the spike might indicate a breach at the crossover joint. Another explanation for the spike is that mud hit the formation earlier than modeled. Blankenship, interview; John Smith (Expert witness), email to Commission staff, January 13, 2011; Internal BP document (BP-HZN-OSC 113091-93). "It was expected that the operational results would differ from these theoretical curves due to a strong out transition zone between the Macondo oil and the kill mud…. The early increase in pressure at around 500 bbls pumped is assumed to be due to a long transition zone and 'roping' of mud. The gradual increase in pressure is expected to be due to mud gradually packing off the Macondo reservoir sand face." Internal BP document (BP-HZN-OSC 113091-93). BP's offshore kill and cement team retained uncertainty: "After pumping 330 bbls of mud into the well, there was a significant deviation from the predicted BOP pressure suggesting a flow path outside of the assumptions and geometries incorporated into the diagnostic model…. Based on the actual kill data divergence from the predicted pressure schedule, multiple flow path options exist for the lower portion of the well, below the 9633.5' MD." Internal BP document (BP-HZN-OSC 113128).

[76] John Smith (Expert witness), email to Commission staff, December 26, 2010. It also appears that the modeling did not account for the volume of the shoe track or the annulus below the float collar. Steve Lewis (Expert witness), email to Commission staff, December 19, 2010.

[77] Internal BP document (Macondo well schematic).

[78] Internal BP document (BP-HZN-BLY 48255, 48259-60).

[79] Internal BP document (BP-HZN-MBI 98759).

[80] *Ibid*; Morey, interview. Connections are designed to withstand the same external pressures as the pipe body. Morey, interview.

[81] Internal BP document (BP-HZN-MBI 98759).

[82] Testimony of Lance John (Weatherford), Hearing before the Deepwater Horizon Joint Investigation Team, July 19, 2010, 243. "We have a thread rep on location which is usually the connection – whatever connection we're running there and also we have our Jam system that we monitor to make up torques and turns. Q. And did you verify that all the connections were to your standard? A. Correct. Well, right away the thread rep is there also looking at the graph as it's being – the connection's being made up." *Ibid.*

[83] Internal Weatherford documents (WFT 38, 43, 49); Steve Lewis (Expert witness), email to Commission staff, January 6, 2011. "I saw nothing in the WFD data or reports that would make me question their statement that all the joints that they made up or checked were made to specification and within normal variable limits." Lewis, email, January 6, 2011.

[84] The logs made available to the investigative team did not contain a record of the connections made up onshore. Those include the reamer shoe, the centralizer subs, the float collar, and the crossover joint. Lewis, email, January 6, 2011.

[85] Testimony of Lance John, 255. "They pulled a connection out of the box. The driller picked up on it. It pulled the connection. It's a wedge-type connection, so when it pulled it out, it sprung back in and damaged the one below it so we laid those out and replaced them with new joints.... We had – one of them was a double, so we had to lay out the double and we put a single and we replaced them with new joints.... The driller, when he slacked off into the box, slacked off too much and the pipe fell to the side and he was straightening it back up and as he was straightening it, is when he pulled it out of the box." *Ibid.*

[86] *Ibid.* That action would have resolved the issue. Morey, interview.

[87] Since the blowout, at least one BP engineer involved with the well has testified that there could have been a breach in the production casing, namely through one or more of the threaded connections between casing joints. "Every joint of casing is screwed together, and there were several joints having a thread that any of those threads could leak." Testimony of Mark Hafle, 77. Transocean's internal investigator does not consider a break in the casing above the float mechanism a possibility because he views that component as being a stronger connection than the "internal guts" of the float assembly. Bill Ambrose (Transocean), interview with Commission staff, November 2, 2010.

[88] Testimony of Nathaniel Chaisson (Halliburton), Hearing before the Deepwater Horizon Joint Investigation Team, August 24, 2010, 432-34.

[89] *Ibid.*

[90] Confidential source, interview.

[91] Testimony of Nathaniel Chaisson, 432-34 ("Phone calls were made to BP in Houston and all of those phone calls and discussions were handled between BP personnel."); Confidential source, interview; Internal BP document (BP-HZN-MBI 129068)(Guide, Morel, and Kaluza considered the possibility of a casing breach).

[92] Internal BP document (BP-HZN-MBI 129068). Clawson was unsure of the meaning of the email, but he did not have a follow-up conversation about it with Morel. Bryan Clawson (Weatherford), interview with Commission staff, October 28, 2010.

[93] Confidential source, interview.

[94] *Ibid.*

[95] After the cement job, the rig crew performed a positive pressure test on the well to test the integrity of the production casing. They pumped 2,500 psi of pressure into the production casing, which held steady for 30 minutes. The fact that the positive pressure test passed makes a casing breach unlikely but does not definitively rule out a breach. The positive pressure test assesses whether the production casing can hold pressure from the inside. It does not test whether the casing can withstand pressure exerted from the outside. The positive pressure test also did not test the casing below the top wiper plug. Finally, a casing breach could have occurred after the positive pressure test was complete.

[96] Smith, email, December 26, 2010. Certain other data that BP employs as the basis for its conclusion that there was shoe track flow is similarly insufficiently sensitive to distinguish a casing breach near the bottom of the production casing from a failure of the shoe track cement. For example, BP argues that the increase (rather than decrease) in drill pipe pressure prior to the blowout indicates that flow came up through the production casing and pushed mud up around the drill pipe. BP, Deepwater Horizon *Accident Investigation Report*, app. G, 221. Hydrocarbons entering through a breach near the bottom of the production casing could be responsible for pushing up the mud.

[97] Internal Weatherford documents (WFT 38, 43, 49). The logs made available to the investigative team did not contain a record of the connections made up onshore. Those include the reamer shoe, the centralizer subs, the float collar, and the crossover joint. Lewis, email, January 6, 2011.

Chapter 4.2

[1] Steve Lewis (Expert witness), interview with Commission staff, September 21, 2010.

[2] *Ibid.*

[3] "In the drilling business it is standard practice to always have multiple barriers in place in the wellbore at any given time.... Only when at least two mechanical barriers are in place, and sufficiently tested, can the drilling fluid be removed as a barrier." Hearing to Review Recent Issues in Offshore Oil and Gas Development, Before the S. Comm. on Energy and Natural Resources, 111th Cong. (May 11, 2010)(statement of F.E. Beck, Texas A&M University). Shell's standard "specifies that all planned well operations must normally be executed under the protection of two independent barriers between the reservoir and the environment...." Joseph Leimkuhler (Shell), letter to Commission staff, September 22, 2010. Indeed, BP's own Drilling and Well Operations Practice manual specifies that (1) "During well construction and maintenance activities, operations shall be conducted with one active barrier and one contingent barrier installed to address critical operational risks and contain the well," and (2) "Prior to breaking containment of any well control equipment such as the removal of a tree, BOP or any component, there shall be two independent mechanical barriers to flow fitted in all wells." Internal BP document (BP-HZN-MBI 130846, 130875).

[4] Gregg Walz (BP), interview with Commission staff, October 6, 2010.

[5] Internal BP document (BP-HZN-MBI 13494).

[6] David Sims (BP), interview with Commission staff, December 14, 2010.

[7] Guide described the Macondo well as a production well with an exploration tail. John Guide (BP), interview with Commission staff, September 17, 2010. So did Walz and Sims. Walz, interview; Sims, interview.

[8] Guide, interview, September 17, 2010.

[9] *Ibid.*

[10] *Ibid.*

[11] Steve Morey (BP), interview with Commission staff, December 22, 2010; Patrick O'Bryan (BP), interview with Commission staff, December 17, 2010.

[12] Testimony of Mark Hafle (BP), Hearing before the Deepwater Horizon Joint Investigation Team, May 28, 2010, 60.

[13] Sims, interview.

[14] Morey, interview.

[15] Internal BP document (BP-HZN-BLY 47277-84).

[16] Internal BP document (BP-HZN-BLY 47280-81).

[17] Technically, the 16-inch pipe is a "liner" rather than a "casing," because it hangs 160 feet below the wellhead. Casing runs all the way up to the wellhead, where it hangs from a "casing hanger." A liner does not run all the way up to the wellhead and instead hangs from a "liner hanger" placed farther down in the well. For simplicity's sake, and because individuals in the oil and gas industry often use the terms interchangeably, this Report nevertheless refers to the 16-inch pipe as a "casing."

[18] Internal BP document (BP-HZN-MBI 98759).

[19] *Ibid.*

[20] *Ibid.*

[21] *Ibid.*

[22] Internal BP document (BP-HZN-MBI 118210).

[23] It is worth noting that some refer to the production casing as a protective casing, in that it protects the production tubing during later production. Guide, interview, September 17, 2010; John Smith (Expert witness), interview with Commission staff, September 7, 2010.

[24] Testimony of Steve Lewis (Expert witness), Hearing before the National Commission, November 9, 2010, 52; Confidential industry expert, interview with Commission staff. On the relief well, 13⅝-inch casing ran all the way to the wellhead to help ensure better pressure integrity for the dynamic kill. Ronnie Sepulvado (BP), interview with Commission staff, August 20, 2010.

[25] Ronnie Sepulvado, interview, August 20, 2010; Confidential industry expert, interview.

[26] Ronnie Sepulvado, interview, August 20, 2010. From the beginning, the 16-inch casing was not supposed to be seated in the high-pressure housing of the wellhead. And the 13⅝-inch casing was not supposed to be tied back. Guide, interview, September 17, 2010.

[27] A 13⅝-inch protective casing would have eliminated the value of the burst disks in the 16-inch casing. Guide, interview, September 17, 2010. Almost every exploration well will have an intermediate long string at 13⅝ inches; Macondo was different because it had a production casing long string. Sims, interview.

[28] Testimony of Mark Bly (BP), Hearing before the National Academy of Engineering, September 26, 2010; Ronnie Sepulvado (BP), interview with Commission staff, October 26, 2010.

[29] Ronnie Sepulvado, interview, August 20, 2010. To the contrary, BP drilling engineer Brian Morel commented, in an email, that BP would "come back to run the tieback" "at a later date." Internal BP document (BP-HZN-MBI 199432).

[30] Testimony of Micah Burgess (Transocean), Hearing before the Deepwater Horizon Joint Investigation Team, May 29, 2010, 117. The *Horizon* crew encountered more problems than usual. E.C. Thomas (Expert witness), interview with Commission staff, October 27, 2010.

[31] Internal BP document (BP-HZN-OSC 1449).

[32] Internal BP document (BP-HZN-OSC 4094).

[33] *Ibid.*; Allen Seraile (Transocean), interview with Commission staff, January 7, 2011.

[34] John Guide (BP), interview with Commission staff, January 19, 2011.

[35] Internal BP document (BP-HZN-OSC 1490); Internal BP document (BP-HZN-OSC 1493).

[36] Guide, interview, September 17, 2010.

[37] Internal BP document (MC 252-1 DDR).

[38] Thomas, interview.

[39] Internal BP document (MC 252-1 DDR).

[40] Internal BP document (BP-HZN-MBI 324854).

[41] Internal BP document (BP-HZN-MBI 109949).

[42] Internal BP document (BP-HZN-MBI 197837).

[43] One BP engineer said, "We have been encountering issues on Macondo and the well design is rapidly changing." Internal BP document (BP-HZN-MBI 110317).

[44] Internal BP document (BP-HZN-MBI 109949).

[45] Guide, interview, January 19, 2010.

[46] Internal BP document (BP-HZN-CEC 8723).

[47] Internal BP document (BP-HZN-MBI 173605); Internal BP document (BP-HZN-CEC 18891).

[48] Internal BP document (BP-HZN-MBI 173605); Internal BP document (BP-HZN-CEC 18891).

[49] Internal Transocean document (TRN-USCG_MMS 11596).

[50] Internal BP document (BP-HZN-MBI 126338); Guide, interview, January 19, 2011.

[51] Internal Transocean document (TRN-USCG_MMS 11597).

[52] Internal BP document (BP-HZN-MBI 126338).

[53] *Ibid.*

[54] *Ibid.*

[55] *Ibid.*

[56] *Ibid.*

[57] Erick Cunningham (BP), interview with Commission staff, January 19, 2011. As discussed in Chapter 4.4, at this meeting the group also considered the merits of using nitrogen foamed cement at the bottom of the Macondo well.

[58] *Ibid.*

[59] *Ibid.*

[60] *Ibid.*

[61] *Ibid.*

[62] *Ibid.*

[63] Confidential source, interview with Commission staff.

[64] *Ibid.*; Walz, interview. BP engineers raised this concern with Gagliano, and Gagliano followed up with other Halliburton personnel to run down the cause of the error. Confidential source, interview.

[65] Confidential source, interview.

[66] Sims, interview; Walz, interview.

[67] Walz, interview.

[68] Internal BP document (BP-HZN-MBI 127267).

[69] *Ibid.*

[70] Internal BP document (BP-HZN-MBI 127266).

[71] Ronnie Sepulvado, interview, August 20, 2010.

[72] Internal BP document (BP-HZN-MBI 255906).

[73] Testimony of Paul Johnson (Transocean), Hearing before the Deepwater Horizon Joint Investigation Team, August 23, 2010, 360.

[74] *Ibid.*

[75] Cunningham, interview.

[76] Confidential source, interview.

[77] It is not clear whether an accurate Halliburton cementing model—with all the correct inputs—was ever run. The engineers turned off the faulty part of the model, manually input logging data, and reran the model. ECD levels went down but were still too high. Cunningham suggested adding a small amount of base oil in front of the spacer fluid in order to lower ECD. Cunningham, interview; Confidential source, interview.

[78] The completion engineers do not appear to have been aware of the final decision at that point. "[N]o one told us what the actual decision was, so we thought y'all were going with the liner until Doris called me for a green light on the MoC." Internal BP document (BP-HZN-MBI 254886).

[79] Internal BP document (BP-HZN-BLY 125443).

[80] Walz, interview.

[81] BP, Deepwater Horizon *Accident Investigation Report* (September 8, 2010), 76.

[82] Testimony of John Guide (BP), Hearing before the Deepwater Horizon Joint Investigation Team, July 22, 2010, 183.

[83] Internal BP document (BP-HZN-MBI 143292).

[84] Testimony of Mark Bly, 306. There was some uncertainty among the team members as to whether a formal management of change process was even required because the original well design had included a long string production casing, the original well design had already been peer-reviewed, and the team had never actually switched the plan in the interim to a liner. Guide, interview, September 17, 2010; Sims, interview. Nevertheless, Sims suggested that the team use the process as a vehicle to capture and record the issues discussed in the previous few days (April 13 to 15) because there had been a great deal of conversation about different options. Sims, interview.

[85] Internal BP document (BP-HZN-MBI 143259).

[86] *Ibid.*

[87] *Ibid.*; Internal BP document (BP-HZN-MBI 143292).

[88] BP, Deepwater Horizon *Accident Investigation Report*, app. O, 1-2; Testimony of Charlie Williams (Shell), Hearing before the National Commission, November 9, 2010, 40; Confidential industry expert, interview.

[89] Anadarko representatives assert, after the blowout, that they would not have installed a long string in the Macondo well, given the cement job. Anadarko representatives, interview with Commission staff, September 29, 2010. The risk of using a long string depends on how reliable the cement job is. According to one industry expert, the cement job here should not have been relied on because of inadequate rock strength, repeated incidents of lost circulation, slow pump rates, small cement volume, and wiper plugs in tapered casing. Confidential industry expert, interview. "If you're concerned that you will lose returns or might lose returns when you're cementing, running the liner is in my view more straightforward on reestablishing the barriers." Testimony of Charlie Williams, 42-43.

[90] Bill Ambrose (Transocean), interview with Commission staff, September 21, 2010.

[91] Confidential industry expert, interview; Bill Ambrose (Transocean), interview with Commission staff, November 2, 2010; Testimony of Tommy Roth (Halliburton), Hearing before the National Academy of Engineering, September 26, 2010; Confidential source, interview with Commission staff. To be sure, industry engineers usually wipe long string production casings with two wiper plugs, whereas it is common practice to wipe a liner with only one wiper plug. In that case, the difference in contamination would be reduced.

[92] Erik B. Nelson and Dominique Guillot, eds., *Well Cementing*, 2nd ed. (Sugar Land, TX: Schlumberger, 2006), 497.

[93] Confidential industry expert, interview. Another reason for this may be wellbore geometry. Mark Bly (BP), interview with Commission staff, September 8, 2010.

[94] Murry Sepulvado (BP), interview with Commission staff, December 10, 2010.

[95] Confidential industry expert, interview.

[96] Testimony of Charlie Williams, 42; Confidential industry expert, interview.

[97] Testimony of Charlie Williams, 42.

[98] Kris Ravi (Halliburton), interview with Commission staff, September 19, 2010; Ambrose, interview, November 2, 2010.

[99] Patrick O'Bryan, interview.

[100] Internal BP document (BP-HZN-OSC 5558). Nonproductive time counts against the performance rating of the rig and its personnel. Internal BP document (BP-HZN-OSC 8995); Internal BP document (BP-HZN-MBI 117622).

[101] Paul Tooms (BP), interview with Commission staff, October 13, 2010; Internal Department of the Interior document (OSC-DWH BOEM-WDC-B06 0001-0007), 133-34.

[102] Internal Department of the Interior document (OSC-DWH BOEM-WDC-B06 0001-0007), 133-34.

[103] Internal Department of Energy document (presentation on mud flow, June 1, 2010); Internal Department of the Interior document (OSC-DWH BOEM-WDC-B06 0001-0007), 133.

[104] Coast Guard official, interview with Commission staff; MMS official, interview with Commission staff; Senior administration official, interview with Commission staff.

[105] Tooms, interview.

[106] Coast Guard official, interview; MMS official, interview; Senior administration official, interview.

[107] Government science advisor, interview with Commission staff; Senior administration official interview; Internal Department of Energy document (presentation on mud flow).

[108] Senior administration official, interview.

[109] Internal Department of the Interior document (OSC-DWH BOEM-WDC-B06 0001-0007), 166-68; Nonpublic BP document (presentation to the Department of Energy, May 25, 2010), 3-5; BP legal team, letter to Commission staff, November 2, 2010, att.

[110] Internal Department of the Interior document (OSC-DWH BOEM-WDC-B06 0001-0007), 135; Doug Suttles (BP), interview with Commission staff, October 13, 2010. One government official told Commission staff that another concern with installing a second BOP was weight: The existing BOP stack was listing at 2 degrees from vertical, and there was a risk that adding weight to its top would lead to further damage or collapse. Senior administration official, interview.

[111] BP, letter.

[112] BP, letter; Richard Lynch (BP), interview with Commission staff, October 13, 2010; Tooms, interview.

[113] Government science advisor, interview with Commission staff.

[114] Government science advisor, interview with Commission staff.

[115] Confidential industry expert, interview.

[116] O'Bryan, interview.

[117] *Ibid.*; Guide, interview, September 17, 2010.

[118] Guide, interview, September 17, 2010.

[119] Morey, interview.

[120] Guide, interview, September 17, 2010; Mike Zanghi (BP), interview with Commission staff, December 15, 2010.

[121] O'Bryan, interview; Internal BP document (BP-HZN-OSC 8007).

[122] Morey, interview.

[123] Walz, interview.

[124] BP, Deepwater Horizon *Accident Investigation Report*, 75.

[125] Confidential industry expert, interview.

Chapter 4.3

[1] Erik Nelson and Dominique Guillot, eds., *Well Cementing*, 2nd ed. (Sugar Land, TX; Schlumberger, 2006), 143, 445.

[2] *Ibid, Well Cementing*, 148.

[3] American Petroleum Institute, *Recommended Practices for Isolating Potential Flow Zones in Well Construction*, 1st ed. (May 2010)("API RP 65 - Part 2 § 4.8.2.1").

[4] Norman Hyne, *Nontechnical Guide to Petroleum Geology, Exploration, Drilling, and Production* (Tulsa, Oklahoma; PennWell, 2004), 297-328.

[5] Nelson and Guillot, eds., *Well Cementing*, 445.

[6] *Ibid*, 445. American Petroleum Institute, *Technical Report on Selection of Centralizers for Primary Cementing Operations* , 1st ed. (May 2008)("API 10TR4 § 2.1"). Poor centralization can also lead to gas flow during the cement job. Gas flow may occur as the cement begins to set. As the cement gels, it no longer transmits the full amount of hydrostatic pressure from the fluids above it in the well. This can allow gas to flow into the cement, weakening it. Halliburton legal team, interview with Commission staff, July 27, 2010. While inadequate centralization can cause both channeling and gas flow, the two effects are different.

Halliburton legal team, interview with Commission staff, August 17, 2010. Testimony of Jesse Gagliano (Halliburton), Hearing before the Deepwater Horizon Joint Investigation Team, August 24, 2010, 416-17.

[7] Nelson and Guillot, eds., *Well Cementing*, 188.

[8] This is a simplified classification of centralizers for the purposes of this section. The API lists seven different types of centralizers falling into three main categories. One expert categorized centralizers as being of two types, either bow spring or rigid, each of which could be placed in two ways, either as a "sub" screw-on or "slip-on" that is secured by stop collars. Steve Lewis (Expert witness), email to Commission staff, October 14, 2010.

[9] Nelson and Guillot, eds., *Well Cementing*, 137.

[10] *Ibid.*, 164.

[11] *Ibid.*, 445.

[12] *Ibid.*, 446.

[13] Testimony of Daniel Oldfather (Weatherford), Hearing before the Deepwater Horizon Joint Investigation Team, October 7, 2010, 44-45.

[14] Testimony of Steve Lewis (Expert witness), Hearing before the National Commission, November 9, 2010, 112-13.

[15] BP, Deepwater Horizon *Accident Investigation Report* (September 8, 2010), 70.

[16] Testimony of Ronnie Sepulvado (BP), Hearing before the Deepwater Horizon Joint Investigation Team, July 20, 2010, 72.

[17] API RP 65 - Part 2 § 4.8.4.

[18] Nelson and Guillot, eds., *Well Cementing*, 148.

[19] The use of an auto-fill tube permits mud to flow up the casing as the casing is lowered. This means "old mud" is also inside of the casing, and not only in the annular space. If the crew circulates bottoms up after converting the float equipment, the old mud that was inside the casing is forced into the annular space. In order to completely purge the system to entirely "new mud," inside the casing and in the annular space, it is necessary to pump new mud "surface to surface." That is, pump a volume of new mud equal to the volume of the casing plus the volume of the annulus.

[20] Nelson and Guillot, eds., *Well Cementing*, 1.

[21] American Petroleum Institute, *Technical Report on Cement Sheath Evaluation*, 2nd ed. (September 2008), 1 ("API 10TR § 2").

[22] API 10TR § 2, 1.

[23] Testimony of Ronnie Sepulvado, 112.

[24] *Ibid.*, 112-13.

[25] John Guide (BP), interview with Commission staff, September 17, 2010; Confidential source, interview with Commission staff.

[26] Lift pressure calculations can provide a rough estimate of the top of cement.

[27] Electric line logs are a type of wire line logs which conduct data to the surface as they are run. Tubing conveyed logs may also be used, which are logs run on drill pipe, tubing work string, or coiled tubing.

[28] "Cement Bond Log," Schlumberger Oilfield Glossary, http://www.glossary.oilfield.slb.com/Display.cfm?Term=cement%20bond%20log.

[29] Nelson and Guillot, eds., *Well Cementing*, 561-63.

[30] *Ibid.*

[31] David Johnson and Kathryne Pile, *Well Logging in Non-technical Language*, 2nd ed. (Tulsa, OK; PennWell, 2004), 196.

[32] Nelson and Guillot, eds., *Well Cementing*, 562.

[33] *Ibid.*, 583-85.

[34] "Cement Bond Log," Schlumberger Oilfield Glossary, http://www.glossary.oilfield.slb.com/Display.cfm?Term=cement%20bond%20log.

[35] Nelson and Guillot, eds., *Well Cementing*, 583.

[36] API 10TR § 10, 75-77.

[37] See, e.g., Nelson and Guillot, *Well Cementing*, 7.

[38] Confidential industry expert, interview with Commission staff.

[39] Testimony of Richard Vargo (Halliburton), Hearing before the National Commission, November 8, 2010, 320.

[40] Nelson and Guillot, eds., *Well Cementing*, 24-25.

[41] American Petroleum Institute, *Recommended Practice for Testing Well Cements 10B-2* (2005)(API RP 10B - Part 2").

[42] Nelson and Guillot, eds., *Well Cementing*, 299-300.

[43] Internal Transocean document (TRN-USCG-MMS 11597)(April 9, 2010).

[44] Internal BP document (BP-HZN-MBI 126338).

[45] *Ibid.*

[46] Internal BP document (BP-HZN-MBI 198602).

[47] Brett Clanton, "New tactic might seal leaking well sooner, BP CEO says," *Houston Chronicle*, May 5, 2010.

[48] Internal Schlumberger document (SLB-EC 1-2); Internal BP document (BP-HZN-MBI 136824)(April 10, 2010); Internal BP document (BP-HZN-MBI 136829)(April 11, 2010); Internal BP document (BP-HZN-MBI 136833)(April 12, 2010); Internal BP document (BP-HZN-MBI 136837)(April 13, 2010); Internal BP document (BP-HZN-MBI 136841)(April 14, 2010); Internal BP document (BP-HZN-MBI 136845)(April 15, 2010). Schlumberger ran a "triple-combo" log that measured formation density, porosity and resistivity, natural gamma radiation, hole size, and fluid temperature; a combinable magnetic resonance (CMR) tool that could estimate the distribution of pore sizes, an elemental capture spectroscopy (ECS) sonde to measure formation lithology, an oil-base microimager (OBMI) tool to visualize the formation, a modular formation dynamics tester (MDT) tool to collect samples and make pressure measurements, and coring tools to take sidewall samples.

[49] Internal Transocean document (TRN-USCG_MMS 11626)(April 16, 2010).

[50] Testimony of John Guide (BP), Hearing before the Deepwater Horizon Joint Investigation Team, July 22, 2010, 150.

[51] Internal BP document (BP-HZN-MBI 129617).

[52] Confidential source, interview with Commission staff.

[53] Internal BP document (BP-HZN-MBI 21304).

[54] Chapter 4.4 explains that BP may originally have chosen to use nitrogen foamed cement for other reasons.

[55] Internal BP document (BP-HZN-MBI 193550).

[56] *Ibid.*

[57] Guide, interview, September 17, 2010.

[58] A number of BP personnel have suggested or stated that APB was a factor in determining TOC at Macondo. David Sims (BP), interview with Commission staff, December 14, 2010; Guide, interview, September 17 2010; Confidential source, interview; Erick Cunningham (BP), interview with Commission staff, January 19, 2011. BP's technical guidance also states that APB may be a factor in determining TOC. Internal BP document (BP-HZN-OSC 8007). However, according to Gregg Walz, the engineering team leader, APB concerns did not drive TOC or cement volume decisions at Macondo. Gregory Walz (BP), interview with Commission staff, October 6, 2010. This view is supported by answers Morel received to questions about APB concerns. Limiting TOC was not given as a method to mitigate APB. Internal BP document (BP-HZN-BLY 66793).

[59] BP, Deepwater Horizon *Accident Investigation Report*, 54.

[60] Internal BP document (BP-HZN-MBI 143300). The Bly report states that the top hydrocarbon zone was 17,788 feet. BP, Deepwater Horizon *Accident Investigation Report*, 54.

[61] Internal BP document (BP-HZN-CEC 20234).

[62] Guide, interview, September 17, 2010.

[63] BP's April 14 Application for Permit to Drill shows TOC at 17,300. Internal BP document (BP-HZN-MBI 23746). The Bly report states that the TOC was at 17,260 feet. BP, Deepwater Horizon *Accident Investigation Report*, 54.

[64] Internal BP document (BP-HZN-MBI 143259).

[65] *Ibid.*

[66] *Ibid.*

[67] *Ibid.*

[68] Internal BP document (BP-HZN-MBI 143291). In that comment, Sims also told Hafle: "Need to change the reviewers to Walz, Reiter, Mix and the approvers to Sims, Sprague, Frazelle."

[69] Internal BP document (BP-HZN-MBI 143292).

[70] Internal BP document (BP-HZN-CEC 22667). BP made this decision out of concern that cement above the float collar might interfere with logging.

[71] Internal Halliburton document (HAL_11002).

[72] Internal BP document (BP-HZN-MBI 143295); Internal BP document (BP-HZN-CEC 22663); BP, Deepwater Horizon *Accident Investigation Report*, 34.

[73] API RP 65 - Part 2 § 4.6.5.2.

[74] API RP 65 - Part 2 § 4.6.5.2.

[75] Internal BP document (BP-HZN-MBI 127537-39); Internal Halliburton document (HAL_11196).

[76] The September 2009 and January 2010 drilling programs called for a **minimum** of 5 bpm. Internal BP document (BP-HZN-CEC 8860)(September 2009 Drilling Program); Internal Transocean document (TRN-HCJ 93593)(January 2010 Drilling Program). Later versions reduced the displacement rate. The April 12 drilling plan called for a minimum of 3 bpm. Internal BP document (BP-HZN-CEC 21265). The April 15 drilling plan called for pump rates of 2 to 4 bpm. Internal BP document (BP-HZN-CEC 17626-27).

[77] Gagliano claimed that Morel made the decision to use foamed cement, though there is conflicting information on this point. Internal BP documents (BP-HZN-BLY 61276); Internal BP document (BP-HZN-BLY 61294); Confidential source, interview.

[78] Internal BP document (BP-HZN-MBI 218202).

[79] *Ibid.*

[80] API 10TR4 § 1; American Petroleum Institute, *Recommended Practice for Centralizer Placement and Stop-collar Testing*, Part 2, 1st ed. (August 2004)("API 10D2 § 4"). API does set forth a formula to calculate standoff ratios, even though it has no specific recommended ratio. API did have a "minimum standard" ratio of 67% standoff for performance of casing bow spring centralizers. But that 67% ratio is simply a means to help manufacturers produce API-quality centralizers. No centralization standoff ratio standards, or a standard for the distance of centralized casing, were found in API RP 10D-2, API Technical Report 10TR4e1 or the post-blowout Recommend Practice 65.

[81] Internal BP document (BP-HZN-MBI 193557). BP cementing expert Cunningham explained that a "distinct permeable zone" could be hydrocarbons or water. For Halliburton, Halliburton engineer Gagliano stated that "you view a percentage standoff and usually shoot for about 70 or above for standoff." Jesse Gagliano (Halliburton), interview with the U.S. House of Representatives Committee on Energy and Commerce, June 11, 2010.

[82] Cunningham, interview.

[83] Internal BP document (BP-HZN-CEC 8858). BP actually set forth three different formulas in this document, but all result in the running of at least 16 centralizers at the eventual total depth of the well. Presumably because every joint to 100 feet above the highest "permeable zone" would be centralized,

modeling would reflect that this was sufficiently "centralized pipe." Internal BP documents (BP-HZN-CEC 8857-58, 8863).

[84] Internal Transocean document (TRN-HCJ 93591).

[85] Internal BP document (BP-HZN-MBI 193557). ETP calls for "centralised pipe" 100 feet above permeable zone. The original 2009 plan called for centralizers every joint to 500 feet above production interval. As 500 feet above the production interval also happens to be 100 feet above the highest permeable zone, the 2009 plan appears to comply with BP's internal guidance. The January 2010 plan, however, calls for centralizers "every other joint" after the first five joints (and the first five joints do not reach 100 feet above the highest permeable zone). If centralizers "every other joint" reflected sufficient standoff in modeling, this would comply. If centralizers "every other joint" did not show sufficient standoff, it would not comply.

[86] Internal BP document (BP-HZN-BLY 66450-51).

[87] *Ibid.*; Bryan Clawson (Weatherford), interview with Commission staff, October 28, 2010. Clawson told the Chief Counsel's team that manufacturing normally takes two to three weeks and that rush jobs can be completed in just seven to 10 days.

[88] Confidential source, interview; Internal BP document (BP-HZN-MBI 126184)(April 12, 2010). No basis was given for this decision. A later cement model was prepared that supported this decision, but this was not available to the BP team when they decided six centralizers would be sufficient. Internal BP document (BP-HZN-MBI 126815). Wells team leader Guide stated that the decision that six centralizers were sufficient was made by Morel or Hafle. John Guide, interview with Commission staff, January 19, 2011.

[89] Testimony of Brett Cocales (BP), Hearing before the Deepwater Horizon Joint Investigation Team, August 27, 2010, 89.

[90] Internal BP document (BP-HZN-MBI 126815-34); BP, Deepwater Horizon *Accident Investigation Report*, 62.

[91] Internal BP document (BP-HZN-MBI 126815-34); BP, Deepwater Horizon *Accident Investigation Report*, 62.

[92] Gagliano, interview.

[93] *Ibid.*

[94] *Ibid.*; Testimony of Jesse Gagliano, 253; Internal BP document (BP-HZN-MBI 127363).

[95] Cunningham, interview.

[96] Testimony of John Guide, July 22, 2010, 87.

[97] Internal BP document (BP-HZN-CEC 22433). Walz emailed Guide to tell him of the decision made in his absence.

[98] *Ibid.* This is what Walz meant when he wrote of his decision, "we needed to be consistent with honoring the model." Testimony of Gregory Walz (BP), Hearing before the Deepwater Horizon Joint Investigation Team, October 7, 2010, part 1, 164.

[99] Testimony of Jesse Gagliano, 129-30; Internal BP document (BP-HZN-BLY 61327).

[100] Internal BP document (BP-HZN-MBI 127363).

[101] Internal Halliburton document (HAL_10648); Gagliano, interview.

[102] Internal BP document (BP-HZN-BLY 61327); Internal Halliburton document (HAL_10604); Internal Halliburton document (HAL_10608); Internal Halliburton document (HAL_10713); Internal Halliburton document (HAL_10717). Besides assuming different numbers of centralizers, the two modeling runs differed in other ways as well; they assumed different standoff values above the centralized casing section and used wellbore survey data in different ways.

[103] Testimony of Brett Cocales, 267; Gagliano, interview. Cocales was not given instructions regarding what type of centralizers to order for the production casing. In a series of phone calls, Clawson told Cocales that that there were 32 centralizers (Cocales remembered 31) that were in BP's inventory. Clawson also told Cocales that these were bow spring centralizers with stop collars. Clawson, interview; Testimony of Brett Cocales, 267. Gagliano's testimony suggests that his cement models drove the decision to specifically use *15* additional centralizers, rather than the capacity of the helicopter. Gagliano, interview. However, Walz stated that he asked Gagliano to run model with 15 additional centralizers after having determined what could be procured. Internal BP document (BP-HZN-BLY 61327).

[104] Testimony of Brett Cocales, 216.

[105] Many BP personnel believed that the Thunder Horse centralizers had integrated stop collars. Guide, interview, September 17, 2010; Confidential source, interview; Internal BP document (BP-HZN-CEC 22433); BP, Deepwater Horizon *Accident Investigation Report*, 63.

[106] Internal BP document (BP-HZN-MBI 127478-80). Gregg Walz stated that he and David Sims looked at the schematics and that both interpreted them as having integrated stop collars. Internal BP document (BP-HZN-BLY 61328).

[107] Internal BP document (BP-HZN-CEC 22433-34).

[108] Testimony of Daniel Oldfather, 6-7.

[109] *Ibid.*, 6-9, 11. Because the BP team thought that centralizers had an integrated design, they did not realize that a box that Weatherford had delivered with the centralizers contained the stop collars. Instead they assumed the box contained equipment and tools. For this reason the team did not consider it a problem that the box was shipped by boat, even though shipping the stop collars by this slower method defeated the purpose of flying the centralizers by helicopter. Internal BP document (BP-HZN-BLY 61328). The accessories' arrival was delayed and did not arrive at the rig until just before running casing. Internal BP document (BP-HZN-CEC 20268).

[110] Testimony of Daniel Oldfather, 9-10; Clawson, interview.

[111] Confidential source, interview.

[112] *Ibid.*

[113] Internal BP document (BP-HZN-MBI 255668).

[114] *Ibid.*

[115] Internal BP document (BP-HZN-CEC 22433).

[116] BP, Deepwater Horizon *Accident Investigation Report*, 63. Guide was concerned about this scenario because it recently happened on another well in the Gulf of Mexico. Testimony of John Guide, July 22, 2010, 373-74. Guide's concern was not unique. The API indicates that "throughout the industry, concerns have been expressed regarding every type of centralizer installation." API 10TR4 § 3.9.

[117] In explaining the decision not to run additional centralizers, Cocales stated that there was concern over the cost involved. He stated to BP investigators that in addition to concern about the centralizers coming off in the well, the centralizers were not installed "due to time required and labor involved." Internal BP document (BP-HZN-BLY 61225). Hafle stated that the installation would have taken even longer, from 12 to 18 hours. Internal BP document (BP-HZN-BLY 61368). However, Walz told investigators that "time or costs were not discussed as a consideration" in a phone call with Guide following the order to cancel the installation of additional centralizers. Internal BP document (BP-HZN-BLY 61328).

[118] Internal BP document (BP-HZN-BLY 61328-29). Guide said that earlier that morning, when everyone believed that the centralizers had integrated stop collars but no one was certain when they would arrive, he stated that the rig crew could stop operations to wait for their arrival. But Guide was unwilling to wait when he learned they had separate stop collars. He explained that the hole can deteriorate while waiting to run casing. Given that the hole had already been open for several days of logging, however, it is unclear how urgent it was to run the production casing.

[119] Testimony of John Guide, July 22, 2010, 312; Internal BP document (BP-HZN-CEC 22433); Internal BP document (BP-HZN-CEC 22666). According to Walz, this type of centralizer had only been used twice before. On one of those two occasions, the rig crew had problems as they were running the casing into the hole. Internal BP document (BP-HZN-MBI 61328); BP, Deepwater Horizon *Accident Investigation Report*, 63.

[120] Internal BP document (BP-HZN-MBI 127087).

[121] Internal Halliburton document (HAL_10648).

[122] Internal BP document (BP-HZN-MBI 128316-17).

[123] Confidential source. Morel may have used the 3-D model to make his centralizer placements, but it appears he did not receive it until after he had made his final centralizer adjustments to the casing tally. The reference to the 3-D model in Morel's correspondence suggests it was only used to argue that 21 centralizers

were unnecessary in a straight hole. Internal BP document (BP-HZNMBI 128430-35); Internal BP document (BP-HZN-MBI 128316).

[124] Internal Halliburton document (HAL_10649); Internal Halliburton document (HAL_10713); Internal BP document (BP-HZN-MBI 127087); Internal BP document (BP-HZN-MBI 128316).

[125] Internal BP document (BP-HZN-CEC 22669).

[126] Testimony of Brett Cocales, 59-60; Testimony of John Guide, October 7, 2010, part 2, 111-12; Testimony of Jesse Gagliano, 264, 284; Gagliano, interview.

[127] Internal BP document (BP-HZN-CEC 22670).

[128] Transocean was not involved in the determination of how many centralizers to use. Testimony of Jimmy Harrell (Transocean), Hearing before the Deepwater Horizon Joint Investigation Team, May 27, 2010, 95.

[129] Gagliano, interview; Internal BP document (BP-HZN-MBI 128489).

[130] Internal BP document (BP-HZN-MBI 128708); BP, Deepwater Horizon *Accident Investigation Report*, 64; Testimony of Richard Vargo, 322.

[131] Testimony of Gregory Walz, 53-54. Walz also stated that he "has a vague recollection that Jesse Gagliano made an off-hand comment about gas flow potential" during that conversation. Internal BP document (BP-HZN-BLY 61332).

[132] Testimony of Greg Walz, 53-54. Walz stated that his discussions with Guide that morning centered around the increased ECD, channeling, the potential for lost circulation and the contingent use of a CBL in that event. There was a brief discussion of the "SEVERE" gas flow potential but that they believed the foamed cement slurry design would protect against gas migration. Internal BP document (BP-HZN-BLY 61332).

[133] Testimony of Jesse Gagliano, 331.

[134] BP, Deepwater Horizon *Accident Investigation Report*, 64; Internal Halliburton document (HAL_10990).

[135] Internal BP document (BP-HZN-MBI 129221); Internal Halliburton document (HAL_11003).

[136] BP admits this. BP, Deepwater Horizon *Accident Investigation Report*, 63.

[137] Internal Halliburton document (HAL_11010); Gagliano, interview.

[138] Internal BP document (BP-HZN-MBI 136936)(April 18); Internal BP document (BP-HZN-MBI 136946).

[139] Testimony of Daniel Oldfather, 45.

[140] Internal BP document (BP-HZN-MBI 198504).

[141] Guide, interview, September 17, 2010.

[142] *Ibid.*

[143] Internal BP document (BP-HZN-MBI 129226).

[144] Internal BP document (BP-HZN-BLY 62079).

[145] Internal BP document (BP-HZN-MBI 137743).

[146] Testimony of Steve Lewis, 91.

[147] Calculation based on Internal BP document (BP-HZN-MBI 129229).

[148] Internal BP document (BP-HZN-MBI 136941). BP's Engineering Technical Practices call for rig personnel to convert the float equipment before the casing reaches the hydrocarbon zones. Internal BP document (BP-HZN-MBI 130799, 130846). However, BP requested a dispensation from MMS from this step in their January 2010 well plan, citing the need to reduce surge pressure given the narrow pore pressure and fracture gradient at Macondo. Internal Transocean document (TRN-HCJ 93610).

[149] Internal BP document (BP-HZN-MBI 136941).

[150] *Ibid.*

[151] *Ibid.*

[152] Internal BP document (BP-HZN-MBI 128966).

[153] Clawson, interview. According to John Guide, BP wanted to confirm the rated pressure of the float collar and check the compressibility of the mud to ensure they were in fact pressuring up against the float collar and not another piece of equipment. They verified the compressibility of the mud, which was consistent with pressures expected when pressuring up against the float collar. Testimony of John Guide, July 22, 2010, 197.

[154] Clawson, interview. Independent experts understand this recommendation to mean 6,800 psi was the highest differential pressure that could safely be applied to the equipment.

[155] Internal BP document (BP-HZN-MBI 21330). If the ports on the auto-fill tube are completely clogged such that there is no circulation, the pressure differential can build up to levels that will cause the brass pins in the collar to fail, pushing the ball through the end of the float equipment. Clawson indicated this would occur if a pressure differential of 1,300 psi existed.

[156] Internal BP document (BP-HZN-MBI 21330).

[157] Internal BP document (BP-HZN-MBI 136941).

[158] Internal BP document (BP-HZN-BLY 61364). Kaluza said Guide gave permission to go to 3,000 psi. Internal BP document (BP-HZN-MBI 129616). Testimony is inconsistent as to who from the rig was calling Guide to discuss problems converting. Both Morel and Kaluza stated they called Guide. However, Guide testified he spoke with Morel regarding float conversion. Testimony of John Guide, July 22, 2010, 197.

[159] Internal BP document (BP-HZN-MBI 129068).

[160] *Ibid.*

[161] Testimony of Nathaniel Chaisson (BP), Hearing before the Deepwater Horizon Joint Investigation Team, August 24, 2010, 432. In addition, Gagliano was informed of the multiple conversion attempts, stating that "I guess the concern was after the fact that it took so much pressure to convert the float that they may have broke something else down or something else happened." Gagliano, interview.

[162] Internal BP document (BP-HZN-MBI 257031).

[163] Testimony of Jesse Gagliano, 287.

[164] Internal BP document (BP-HZN-MBI 137367).

[165] Halliburton cementer Nathaniel Chaisson testified before the U.S. Coast Guard/BOEMRE Joint Investigation "concern was shown by many people on the rig floor." Testimony of Nathaniel Chaisson, 432.

[166] Confidential source, interview.

[167] *Ibid.*

[168] Testimony of Brett Cocales, 70-71. Maxie Doyle of M-I SWACO reviewed the modeling with John Guide and could not estimate pressures as low as those seen on the drill pipe. Doyle later emailed Brett Cocales, John Guide, Mark Hafle, Brian Morel, and Gregg Walz and informed them "I have had several individuals double check and critique[] my inputs and still cannot explain the difference." Internal BP document (BP-HZN-MBI 129104).

[169] Testimony of Brett Cocales, 72; Internal BP document (BP-HZN-MBI 137367); Internal BP document (BP-HZN-MBI 136941).

[170] Internal BP document (BP-HZN-MBI 136941); Internal BP document (BP-HZN-BLY 61355).

[171] Internal BP document (BP-HZN-MBI 137367).

[172] Internal BP document (BP-HZN-MBI 137367).

[173] *Ibid.* Another source indicated 390 psi. Internal BP document (BP-HZN-MBI 136941); Internal BP document (BP-HZN-MBI 129616).

[174] Internal BP document (BP-HZN-MBI 137367).

[175] *Ibid.*

[176] Internal BP document (BP-HZN-MBI 129221). MMS approved a regulatory dispensation that modified the standard testing regime for the diverter sealing element at Macondo. Internal BP document (BP-HZN-

OSC 512). The Chief Counsel's team has not found evidence suggesting that this dispensation affected the ability of the diverter sealing element.

[177] Internal BP document (BP-HZN-MBI 21304). The rig crew closed the lower annular and pumped down the drill pipe, taking returns up the choke and kill lines. They then shut off the pumps and monitored the flow line for losses. The rig crew determined the diverter was closed. Internal BP document (BP-HZN-OSC 4027); Internal BP document (BP-HZN-MBI 136942); Internal BP document (BP-HZN-MBI 21304).

[178] Testimony of Nathaniel Chaisson, 432. Morel was similarly unsure how they established circulation. Internal BP document (BP-HZN-MBI 129068).

[179] Confidential source, interview. The engineers thought upcoming pressure tests after cementing would confirm the integrity of the system. During a positive pressure test, additional fluid is pumped into the well, which is then sealed and monitored for pressure changes. Constant pressure confirms the casing, seal assembly, and BOP do not contain holes or leaks and can therefore hold pressure. But a positive pressure test does not confirm the integrity of the casing below the top wiper plug. On April 20 between 10:30 a.m. and noon, the crew conducted two positive pressure tests, both of which indicated the well was holding pressure. A negative pressure test similarly tests the well system for leaks, but the negative pressure test at Macondo did not indicate well integrity.

[180] Testimony of Brett Cocales, 72-73.

[181] Confidential source, interview.

[182] Internal BP document (BP-HZN-MBI 129616).

[183] API, *Recommended Practices for Isolated Potential Flow Zones During Well Construction,* part 2, 1st ed. (May 2010), 65 ("API RP 65 § 4.8.4"). The relevant portion of the second edition (§ 5.8.7) is not as prescriptive.

[184] Gagliano, interview.

[185] Steve Lewis (Expert witness), email to Commission staff, November 19, 2010. The precise calculation is 2,757.2 barrels.

[186] Testimony of Jesse Gagliano, 341-342. The actual pre-cement circulation time took roughly half an hour. *Ibid.*, 343.

[187] Internal BP document (BP-HZN-CEC 8860); Internal Transocean document (TRN-HCJ 93593).

[188] Internal BP document (BP-HZN-CEC 21445).

[189] Internal BP document (BP-HZN-CEC 21265). This pipe capacity equals 883.82 barrels. Internal BP document (BP-HZN-CEC 21445).

[190] Internal BP document (BP-HZN-CEC 17626).

[191] Testimony of John Guide, July 22, 2010, 150.

[192] BP legal team, interview with Commission staff, August 9, 2010.

[193] Testimony of Gregory Walz, 10-11.

[194] *Ibid.*, 11.

[195] Testimony of John Guide, July 22, 2010, 150; Guide, interview, September 17, 2010.

[196] Testimony of Nathaniel Chaisson, 437-38.

[197] Internal BP document (BP-HZN-CEC 21448). This document was prepared on April 17 by BP well site leaders Don Vidrine, Bob Kaluza, BP engineer Brian Morel, Halliburton cementer Nathaniel Chaisson, Halliburton foam team leader Paul Anderson, and Terry Leblue of Blackhawk. Testimony of Nathaniel Chaisson, 436.

[198] Testimony of Nathaniel Chaisson, 436.

[199] Sperry-Sun data, April 19, 2010, 16:15 – 16:30. When this circulation was established, the rig crew believed the float equipment had converted.

[200] Internal BP document (BP-HZN-CEC 21448). Halliburton concluded 346 barrels were circulated prior to cementing. Sperry-Sun data, April 20, 2010, 19:20. An independent expert calculated a circulation of

approximately 352.25 bbl; 350 bbl represents the approximate average, 349.125, of these two figures. Steve Lewis (Expert witness), email to Commission staff, November 17, 2010.

[201] Internal BP document (BP-HZN-CEC 21448).

[202] This volume is also significantly lower than the 4,140 bbl, or 1.5 annular volume, recommended by API. The Chief Counsel's team disagrees with BP witness testimony indicating the volume pumped at Macondo is in accord with the API RP 65. Testimony of Brett Cocales, 209.

[203] Internal Halliburton document (HAL_11196-97).

[204] Commission calculation based on internal Halliburton document (HAL_11196-97).

[205] Commission calculation based on internal Halliburton document (HAL_10994); Lewis, email, November 19, 2010.

[206] API RP 65 §4.8.4.

[207] Commission calculation based on internal Halliburton document (HAL_10994); Steve Lewis, email, November 19, 2010.

[208] Internal BP document (BP-HZN-CEC 8860).

[209] Internal Transocean document (TRN-HCJ 93593).

[210] Calculated based on internal BP document (BP-HZN-CEC 21445). Using the internal volume calculated by independent experts, 900 barrels, 1.5 × pipe volume would be 1,350 bbl.

[211] Internal BP document (BP-HZN-CEC 21265).

[212] Internal BP document (BP-HZN-CEC 21445). Independent experts calculated total internal volume at 900 barrels.

[213] Internal BP document (BP-HZN-CEC 17626). Independent experts calculated total internal volume at 900 barrels.

[214] Internal BP document (BP-HZN-CEC 21445).

[215] Internal BP document (BP-HZN-CEC 21448).

[216] Internal BP document (BP-HZN-MBI 137364).

[217] *Ibid.*

[218] Internal Halliburton document (HAL_11196-97).

[219] Internal BP document (BP-HZN-MBI 137368).

[220] 39 barrels unfoamed, 48 bbl after foaming. Internal Transocean document (TRN-USCG_MMS 11640); Internal BP document (BP-HZN-MBI 137365)

[221] Internal BP document (BP-HZN-MBI 137368)

[222] Internal BP document (BP-HZN-MBI 137365).

[223] Internal BP document (BP-HZN-MBI 137365).

[224] Internal BP document (BP-HZN-MBI 129068); Internal BP document (BP-HZN-MBI 137369).

[225] Internal BP document (BP-HZN-MBI 137370).

[226] Internal BP document (BP-HZN-MBI 139592).

[227] Internal BP document (BP-HZN-MBI 21305).

[228] *Ibid.*

[229] Cathleenia Willis (Sperry Drilling), interview with Commission staff, October 21, 2010; Testimony of Nathaniel Chaisson, 411-12.

[230] Internal BP document (BP-HZN-MBI 137370).

[231] Guide, interview, September 17, 2010.

[232] Internal BP document (BP-HZN-MBI 137370); Internal BP document (BP-HZN-MBI 129052).

[233] Internal Halliburton document (HAL_11208).

[234] Internal BP document (BP-HZN-MBI 137370); Internal Halliburton document (HAL_11208).

[235] Internal BP document (BP-HZN-MBI 129052).

[236] Internal BP document (BP-HZN-MBI 129141).

[237] Internal BP document (BP-HZN-MBI 129055).

[238] Testimony of John Guide, July 22, 2010, 321-22.

[239] Testimony of Steve Lewis, 110-11.

[240] Testimony of Vincent Tabler (Halliburton), Hearing before the Joint Investigation Team, August 25, 2010, 21-22.

[241] Confidential source, interview; Internal BP document (BP-HZN-CEC 21449).

[242] Internal BP document (BP-HZN-CEC 20234); Testimony of Vincent Tabler, 21-22.

[243] Testimony of Vincent Tabler, 22, 33-34.

[244] Testimony of Vincent Tabler, 21-22; Confidential source, interview; Internal BP document (BP-HZN-CEC 21449); Internal BP document (BP-HZN-CEC 20234).

[245] Internal BP document (BP-HZN-MBI 136947).

[246] Internal BP document (BP-HZN-MBI 143304); Internal BP document (BP-HZN-CEC 21665).

[247] BP's decision tree is based on the MMS regulation about cement bond logs. 30 C.F.R. § 250.428 states that if "you encounter…lost returns, cement channeling, or failure of equipment…[t]hen you must (1) Pressure test the casing shoe; (2) Run a temperature survey; (3) Run a cement bond log; or (4) Use a combination of these techniques." BP's decision tree meets that regulation and was apparently designed in light of it. Gagliano, interview. However, the decision tree violates another provision of the MMS regulations. 30 C.F.R. § 250.421 requires that cement be at least 500 feet above a pay zone. Here, BP's decision tree permitted the cement to be as little as 100 feet above the pay zone.

[248] Randy Ezell (Transocean), interview with Commission staff, September 16, 2010. Gagliano agreed with this focus, stating that "it was critical that we design the cement job to have full returns because of the fact that if we lost returns during the cement job, they would have to run a cement bond log to see where the top of the cement was and then potentially remediation would be required." Gagliano, interview.

[249] Internal BP document (BP HZN CEC 21288); Internal BP document (BP-HZN-MBI 255705).

[250] Testimony of Ronnie Sepulvado, 21; Internal BP document (BP-HZN-MBI 126867).

[251] Confidential source, interview with Commission staff.

[252] Internal BP document (BP-HZN-MBI 127104).

[253] Internal BP document (BP-HZN-MBI 126867).

[254] Internal BP document (BP-HZN-MBI 127104).

[255] Internal Schlumberger document (SLB-EC 02); Internal Transocean document (TRN-USCG_MMS 30414)(April 18, 2010); Internal Transocean document (TRN-USCG_MMS 30422)(April 19, 2010).

[256] Testimony of Vincent Tabler, 23.

[257] Testimony of John Guide, July 22, 2010, 43-44. According to Walz, who was on the call, Guide noted that the indications that they had met their plan and that the CBL had only been a contingency. "Guide concluded that there was no reason to run the CBL and [everyone] on the call agreed." Internal BP document (BP-HZN-BLY 61333).

[258] Internal Schlumberger document (SLB-EC 02).

[259] Confidential source, interview with Commission staff.

[260] Testimony of Mark Bly (BP), Hearing before the National Commission, November 8, 2010, 326; Testimony of Bill Ambrose (Transocean), Hearing before the National Commission, November 8, 2010, 326; Testimony of Richard Vargo, 326.

[261] Greg Meeker (USGS), interview with Commission staff, January 27, 2011.

[262] Craig Gardner (Chevron), letter to Commission staff, October 26, 2010.

[263] Testimony of Richard Vargo, 331; Testimony of Jesse Gagliano, 321, 337.

[264] Testimony of Richard Vargo, 329, 355.

[265] Many would dispute the assumption the model is accurate. Testimony of John Guide, July 22, 2010, 275. Gagliano stated that BP "questioned the validity of the model." Testimony of Jesse Gagliano, 386. BP's cementing specialist Erick Cunningham stated that there are limitations to all such models and that their results always require the application of sound engineering judgment. Cunningham, interview. Moreover, had BP run the cement job on the basis of the 21-centralizer design April 18 model, BP would apparently have *relied* on the model's inaccuracy. That model predicted that the final two barrels of the cement job would have exceeded the fracture gradient. Internal BP document (BP-HZN-BLY 61327). For that reason, Walz alluded to disagreement over the accuracy of the model. Internal BP document (BP-HZN-CEC 22433).

[266] BP, Deepwater Horizon *Accident Investigation Report*, 65.

[267] *Ibid*, 35. Internal BP document (BP-HZN-CEC 8858); Internal BP document (BP-HZN-CEC 8858); Internal Transocean document (TRN-HCJ 9351); Internal Transocean document (TRN-HCJ 93526).

[268] One of BP's representatives stated that Halliburton's cement modeling was little more than a marketing tool.

[269] Internal BP document (BP-HZN-MBI 136849).

[270] Testimony of John Guide, July 22, 2010, 289.

[271] Halliburton observes circulating full bottoms up is a Halliburton "best practice." Testimony of Jesse Gagliano, 263.

[272] BP, Deepwater Horizon *Accident Investigation Report*, 66.

[273] Confidential source, interview.

[274] Internal BP document (BP-HZN-MBI 143295); Internal BP document (BP-HZN-CEC 22663); *see also* Guide, interview, September 17, 2010; Internal BP document (BP-HZN-CEC 22666). In the April 14 meeting discussing the design of the cement job however, neither BP nor Halliburton raised concerns about the volume of cement to be used. Internal BP document (BP-HZN-BLY 61326).

[275] BP, Deepwater Horizon *Accident Investigation Report*, 67. At the same time, BP's cementing engineer pointed out that there was little additional volume that could have been added. The TOC was reasonably close to the previous casing string and the entire annular volume was nearly full of cement. Nevertheless, the design would have been different had the well constraints been different. Cunningham, interview.

[276] Gagliano indicated that BP told him to keep ECD below 14.7 ppg. In order to do that the pump rate needed to be 4 bbl per minute or less. Testimony of Jesse Gagliano, 282; Testimony of Richard Vargo, 345-46. One expert estimates the cement flow rate could have been up to twice as fast as the rate pumped at Macondo. Steve Lewis (Expert witness), email to Commission staff, October 21, 2010.

[277] Testimony of Jesse Gagliano, 283-284.

[278] Confidential sources, interviews with Commission staff; Testimony of Tommy Roth (Halliburton), presentation to the National Academy of Engineering, September 26, 2010.

[279] Internal BP document (BP-HZN-MBI 117537).

[280] Guide, interview, September 17, 2010.

[281] *Ibid.*; Internal BP document (BP-HZN-CEC 21261).

[282] Testimony of Jesse Gagliano, 259.

[283] Both BP and Transocean personnel testified that additional pressure and multiple attempts at conversion are not unusual, particularly where equipment is clogged with debris. John Guide had previously experienced cuttings getting stuck between the ball and the tube, and believed this required applying

additional pressure. Testimony of John Guide, July 22, 2010, 147-48. Transocean offshore installation manager (OIM) Jimmy Harrell stated he was not concerned with additional pressure and that it is not usual that several attempts at higher pressures were needed to convert. Testimony of Jimmy Harrell, 85, 122. And Transocean senior toolpusher Randy Ezell also stated that the applied pressures were not too high due to debris that needed to be cleared. Ezell, interview. Another failure mode could be created by interference of the diverter ball, though this seems unlikely. After the casing is installed, the rig crew drops a 1.625-inch brass ball to close the hole, which closes the diverter gate and seals the inside of the casing from the annular space. The ball eventually lands on the float equipment. Email communications indicate the rig crew considered this scenario. It appears unlikely the diverter ball prevented conversion, as even with the diverter ball there would be more than 2 inches of space for fluid to circulate around the ball and through the float equipment. Internal BP document (BP-HZN-MBI 127277). BP's post-blowout tests simulated the diverter ball on top of the float equipment. BP tests where the Allamon ball was on top of the float collar established circulation at 3,210 psi and the float equipment converted at 11.5 bpm in the subsequent flow surge. Internal BP document (BP-HZN-BLY 62214-15).

[284] Internal BP document (BP-HZN-CEC 8860).

[285] Internal Transocean document (TRN-HCJ 93593).

[286] Internal BP document (BP-HZN-CEC 21265).

[287] Internal BP document (BP-HZN-CEC 17626).

[288] Internal BP document (BP-HZN-MBI 129226-32).

[289] Commission calculation based on internal BP document (BP-HZN-MBI 129229); Lewis, email, October 21, 2010. Evidence has indicated mud weight as both 14.0 or 14.17 ppg. For purposes of this calculation, an independent expert used a mud weight of 14.1 ppg. A mud weight in this range would require a flow rate of approximately 6 bpm to create a differential pressure of approximately 600 psi to convert the float equipment.

[290] Lewis, email, October 21, 2010; Testimony of Steve Lewis, 95.

[291] Testimony of Steve Lewis, 93.

[292] Internal BP document (BP-HZN-MBI 136941).

[293] Testimony of Steve Lewis, 96; BP, Deepwater Horizon *Accident Investigation Report*, 70.

[294] Commission calculation based on internal BP document (BP-HZN-MBI 129229).

[295] Stress Engineering conducted performance tests including flow endurance tests, steady-state flow conversion tests, flow-surge conversion tests, flow-surge tests on already converted float equipment, and mechanical failure tests of auto-fill tubes. Internal BP document (BP-HZN-BLY 62187).

[296] Internal BP document (BP-HZN-BLY 62156).

[297] Internal BP document (BP-HZN-MBI 136941).

[298] Internal BP document (BP-HZN-BLY 62087). Stress Engineering could not simulate field conditions of the surge accurately enough for results to be conclusive. Internal BP document (BP-HZN-BLY 62066).

[299] Testimony of Steve Lewis, 112-13.

[300] BP, Deepwater Horizon *Accident Investigation Report*, 71. So does wells team leader John Guide, who stated that the team could have been fooled that the floats held because the u-tube pressure during the check was so low. Guide, interview, January 19, 2011.

[301] Internal Halliburton document (HAL_11004). The Chief Counsel's team's expert calculated an even lower pressure differential. Based on figures in the final cement plan, fluids inside the casing and in the annulus may have been in almost perfect balance with only a 0.8 psi differential. Steve Lewis (Expert witness), email to Commission staff, December 14, 2010.

[302] Lewis, email, December 14, 2010.

[303] Testimony of Daniel Oldfather, 49.

[304] Testimony of Steve Lewis, 110.

[305] Internal BP document (BP-HZN-MBI 137753).

[306] Internal BP document (BP-HZN-MBI 21304).

[307] Ronnie Sepulvado (BP), interview with Commission staff, August 20, 2010.

[308] An obstruction in the reamer shoe could cause pressure to be the same above and below the auto-fill tube, which would mean there would be insufficient differential pressure to dislodge the tube.

[309] Testimony of Richard Sears, Hearing before the National Commission, November 9, 2010, 324.

[310] Internal BP document (BP-HZN-MBI 21330).

[311] According to BP's John Guide, he and Brian Morel did not discuss this possibility. Testimony of John Guide, October 7, 2010, 179-80.

[312] Testimony of Steve Lewis, 99; Clawson, interview.

[313] BP, Deepwater Horizon *Accident Investigation Report*, 70.

[314] Testimony of Steven Robinson (BP), Hearing before the Joint Investigation Team, December 8, 2010, part 1, 39-40; Guide, interview, September 17, 2010. There is some disagreement as to whether float equipment constitutes a barrier. An expert retained by the Commission considers float equipment a barrier, but only to its pressure rating. Steve Lewis (Expert witness), interview, September 15, 2010.

[315] Clawson, interview.

[316] API RP 65 - Part 2 §4.4.3.

[317] API RP 65 - Part 2 §3.4.

[318] Testimony of John Guide, July 22, 2010, 289.

[319] Ezell, interview; Testimony of Jimmy Harrell, 95.

[320] Internal BP document (BP-HZN-CEC 22670).

[321] Joseph Leimkuhler (Shell), letter to Commission staff, September 22, 2010.

[322] BP's own report appears to agree that better risk identification and assessment would have helped avoid the accident. The Bly report states that a "formal risk assessment might have enabled the BP Macondo well team to identify further mitigation options to address risks...."BP, Deepwater Horizon *Accident Investigation Report*, 36; Testimony of Mark Bly, 345.

[323] Internal BP document (BP-HZN-CEC 17626)(April 15 Drilling Program).

[324] Testimony of Steve Lewis, 103; Steve Lewis, email, October 31, 2010.

[325] Clawson, interview.

[326] *Ibid.* BP would have known of the need for 7-inch centralizers at the time they decided to run a 9⅞-inch liner. That liner was specified nearly a week earlier in the March 25 APD that BP filed to MMS. Steve Lewis (Expert witness), email to Commission staff, January 1, 2011; Internal BP document (BP-HZN-MBI 23666). When asked by BP investigators why he did not order additional centralizers at the time the APD was submitted, Walz stated that he could not recall any specific discussion about centralizers but thought he "had a conversation with Brian Morel about ordering what they normally ordered." Internal BP document (BP-HZN-BLY 61329). Morel promptly ordered other 7-inch casing equipment but not centralizers. On March 25, for instance, he asked a BP materials coordinator, "Can you have (2) cross-over made for 7" x 9-7/8" with a rush on it (2 weeks or less)?" Internal BP document (BP-HZN-BLY 64818).

[327] Confidential source, interview. The "7-10" range was generated within the BP drilling team based on a rough estimate for what was needed to cover the pay zone given hole conditions. But it is unclear why the goal of the well plan or the well conditions would have been any different than when the original plan was designed.

[328] Weatherford sales representative Bryan Clawson indicated that in the case of BP's final production casing, Weatherford would have been able to manufacture additional centralizers in time, had they been promptly asked. Clawson, interview. Other BP rig well teams faced simultaneous centralizer needs that Weatherford could not fill, and approached other suppliers. Internal BP document (BP-HZN-MBI 252278).

[329] The Bly report states that the Macondo team "erroneously believed they had received the wrong centralizers." BP, Deepwater Horizon *Accident Investigation Report*, 63. This statement is ambiguous. The well team "erroneously believed" that it had ordered one-piece centralizers and therefore did not expect to

receive centralizers with separate stop collars. But the team was correct to conclude that the centralizers Weatherford shipped were not, in fact, one-piece centralizers.

[330] Internal BP document (BP-HZN-MBI 255906). The "flying by the seat of our pants" quote was apparently made by well site leader Vidrine, who was concerned that with all of the special deliveries, crucial pieces might be missing for operations. Guide believed that the logistics changes were making things much more difficult on the well site leaders. Guide, interview, January 19, 2011.

[331] BP, Deepwater Horizon *Accident Investigation Report*, 63. By contrast, BP had decided to use a long string using a management of change process—and that management of change process had been based on a 70% standoff ratio. Internal BP document (BP-HZN-BLY 61198).

[332] Testimony of Daniel Oldfather, 9-15. Oldfather had installed centralizers on 10 different occasions, and at no time had centralizers gotten hung up in the well. He had heard of other jobs, with different types of centralizers, where the centralizers had been damaged. Testimony of Daniel Oldfather, 67, 72. In his defense, Guide notes that his personal experience is based on running hundreds of casing strings. Guide, interview, January 19, 2010.

[333] BP, Deepwater Horizon *Accident Investigation Report*, 63.

[334] Testimony of Jesse Gagliano, 295; Testimony of Gregg Walz, 71.

[335] BP, Deepwater Horizon *Accident Investigation Report*, 64. As discussed, however, the cement modeling was rerun on Halliburton's initiative.

[336] Testimony of John Guide, July 22, 2010, 275.

[337] Internal BP document (BP-HZN-CEC 22666).

[338] Testimony of Daniel Oldfather, 12.

[339] *Ibid.*, 57.

[340] *Ibid.*, 14.

[341] Gagliano, interview.

[342] *Ibid.*

[343] Testimony of Jesse Gagliano, 259; Internal BP document (BP-HZN-MBI 128489).

[344] If the rig crew believed the gauge was inaccurate, they could have compared pressure readings from that gauge to other sources of drill pipe pressure readings on the rig. Rigs typically have several different drill pipe pressure gauges, including electronic and analog gauges. The rig crew could have also observed drill pipe pressure on gauges at the cementing unit. Darryl Bourgoyne, email to Commission staff, January 1, 2011.

[345] Testimony of Steve Lewis, 105. However, applying back pressure can be undesirable for other reasons. Nelson and Guillot, eds., *Well Cementing*, 366.

[346] Ronnie Sepulvado (BP), interview with Commission staff, September 1, 2010.

[347] Internal BP document (BP-HZN-MBI 193558).

[348] *Ibid.*

[349] Guide, interview, September 17, 2010.

[350] When asked by BP investigators if he was aware of BP's technical guidance regarding lift pressures, Walz stated that at the time he "didn't make the connection" but now "understands that the mud weight differentials were not high enough to get a valid confirmation of good cement placement. Internal BP document (BP-HZN-BLY 61332).

[351] Confidential industry experts, interview with Commission staff.

[352] Steve Lewis (Expert witness), interview with Commission staff, September 21, 2010.

[353] Richard Sears (Expert witness), email to Fred Bartlit, September 20, 2010; Confidential industry experts, interviews.

[354] BP, Deepwater Horizon *Accident Investigation Report*, 36. BP investigator Mark Bly elaborated that using those criteria, "in hindsight, those could have caused the team to think a bit more carefully about it." Testimony of Mark Bly, 202-03, 375.

[355] Internal BP document (BP-HZN-MBI 195582).

[356] See, e.g., Steve Lewis (Expert witness), email to Commission staff, October 13, 2010. Not surprisingly, Halliburton has stated since the blowout that a cementing operation cannot be successfully concluded without performing a cement bond log. Testimony of Richard Vargo, 318-19, 321.

[357] Gregg Walz, interview with Commission staff, October 16, 2010.

[358] Gagliano, interview; Testimony of Jesse Gagliano, 270-71; Nelson and Guillot, eds., *Well Cementing*, 583.

[359] See, e.g., Tommy Roth (Halliburton), House Energy and Commerce Committee Staff Briefing, June 3, 2010; Testimony of Tommy Roth, presentation to NAE; Testimony of Richard Vargo, 325.

[360] Testimony of Gregg Walz, 53-54.

[361] Gagliano, interview.

[362] *Ibid.*

[363] Testimony of Jesse Gagliano, 335, 360.

[364] Gagliano interview. Gagliano stated that "I was never directly involved with those conversations. That's usually a decision made by BP. I was in a room when some of those conversations took place, but they weren't actually talking directly to me. So I have really no decision in that." *Ibid.*

[365] Testimony of Nathaniel Chaisson, 432.

[366] Testimony of Jesse Gagliano, 249-50; Testimony of Nathaniel Chaisson, 442.

[367] Internal Halliburton document (HAL_11010). One Macondo well site leader, for instance, did not know that the jagged lines indicated channeling. The report did indicate a "SEVERE" potential for gas flow, which while not an indication of channeling, can be a product of insufficient centralization.

[368] Testimony of Ronnie Sepulvado, 133; Murry Sepulvado, interview with Commission staff, December 10, 2010.

[369] Testimony of Nathaniel Chaisson, 440. In addition, the Halliburton post-job report indicated that the TOC was at 17,300 feet, where it would have been expected without any channeling. Internal Halliburton document (HAL_11195). However, Halliburton has indicated that this figure is a "volumetric calculation" designed simply to reflect full returns and was not an "engineering analysis." Halliburton legal team, interview with Commission staff, September 18, 2010.

Chapter 4.4

[1] Erik B. Nelson and Dominique Guillot, eds., *Well Cementing*, 2nd ed. (Sugar Land, TX: Schlumberger, 2006), 256.

[2] *Ibid*, 256. "Pounds per gallon" (ppg) units are used here for consistency rather than the more technically precise "pounds-mass per gallon" (lbm/gal).

[3] Jean de Rozieres and Remi Ferriere, "Foamed Cement Characterization Under Downhole Conditions and Its Impact on Job Design," *SPE Production Engineering* (August 1991), 298.

[4] S.L. Pickett and S.W. Cole, "Foamed Cementing Technique for Liners Yields Cost-Effective Results," *Society of Petroleum Engineers 27679* (1994), 521-24. "Failure to stabilize foam will...cause the [nitrogen] bubbles to coalesce and form interconnected flow paths." *Ibid.*, 523-24.

[5] D.S. Kulakofsky, P.G. Creel, and D.L. Kellum, "Techniques for Planning and Execution to Improve Foam Cement Job Performance," *Society of Petroleum Engineers 15519* (1986), 1-2.

[6] Michael T. Olanson, "Application of Foam Cements in Alberta," *Journal of Canadian Petroleum Technology* (October 1985), 51.

[7] N.R. Loeffler, "Foamed Cement: A Second Generation," *Society of Petroleum Engineers* 12592 (1984), 2. "[A]ccurate foam slurry design as confirmed by laboratory testing is an indispensable prerequisite to effective placement design engineering." *Ibid.*

[8] American Petroleum Institute, *Recommended Practice on Preparation and Testing of Foamed Cement Slurries at Atmospheric Pressure 10b-4* (July 2004)("API RP 10b-4").

[9] *Ibid.*, § 9.3.4.

[10] BP and Halliburton used foamed cement on the 28-inch casing and the 22-inch casing. Internal BP document (BP-HZN-OSC 603); Internal BP document (BP-HZN-MBI 14118); Internal BP document (BP-HZN-MBI 14193).

[11] Internal Halliburton document (GOM foamed cement jobs).

[12] Erick Cunningham (BP), interview with Commission staff, January 17, 2011.

[13] *Ibid.*; Internal BP document (BP-HZN-BLY 61276).

[14] Internal BP document (BP-HZN-MBI 218202).

[15] Jesse Gagliano (Halliburton), interview with Commission staff, September 10, 2010.

[16] *Ibid.*

[17] The February recipe included more retarder than the final recipe because in February BP expected Macondo to be a hotter well than it eventually proved to be. High well temperatures increase the speed at which cement cures. And if cement cures prematurely, it may become impossible to pump before it reaches the bottom of the well. One way to solve this problem is by adding retarder to the cement recipe.

[18] Internal Halliburton document (HAL_DOJ 61-62, 67-68).

[19] Confidential industry expert, interview with Commission staff.

[20] Internal BP document (BP-HZN-MBI 109218, 109223-24).

[21] Internal BP document (BP-HZN-MBI 117603).

[22] *Ibid.*

[23] Gagliano, interview; Internal Halliburton document (HAL DOJ 35).

[24] Gagliano, interview.

[25] Internal Halliburton document (HAL DOJ 35-36).

[26] Halliburton contends that its lab personnel performed this April 13 test improperly. Donald Godwin (Halliburton), letter to Commission staff, November 18, 2010.

[27] Internal BP document (BP-HZN-OSC 6224).

[28] Internal BP document (BP-HZN-MBI 128542).

[29] Internal BP document (BP-HZN-BLY 47095).

[30] *Ibid.*

[31] Internal BP document (BP-HZN-OSC 6224).

[32] Internal Halliburton document (HAL DOJ 42-43).

[33] API RP 10b-4 § 7.2(j).

[34] Internal Halliburton document (HAL DOJ 43). Cement laboratories routinely process samples around the clock in response to the intense time pressures in the drilling industry.

[35] Donald Godwin (Halliburton), letter to Commission staff, January 7, 2011, 2.

[36] Halliburton document (presentation to the National Academy of Engineers, September 26, 2010), 7. "The foam slurry was transferred to a stability test cell and cured for 48 hours." *Ibid.*

[37] Internal BP document (BP-HZN-MBI 136947).

[38] Testimony of Richard Vargo (Halliburton), Hearing before the National Commission, November 8, 2010, 367-68.

[39] Donald Godwin (Halliburton), letter to Commission staff, January 10, 2011, 1; Godwin, letter, January 7, 2011, 2.

[40] Godwin, letter, January 10, 2011, 2.

[41] Godwin, letter, January 7, 2011, 2.

[42] Internal BP document (BP-HZN-MBI 171151).

[43] BP, Deepwater Horizon *Accident Investigation Report* (September 8, 2010), 55-60, 66-68.

[44] BP, Deepwater Horizon *Accident Investigation Report*, 58. At the time BP issued its report, it was only aware of the results of the final April foam stability test conducted by Halliburton on or about April 18. BP apparently had not yet realized that its personnel had also received testing data from Halliburton on March 8, let alone that Halliburton had not reported the results of two foam stability tests to BP at all.

[45] *Ibid*, app. K.

[46] *Ibid*, 59.

[47] *Ibid*, 60.

[48] Craig Gardner (Chevron), letter to Commission staff, October 26, 2010.

[49] Donald Godwin (Halliburton), statement to Commission staff, October 12, 2010.

[50] Gardner, letter, 2.

[51] Confidential industry experts, interviews with Commission staff.

[52] Testimony of Richard Vargo, 360-61.

[53] Godwin, letter, November 18, 2010.

[54] Fred Bartlit, letter to Donald Godwin, November 23, 2010.

[55] American Petroleum Institute, Recommended Practice for Testing Well Cements 10b-2 (January 2007)("API RP 10b-2").

[56] Bartlit, letter, November 23, 2010.

[57] Confidential industry experts, interviews.

[58] Nelson and Guillot, eds., *Well Cementing*, 257.

[59] Internal BP document (BP-HZN-MBI 171151).

[60] Testimony of Richard Vargo, 362.

[61] Press Release, Halliburton, Halliburton Comments on National Commission Cement Testing, October 28, 2010; Godwin, letter, November 18, 2010.

[62] Testimony of Richard Vargo, 363. Halliburton expert Richard Vargo admitted that "based on those [February] results, I would not have, at that time on February 17th, chosen to run that [design] in the well." *Ibid*.

[63] *Ibid.*, 366.

[64] Godwin, letter, November 18, 2010; Donald Godwin (Halliburton), letter to Commission staff, December 9, 2010 (citing HAL DOJ 35, 41, 44).

[65] If the latter is true, the Chief Counsel believes that Halliburton personnel may have been motivated in part by the fact that a slurry redesign could have required Halliburton to discard the dry blend that was on the *Deepwater Horizon* and deliver a new dry blend to the rig. That process might have cost time and money—especially if Halliburton first realized the problem just before pumping the job.

[66] Internal BP document (BP-HZN-MBI 130830); Internal BP document (BP-HZN-OSC 6225).

[67] Internal BP document (BP-HZN-MBI 117603); Internal BP document (BP-HZN-MBI 6225).

[68] Internal BP document (BP-HZN-MBI 255509).

[69] Internal BP document (BP-HZN-MBI 128542).

[70] Internal BP document (BP-HZN-BLY 47093-95).

[71] Internal BP document (BP-HZN-MBI 257237).

[72] Internal BP document (BP-HZN-MBI 218202).

[73] Internal BP document (BP-HZN-OSC 6224).

[74] BP, Deepwater Horizon *Accident Investigation Report*, 68. "The investigation team saw no evidence that the BP Macondo well team confirmed that all relevant lab test results had been obtained and considered by the Halliburton in-house cementing engineer before cement placement proceeded." *Ibid.*

[75] Testimony of Mark Bly (BP), Hearing before the National Commission, November 8, 2010, 388.

[76] BP's own report admits that "stronger quality assurance by the BP Macondo well team might have identified potential flaws and weaknesses in the slurry design." BP, Deepwater Horizon *Accident Investigation Report*, 68.

[77] Internal BP document (BP-HZN-MBI 110151).

[78] John Guide (BP), interview with Commission staff, January 19, 2011.

[79] Cunningham, interview.

[80] *Ibid.*

[81] Internal BP document (BP-HZN-OSC 9337).

Chapter 4.5

[1] BP, Deepwater Horizon *Accident Investigation Report* (September 8, 2010), 72.

[2] 30 C.F.R. § 250.1721.

[3] BP repeated the explanation in its internal investigation report, framing certain steps as "required," without critically analyzing or questioning them. BP, Deepwater Horizon *Accident Investigation Report*, 72. "To install the lockdown sleeve, 100,000 lbs of weight was required on the running tool. This required approximately 3,000 ft. of drill pipe to be run below the running tool. Allowing for this length of drill pipe determined the final cement plug setting depth. In turn, the cement plug setting depth 3,000 ft. below the wellhead influenced the differential pressure created during the negative-pressure test." *Ibid.*

[4] John Guide (BP), interview with Commission staff, January 19, 2011; Gregory Walz (BP), interview with Commission staff, October 6, 2010; Confidential source, interview with Commission staff; Testimony of Ross Skidmore (Swift), Hearing before the Deepwater Horizon Joint Investigation Team, July 20, 2010, 61.

[5] Testimony of Ronnie Sepulvado (BP), Hearing before the Deepwater Horizon Joint Investigation Team, July 20, 2010, 145; BP, Deepwater Horizon *Accident Investigation Report*, 72; Confidential source, interview; Guide, interview, September 17, 2010.

[6] Skidmore, interview; Testimony of Ross Skidmore, 248-49; Walz, interview; Ronnie Sepulvado, interview; David Sims (BP), interview with Commission staff, December 14, 2010; Testimony of John Guide (BP), Hearing before the Deepwater Horizon Joint Investigation Team, July 22, 2010, 299.

[7] Guide, interview, September 17, 2010; Ronnie Sepulvado, interview; Testimony of Mark Bly (BP), Hearing before the National Commission, November 8, 2010, 210, 310. A February 11 email discussing an earlier surface plug at Macondo suggests that even that plug "was set in seawater to prevent contamination." Internal BP document (BP-HZN-MBI 196335).

[8] Ronnie Sepulvado, interview. BP explained this in a filing with the MMS. Internal BP document (BP-HZN-MBI 127909). "The requested surface plug depth deviation is for minimizing the chance for damaging the LDS sealing area." *Ibid.*

[9] BP's temporary abandonments of previous wells bear this out, as do interviews with several BP engineers. Internal BP document (BP-HZN-OSC 6083); Internal BP document (BP-HZN-OSC 6158); Murry Sepulvado (BP), interview with Commission staff, December 10, 2010; Ross Skidmore (Swift), interview with

Commission staff, December 21, 2010; John Guide (BP), interview with Commission staff, September 17, 2010.

[10] Ronnie Sepulvado (BP), interview with Commission staff, October 26, 2010; Merrick Kelley (BP), interview with Commission staff, October 22, 2010.

[11] Murry Sepulvado, interview; Skidmore, interview; Guide, interview, September 17, 2010; Kelley, interview.

[12] Kelley, interview; Internal BP document (BP-HZN-OSC 112833); Internal BP document (BP-HZN-OSC 113043).

[13] Internal BP document (BP-HZN-MBI 119062).

[14] *Ibid.*

[15] *Ibid.*

[16] Internal BP document (BP-HZN-MBI 100448).

[17] *Ibid.*

[18] Internal BP document (BP-HZN-MBI 100447).

[19] Internal BP document (BP-HZN-MBI 100445-46). "5.5 rig days at $400,000 = $2,200,000." *Ibid.*

[20] Internal BP document (BP-HZN-MBI 100446).

[21] In early January, Hafle inquired: "I was told the LDS could be installed off critical path with an intervention type vessel. Is this possible?" Kelley responded: "We are building the necessary components to convert the Thunderhorse open water tool for use on a H-4 wellhead at present. We plan to use this tool for the Isabela LDS mid this year. Not sure how successful we will be but will have to rely on the rig as a contingency if necessary." Internal BP document (BP-HZN-MBI 100447).

[22] Internal BP document (BP-HZN-MBI 100446). "For the open water cost of installing the LDS...$480,000." *Ibid.* "If you do it with the rig at the time of the original drilling...$600,000." *Ibid.*

[23] *Ibid.* "The risk you are exposed to is if the LDS does not land out correctly or the nose seals are damaged and the LDS has to be retrieved (can happen but has not happened to us in the past 5 years)." *Ibid.*

[24] *Ibid.*

[25] Internal BP document (BP-HZN-MBI 127907).

[26] Internal BP document (BP-HZN-MBI 119061).

[27] *Ibid.*

[28] Internal BP document (BP-HZN-MBI 199123).

[29] Internal BP document (BP-HZN-MBI 126145).

[30] *Ibid.*

[31] Internal BP document (BP-HZN-MBI 199123).

[32] Internal BP document (BP-HZN-MBI 126333).

[33] *Ibid.* The next day, Morel forwarded Kelley's email to Gregg Walz. *Ibid.*

[34] Internal BP document (BP-HZN-MBI 126450).

[35] Internal BP document (BP-HZN-MBI 126495).

[36] BP's internal investigator acknowledges that there were multiple changes in plan. Testimony of Mark Bly, 305.

[37] Internal BP document (BP-HZN-MBI 199122).

[38] Internal BP document (BP-HZN-MBI 126180); Internal BP document (BP-HZN-MBI 126189-90).

[39] Internal BP document (BP-HZN-MBI 126585-86).

[40] Internal BP document (BP-HZN-MBI 126982). BP has asked the Commission to disregard this procedure because, among other reasons, it was an email sent to a rig clerk (rather than the engineering team).

However, it was also sent to one of the well site leaders on the rig, Ronnie Sepulvado, and Morel specifically asked for feedback on the procedure.

[41] Internal BP document (BP-HZN-MBI 126982).

[42] *Ibid.*

[43] Internal BP document (BP-HZN-BLY 61203).

[44] Internal BP document (BP-HZN-CEC 21281, 21288). Transocean has claimed that conducting a negative pressure test on the shoe track rather than the cement plug is unusual. In the January 2010 abandonment of the Kodiak well, for instance, the negative pressure test was not done until the cement plug was set. Given the risk factors associated with the production casing cement job, it was prudent for BP to conduct a negative pressure test before setting the cement plug. It is unclear, however, whether that was the reason that BP switched the order of the negative pressure test.

[45] Internal BP document (BP-HZN-CEC 21281-88).

[46] Internal BP document (BP-HZN-CEC 20263); Guide, interview, January 19, 2011.

[47] Internal BP document (BP-HZN-CEC 21281-88). Morel was not the only one who thought that MMS might not approve it. Drilling engineer Mark Hafle emailed Morel and told him, "I really don't think MMS will approve deep surface plug. We'll see." Internal BP document (BP-HZN-MBI 127489). BP's concern that MMS might not have approved such a deep surface plug suggests that BP may have recognized the risks attendant to setting such a deep plug.

[48] Internal BP document (BP-HZN-MBI 127907). The APM said that the test would be conducted with "seawater gradient equivalent"; base oil is understood to be such an equivalent. This change, of course, also allowed the rig crew the flexibility to use seawater instead of base oil. Murry Sepulvado, interview.

[49] Internal BP document (BP-HZN-BLY 69066-67); Internal BP document (BP-HZN-MBI 127906).

[50] Proponents of this interpretation point to the very plain order provided on the APM: "1. Negative Test...2. TIH [Trip in Hole] with a 3-1/2" stinger to 8367; 3. Displace to seawater." Internal BP document (BP-HZN-OSC 1436). They note that procedures are meant to be followed sequentially and, as an aside, state that if a different order were intended, this procedure was very poorly written. They also note that the order is in keeping with the April 15 well plan (which called for a negative pressure test at the wellhead), circulated the same day that the APM was prepared. Finally, they note that although a negative pressure test may have made most sense at a depth of 8,367 feet to mimic the underbalanced condition of an abandoned well, negative pressure tests could be conducted at higher depths in a staged testing process—as the BP engineering team in fact planned. Internal BP document (BP-HZN-BLY 61379-80).

[51] BP drilling engineer Brett Cocales and Wells Team Leader John Guide read the second and third steps as subsets of the negative pressure test. Testimony of John Guide, 329-331; Guide, interview, January 19, 2011. Cocales has testified that there would otherwise be no way to know at what depth to conduct the negative pressure test. Testimony of Brett Cocales (BP), Hearing before the Deepwater Horizon Joint Investigation Team, August 27, 2010, 128-29. Moreover, they believe that 8,367 feet is the only depth at which a negative pressure test would make sense, as that is the only depth at which the test would simulate the temporary abandonment of the well. *Ibid.*, 129.

[52] There may not be much significance beyond that. Conducting the negative pressure test at the wellhead rather than at 8,367 feet would probably not have prevented the blowout. In such a test, the pressure buildup, if any, would have been much more subtle than the 1,400 psi seen on the drill pipe. Given that the rig crew proceeded with abandonment despite observing 1,400 psi on the drill pipe, it is unlikely that they would have stopped the operation upon observing even smaller pressures on the drill pipe. John Smith (Expert witness), interview with Commission staff, October 22, 2010. In addition, there is no reason to think that MMS would have rejected a permit that explicitly stated the negative pressure test would take place at a depth of 8,367 feet. Frank Patton (MMS), interview with Commission staff, October 1, 2010.

[53] Internal BP document (BP-HZN-CEC 20273).

[54] Internal BP document (BP-HZN-BLY 61379-80); Internal BP document (BP-HZN-BLY 20087).

[55] Walz, interview.

[56] Internal BP document (BP-HZN-BLY 61380).

[57] Internal BP document (BP-HZN-BLY 70087).

[58] Guide, interview, January 19, 2011.

[59] Morel sent a second Ops Note within 10 minutes after the first one. It added details and did not change the order of the procedures. Internal BP document (BP-HZN-MBI 195579-80).

[60] Testimony of Bill Ambrose (Transocean), Hearing before the National Commission, November 8, 2010, 315.

[61] This would not be a change if one accepted the interpretation that the negative pressure test in the April 16 APM would follow displacement. Internal BP document (BP-HZN-MBI 195579-80).

[62] *Ibid.*

[63] The Ops Note says, "With seawater in kill close annular and do a negative test ~2350 psi differential." Internal BP document (BP-HZN-MBI 139532). The 2,350 psi is not an expected drill pipe pressure. It describes the reduction in pressure at the bottom of the well by replacing the mud with water in the drill pipe and drill pipe-casing annulus down to 8,367 feet. Although the initial drill pipe pressure at the beginning of the negative pressure test was 2,325 psi, an independent expert has stated that the expected pressure at the beginning of the negative pressure test should have been 1,610 psi. John Smith, *Review of Operational Data Preceding Explosion on Deepwater Horizon in MC252* (July 1, 2010), 9.

[64] According to Mark Hafle, John Guide made a conscious decision not to notify the MMS of the change. Internal BP document (BP-HZN-BLY 61380). For his part, Guide said he had difficulty remembering any discussion or decision regarding notifying MMS, but agreed that MMS did not need to be notified in this situation. Guide believed that the plan had always been to conduct the negative pressure test at 8,367 feet. Guide, interview, January 19, 2011. Gregg Walz suggested that Hafle may have been the one to make the decision not to notify MMS of the change. Walz, interview. This assertion seems to be in tension with the fact that Hafle called Kaluza on April 20 out of concern that the new procedure departed from the one approved by MMS. Internal BP document (BP-HZN-CEC 20185, 200, 204).

[65] According to his supervisor Gregg Walz, Hafle probably did not seek permission from MMS for deviating from the APM because the deviation resulted in a stricter test. Walz, interview. Conducting a negative pressure test at 8,367 feet would underbalance the well to a much greater degree than conducting a negative pressure test at the wellhead. However, displacing so much mud with seawater to conduct the test still entailed a level of risk.

[66] Internal BP document (BP-HZN-OSC 6158). As at Macondo, this well altered the procedure to shift from (a) conduct negative pressure test, set plug in mud, then displace to seawater, to (b) partially displace, conduct negative pressure test, finish displacement, set plug in seawater. This application also stated that the reason for combining the displacement and negative pressure test was "to improve rig efficiency." *Ibid.*

[67] Internal BP document (BP-HZN-CEC 20185, 20200, 20204).

[68] This may have been so if rig personnel had conducted a negative pressure test before displacing the well. Prior to displacement, there is neither spacer nor drill pipe in the wellbore. So there would have been no spacer in the well to potentially clog the kill line. And the crew would have sealed the well using the blind shear rams—not the annular preventer—eliminating any suggestion of a "bladder effect." Darryl Bourgoyne (Expert witness), interview with Commission staff, January 26, 2010; Darryl Bourgoyne (Expert witness), email to Commission staff, January 26, 2010.

[69] BP representatives agree that displacing 8,367 feet of mud from the wellbore increased the underbalance but do not agree that the severity of the underbalance increased risk. "There's no reason at all to believe that increases the risk." Testimony of Mark Bly, 209, 215-17. BP is wrong. Greater underbalance in the well places greater stress on the bottomhole cement. More broadly, there was no need to create an underbalance in the first place or to create one before putting more barriers in place.

[70] John Smith (Expert witness), interview with Commission staff, September 7, 2010; Murry Sepulvado, interview; John Guide, interview, January 19, 2011.

[71] The depth of the surface cement plug at Macondo was extremely unusual—perhaps one-of-a-kind. Anadarko representatives, interview with Commission staff, September 29, 2010; Testimony of Richard Vargo (Halliburton), Hearing before the National Commission, November 8, 2010, 212 (has set thousands of plugs in his career and has "never seen it set this deep before"); Walz, interview (cannot recall setting one that deep before); Bill Ambrose (Transocean), interview with Commission staff, November 2, 2010 (unusually deep); Murry Sepulvado, interview (never seen one this deep before); Ronnie Sepulvado, interview (deeper than he had ever seen); Confidential source, interview (deepest in experience); Randy Ezell (Transocean), interview with Commission staff, September 16, 2010 (unusual); Testimony of John Guide, 298 ("it was deeper than normal"); Testimony of Ross Skidmore, 60 (a lot deeper than he had probably seen before); Testimony of Leo Lindner (M-I SWACO), Hearing before the Deepwater Horizon Joint Investigation Team, July 19, 2010, 316 ("much further down than usual"); Allen Seraile (Transocean),

interview with Commission staff, January 7, 2011 (unusual, hadn't done that before); Transocean legal team, interview with Commission staff, September 21, 2010 (poll of 25 Transocean rigs revealed average displacement depth of 150 feet below mudline).

[72] Testimony of Mark Bly, 308.

[73] BP representatives admit this too. *Ibid.*, 312-13.

[74] BP representatives acknowledge this point as well. *Ibid*, 308-09; Murry Sepulvado, interview.

[75] Anadarko representatives, interview; Confidential industry expert, interview with Commission staff; Testimony of Merrick Kelley (BP), Hearing before the Deepwater Horizon Joint Investigation Team, August 27, 2010, 298; Testimony of Ross Skidmore, 250, 259; Steve Lewis (Expert witness), interview, October 20, 2010.

[76] At least one major operator typically sets the lockdown sleeve prior to the surface plug. Confidential industry expert, interview. At least one BP well site leader has done so in the past. Murry Sepulvado, interview. But BP wells team leader John Guide stated that, of the 17 lockdown sleeves he has set in his career, he set all of them last, and doing so was BP's standard practice. Guide, interview, September 17, 2010. It is worth noting that all of those 17 lockdown sleeves were a different model than the one set at Macondo and did not require downward setting force. Guide, interview, January 19, 2011.

[77] BP engineers state that they needed to set the lockdown sleeve last to avoid damaging it. But BP could have managed the risk of damage in other ways and has done so in the past. Kelley, interview. Alternative precautions include installing a seat protector and running pipe more carefully. *Ibid.*

[78] BP, Deepwater Horizon *Accident Investigation Report*, 72; Confidential industry expert, interview; Internal BP document (BP-HZN-MBI 316347)("Often times this ring is removed."); Internal BP document (BP-HZN-BLY 61330)("the OD lock-down ring on casing hanger seal assembly was optional and not being run routinely by the E&A team in all wells...it was a common practice").

[79] Joint Industry Task Force to Address Offshore Operating Procedures and Equipment, "Recommendations for Improving Offshore Safety," May 17, 2010, 4.

[80] Internal BP document (BP-HZN-MBI 196718).

[81] Internal BP document (BP-HZN-MBI 119061).

[82] Internal BP document (BP-HZN-MBI 126145); Internal BP document (BP-HZN-CEC 21268).

[83] Skidmore, interview.

[84] Internal BP document (BP-HZN-MBI 126928).

[85] Skidmore, interview. Skidmore approached Morel or Vidrine, or both. *Ibid.*

[86] Internal BP document (BP-HZN-MBI 199250). "DRIL-QUIP recommends running 100,000 lb of weight below the Running Tool.... Weight above the Running Tool can be substituted for weight below the Running Tool." *Ibid.*

[87] Kelley, interview.

[88] Dril-Quip legal team, email to Commission staff, December 27, 2010; Internal BP document (BP-HZN-MBI 44875). "DRIL QUIP recommends running 100,000 lb of weight below the Running Tool.... Weight above the Running Tool can be substituted for weight below the Running Tool." Internal BP document (BP-HZN-MBI 44875).

[89] Internal BP document (BP-HZN-MBI 44858).

[90] Internal BP document (MC 129 #3 APM 82712).

[91] Internal BP document (BP-HZN-OSC 6118).

[92] *Ibid.*

[93] Kelley, interview.

[94] *Ibid.* (recommended 1,250 to 1,350 feet below the mudline).

[95] Internal BP document (BP-HZN-MBI 196174). "MU & GIH w/ lockdown sleeve on 6-5/8" landing string w/ drill collars below." *Ibid.* On November 12, 2009, BP engineer Tippetts told Dril-Quip representative

Patterson, "We will be using [drill collars] for the tailpipe on Macondo." Internal BP document (BP-HZN-MBI 119061).

[96] Internal BP document (BP-HZN-MBI 196174). On February 22, Tippetts asked BP operations engineer Brett Cocales whether the *Horizon* had the equipment necessary to run the planned drill collars. Cocales informed Tippetts the next day that the *Horizon* did not have the requested equipment and that Tippetts should plan to get it from a supplier. Internal BP document (BP-HZN-MBI 196720).

[97] Internal BP document (BP-HZN-MBI 196719).

[98] In the April 14 forward plan, BP lists the supply vessel *M/V Hilda Lab* as being en route to the rig with nine 6½-inch drill collars on board. Internal BP document (BP-HZN-MBI 199282).

[99] Internal BP document (BP-HZN-MBI 252071). "I keep coming back to sequence of setting casing, set the wear bushing, do the T&A work, pull the bushing, pick drill collars and RIH to set the lock down sleeve."*Ibid.*

[100] Internal BP document (BP-HZN-CEC 21268).

[101] Internal BP document (BP-HZN-MBI 199229, 199239)(~18 6½-inch range 2 drill collars from Alice Chalmers listed for use in tailpipe; "The tail pipe consisting of HT 55 Drill Pipe & 6 ½" drill collars will be used for achieving the required weight down for the Lock Down Sleeve Installation").

[102] Internal BP document (BP-HZN-MBI 199282).

[103] Guide, interview, January 19, 2010. Guide stated that the suggestion to use heavyweight drill pipe instead of drill collars came from Transocean senior toolpusher Randy Ezell. *Ibid.*

[104] According to Guide, making up drill collars would not necessarily *add* time because the drill collars could be made up offline. *Ibid.*

[105] On this, all parties (including BP) agree. Testimony of Mark Bly, 211, 213; Testimony of Richard Vargo, 211; Testimony of Bill Ambrose, 214.

[106] Confidential industry expert, interview; Murry Sepulvado, interview; Steve Lewis (Expert witness), interview with Commission staff, October 29, 2010. Chemicals in oil-based mud can cause more disruption to the physical properties of cement, which is water-based, than seawater. Lewis, interview, October 29, 2010.

[107] Internal BP document (BP-HZN-MBI 196335).

[108] Lewis, interview, October 29, 2010; Testimony of Charlie Williams (Shell), Hearing before the National Commission, November 9, 2010, 45.

[109] BP's internal investigator acknowledges that the company could set mechanical plugs. Testimony of Mark Bly, 310. Many in the industry believe mechanical plugs should be incorporated into routine well design. Confidential industry experts, interview with Commission staff; Murry Sepulvado, interview; Steve Lewis (Expert witness), email to Commission staff, September 20, 2010; Confidential industry expert, interview.

[110] Testimony of Charlie Williams, 44.

[111] An earlier surface cement plug at Macondo appeared to have been "set in seawater to prevent contamination," but when the rig returned to resume drilling, they found that "the surface plug was not hard" because of "a cement/water [contamination] issue." That earlier cement plug had been set using a parabow (a metal retainer), which held the cement in place but presented separate complications. Internal BP document (BP-HZN-MBI 196335).

[112] *Ibid.*

[113] *Ibid.*

[114] *Ibid.*

[115] Testimony of Mark Bly, 213. "There's engineering choices that you make, and I think setting it in mud is something that happens sometimes and sometimes people choose to set them in seawater." *Ibid.*

[116] Testimony of Mark Bly, 214. "It's an engineering tradeoff decision." *Ibid.*

[117] Walz, interview; Internal BP document (BP-HZN-OSC 6083), Internal BP document (BP-HZN-OSC 6158).

[118] Internal BP document (BP-HZN-MBI 126928).

[119] These include the Halliburton Fas Drill (drillable bridge plug), Baker Hughes GT plug (retrievable bridge plug), and parabow (retrievable retainer). Internal BP document (BP-HZN-OSC 6083); Internal BP document (BP-HZN-OSC 6158).

[120] Notably, BP used GT plugs in at least two other wells in 2010, including MC 822 #5 and MC 877 #22. Internal BP document (BP-HZN-MBI 198612). BP engineers also considered running a Halliburton Fas Drill plug. Internal Transocean document (TRN-HCJ 93590); Internal BP document (BP-HZN-OSC 112974)("We are considering a Fas-Drill retainer for the TA plug, vs a GT plug in the 9-7/8" casing.").

[121] Internal BP document (BP-HZN-MBI 198666).

[122] Internal BP document (BP-HZN-MBI 198602). "Thanks for confirming the decision to use the GT Plug for abandonment." *Ibid.* In late February, BP set up a meeting where Morel and Hafle would "go over the necessary Macondo data and needs" and Baker Hughes representative Mark Plante would then make a presentation about the GT plug and procedure. Internal BP document (BP-HZN-MBI 196642). The meeting was originally scheduled for March 3, then postponed "probably until the week of March 8th" because Morel and Hafle were "very busy," and finally scheduled for March 10. *Ibid*; Internal BP document (BP-HZN-MBI 196652); Internal BP document (BP-HZN-MBI 196643). In April, Morel indicated that there was "[s]till some discussion on" whether to use the GT plug and that John Guide would be following up. Internal BP document (BP-HZN-MBI 198762). According to Guide, he was not involved in the decision of whether to use a GT plug but has stated that Morel would have known that Guide opposes setting GT plugs. Guide, interview, January 19, 2011.

[123] Internal BP document (BP-HZN-MBI 198666); Internal BP document (BP-HZN-MBI 198910). "Met with Baker today on this issue, and they are going to come back with a proposal based on fixed and firm commitment from BP for the use of these plugs on a longer term basis. Details are TBD, but we talked about some high level options and quantities based on your feedback for upcoming/ongoing work at Macondo, Atlantis and TH." Internal BP document (BP-HZN-MBI 198910).

[124] Internal BP document (BP-HZN-MBI 199275).

[125] Internal BP document (BP-HZN-MBI 129149-51).

[126] *Ibid.*

[127] Internal BP document (BP-HZN-MBI 128957). "We are at the end of the week and our district would like to know if there is going to be any decision made soon on whether or not the GT packer is going to be run and if so. When it is going to be called out. We just want to make sure you are covered in case something comes up in the weekend." *Ibid.*

[128] Internal BP document (BP-HZN-MBI 128957).

[129] Internal BP document (BP-HZN-MBI 198919).

[130] Internal BP document (BP-HZN-CEC 21269).

[131] Internal BP document (BP-HZN-MBI 128957).

[132] Internal BP document (BP-HZN-MBI 128959).

[133] Internal BP document (BP-HZN-MBI 129145).

[134] This comment was made on April 14 to the BP engineer negotiating the Baker Hughes contract. Internal BP document (BP-HZN-BLY 68031).

[135] *Ibid.*

[136] Internal BP document (BP-HZN-MBI 251262); Confidential source, interview; Walz, interview; Guide, interview, January 19, 2011; Internal BP document (BP-HZN-OSC 6083); Internal BP document (BP-HZN-OSC 6158); Internal BP document (BP-HZN-MBI 252171).

[137] Testimony of Steve Lewis (Expert witness), Hearing before the National Commission, November 9, 2010, 53-55.

[138] Confidential industry expert, interview with Commission staff; Steve Lewis (Expert witness), interview with Commission staff, September 28, 2010.

[139] BP's Drilling and Well Operations Practice manual discusses the use of kill weight fluid as a barrier before breaking containment but discusses it as a replacement for (instead of addition to) one of the two required mechanical barriers. Internal BP document (BP-HZN-MBI 130875).

[140] Confidential industry expert, interview with Commission staff; Testimony of Charlie Williams, 45.

[141] As far as after the displacement and abandonment of the well, mud would not remain a barrier indefinitely. Over time, when left static in the wellbore and not circulated, mud suffers from barite fallout and loses its integrity. Therefore, "mud can only be considered a temporary barrier with a restricted life span dependent on the mud weight and temperature." Joseph Leimkuhler (Shell), letter to Commission staff, September 22, 2010.

[142] BP concedes this. Testimony of Mark Bly, 311, 314.

[143] Internal BP document (BP-HZN-OSC 6083); Internal BP document (BP-HZN-OSC 6158).

[144] Testimony of Charlie Williams, 43-44 (Shell sets three to five plugs); Ezell, interview (normally will set several plugs in addition to the bottomhole cement); Seraile, interview (normally will set two or three plugs before displacing).

[145] Guide, interview, January 19, 2011 (does not recall anyone on the Macondo team suggesting that they should run more than one plug). This may be because setting multiple, intermediate plugs can complicate later re-entry and completion of the well, since retrieving or drilling out the plugs would take time and could disperse debris in the well. *Ibid.* Nevertheless, BP appears to have addressed or accepted these complications in other wells where they have set numerous plugs. Internal BP document (BP-HZN-OSC 6083); Internal BP document (BP-HZN-OSC 6158).

[146] Darryl Bourgoyne (Expert witness), interview with Commission staff, September 10, 2010; Steve Lewis (Expert witness), interview with Commission staff, September 21, 2010.

[147] Lewis, interview, September 21, 2010. Performing a displacement with the BOP closed can involve some minor encumbrances, including wear of the choke, kill, and boost lines, added time, and greater mixing of fluids in the wellbore. *Ibid.*

[148] *Ibid.*

[149] Joint Industry Task Force to Address Offshore Operating Procedures and Equipment, "Recommendations for Improving Offshore Safety," 4.

[150] Testimony of Steve Lewis, 63; Lewis, interview, September 28, 2010. BP wells team leader John Guide stated that temporary abandonment procedures are historically written at the end of a well, not incorporated into the initial drilling program, because the final dimensions of the well are not yet clear. Guide, interview, January 19, 2011. This reasoning is unpersuasive. Many aspects of the well—such as the precise pore pressures of yet-undrilled formations—are not yet clear, but operators still create a casing and drilling fluids program to guide well operations. The engineers then revise those programs as additional information becomes available.

[151] Guide, interview, September 17, 2010. Indeed, because BP recognized Macondo's production potential early on (from the seismic imaging), it involved completion engineers in the well design process from the very beginning. Sims, interview.

[152] Internal BP document (BP-HZN-MBI 180439); Internal BP document (BP-HZN-CEC 8712).

[153] Internal BP document (BP-HZN-CEC 8892).

[154] Internal BP document (BP-HZN-MBI 126338).

[155] Guide, interview, January 19, 2011.

[156] Internal BP document (BP-HZN-MBI 199123).

[157] Walz's acknowledgment came in the context of the centralizer decision. Internal BP document (BP-HZN-CEC 22662). This was not a first for the Macondo team. In early March, onshore engineer Brett Cocales sent an email to the rig's well site leader Earl Lee canceling the conversion of the float equipment on the 16-inch casing. The rig converted the float equipment anyway. This was because Lee did not see Cocales' email until after the casing had been set and cemented. After learning of the mix-up, Cocales wrote, "I understand. We will work on getting you guys any changes in the future sooner so you will have time to review." Internal BP document (BP-HZN-MBI 213550-51).

[158] Testimony of Steve Lewis, 63.

[159] A temporary abandonment procedure should be "designed with the same degree of rigor" as the initial well design. Changes in the procedure should be treated with similar rigor: "if you change one cog...you have to consider whether or not it meshes with the others." Testimony of Steve Lewis, 63-64. The Macondo team

does not appear to have clearly understood whether they should have followed BP's management of change process when changing the temporary abandonment procedures. Macondo team managers David Sims and John Guide stated that changes in the lockdown sleeve setting procedures would not, as a general rule, have required a management of change process. Sims, interview; Guide, interview, September 17, 2010. But BP's own Macondo lockdown sleeve setting procedure appears to set down in writing just such a general rule: "Any deviation, exception or addition to this procedure must be approved by BP or designated representative. BP MOC procedures must be completed prior to implementing any procedural change." Internal BP document (BP-HZN-MBI 199226).

[160] Internal BP document (BP-HZN-MBI 127489).

[161] *Ibid.*

[162] Guide, interview, January 19, 2011.

[163] Internal BP document (BP-HZN-BLY 61203).

[164] Murry Sepulvado, interview.

[165] Internal BP document (BP-HZN-BLY 61361).

[166] Testimony of Steve Lewis, 59; Ronnie Sepulvado, interview.

[167] Confidential source, interview; Walz, interview; Kelley, interview; Internal BP document (BP-HZN-OSC 112884). Before planning the type of pipe, a BP engineer asks "How much pipe is already on the rig that can be used to weight the LIT/LDS?" Internal BP document (BP-HZN-OSC 112884).

[168] Guide, interview, January 19, 2010. Guide stated that the suggestion to use heavyweight drill pipe instead of drill collars came from Transocean senior toolpusher Randy Ezell. *Ibid.*

[169] Internal BP document (BP-HZN-MBI 199236).

[170] According to wells team leader John Guide, temporary abandonment procedures were very standard: displace the wellbore to seawater, negative pressure test, and set a plug. But Guide was speaking only about his own experience. Guide, interview, January 19, 2011. A review of past BP temporary abandonment procedures shows great variation. Internal BP document (BP-HZN-OSC 6083); Internal BP document (BP-HZN-OSC 6158).

[171] Internal BP document (BP-HZN-MBI 130875); Internal BP document (BP-HZN-OSC 7918).

[172] Internal BP document (BP-HZN-OSC 112993).

[173] Internal BP document (BP-HZN-BLY 47094) ("Anyone know if there is any requirements in the MMS regs for a negative test, can't find any specifics?"); Internal BP document (BP-HZN-OSC 112888)("Regs for Temp Abandonment"); Internal BP document (BP-HZN-MBI 128655)("If anyone else has any ideas of where something else might be let me know.").

Chapter 4.6

[1] BP, Deepwater Horizon *Accident Investigation Report* (September 8, 2010), 82; John Smith, *Review of Operational Data Preceding Explosion on Deepwater Horizon in MC252* (July 1, 2010), 17, 25; John Smith (Expert witness), interview with Commission staff, September 7, 2010; Testimony of Mark Bly (BP), Hearing before the National Commission, November 8, 2010, 285; Testimony of Bill Ambrose (Transocean), Hearing before the National Commission, November 8, 2010, 280; Testimony of Richard Vargo (Halliburton), Hearing before the National Commission, November 8, 2010, 285.

[2] Internal BP document (BP-HZN-MBI 136947).

[3] Internal BP document (BP-HZN-MBI 136948); BP, Deepwater Horizon *Accident Investigation Report*, 82.

[4] BP, Deepwater Horizon *Accident Investigation Report*, 82; Smith, *Review of Operational Data Preceding Explosion on Deepwater Horizon in MC252*, 17; Smith, interview, September 7, 2010; Testimony of Mark Bly, 285; Testimony of Bill Ambrose, 280; Testimony of Richard Vargo, 285. Even during the negative pressure test, a leak in the shoe track cement cannot be identified unless some other component of the casing system, such as the float valve equipment, also leaks.

[5] Testimony of John Smith (Expert witness), Hearing before the Deepwater Horizon Joint Investigation Team, July 23, 2010, 265-67.

[6] BP well site leader John Guide said the test was designed "to see if the float equipment and the cement – actually the cement...inside of the casing is holding, [a]nd also the casing itself" and agreed that the negative test is the last evaluative test performed on a well before the BOP is pulled and the rig is demobilized. Testimony of John Guide (BP), Hearing before the Deepwater Horizon Joint Investigation Team, July 22, 2010, 137-38. Daun Winslow, Transocean General Manager for the Gulf of Mexico region, said, "A negative test...is very important to understand that your barriers are in place and they...work and they hold prior to displacing the seawater and removing the blowout preventers from the wellhead. It's very important." Testimony of Daun Winslow (Transocean), Hearing before the Deepwater Horizon Joint Investigation Team, August 24, 2010, 209.

[7] Testimony of Mark Bly, 204, 326; Testimony of Bill Ambrose, 204, 326; Testimony of Richard Vargo, 204, 326.

[8] Internal BP document (BP-HZN-OSC 1438).

[9] Internal BP document (BP-HZN-CEC 20190, 20204).

[10] Internal BP document (BP-HZN-CEC 20200-02).

[11] Testimony of Leo Lindner (M-I SWACO), Hearing before the Deepwater Horizon Joint Investigation Team, July 19, 2010, 273.

[12] Transocean contends that the choke and kill line were not fully displaced of mud. According to their calculations, which they have not shared with the Chief Counsel, the kill line had 22 barrels of mud remaining in it. Bill Ambrose (Transocean), interview with Commission staff, September 21, 2010. Dr. John Smith, an independent expert, has stated that both the volumes pumped and pressures after displacement indicate that the kill line was fully displaced with seawater. John Smith (Expert witness), email to Commission staff, October 3, 2010.

[13] Smith, *Review of Operational Data Preceding Explosion on Deepwater Horizon in MC252*, 9; Internal Halliburton document (HAL_48974).

[14] BP, Deepwater Horizon *Accident Investigation Report*, 83 and app. Q, 4; Internal BP document (BP-HZN-BLY 47100-02); Testimony of Greg Meche (M-I SWACO), Hearing before the Deepwater Horizon Joint Investigation Team, May 28, 2010, 215.

[15] Smith, *Review of Operational Data Preceding Explosion on Deepwater Horizon in MC252*, 18; Testimony of Leo Lindner, 287. Dr. Smith writes that "[a] common industry practice to minimize this occurrence is to use an unweighted, viscous spacer to follow a dense fluid that is being displaced up the annulus." Smith, *Review of Operational Data Preceding Explosion on Deepwater Horizon in MC252*, 18. Leo Lindner, the M-I SWACO engineer, stated that despite having seawater and spacer mixing, you could still have a good negative pressure test. However, he went on to say that "ideally" you would have all the spacer above the annular preventer. Testimony of Leo Lindner, 288.

[16] Internal BP document (BP-HZN-MBI 129256). M-I SWACO wrote, "I do not know the exact [stinger] tool that will be used but if there are any small restrictions in the assembly [setting up] this would be a risk." *Ibid.*

[17] Internal BP document (BP-HZN-BLY 47100-01). The BP well site leader, Murry Sepulvado, stated that the shoreside team had supposedly tested the spacer within hours after its use was suggested. Murry Sepulvado (BP), interview with Commission staff, December 10, 2010. However, BP's own investigation could find no evidence of such a test. "This material is sold by M-I SWACO for lost circulation and has no history or testing for use as a spacer. No evidence of compatibility testing could be found for the Macondo well." BP, Deepwater Horizon *Accident Investigation* Report, app. Q, 6. And although M-I SWACO recognized the possibility that the lost circulation materials presented certain risks, their communications suggested they had assumed rather than tested their compatibility as a spacer. "We do not feel there would be any restriction that would cause the FORM A SQUEEZE to set up and without [an additive in the FORM A SET] there is no cross linking agent to cause it to set up." Internal BP document (BP-HZN-MBI 129256). Although BP's subsequent investigation agreed that it was unlikely that cross-linking caused the blockage, it also found that cross-linking was not the only risk. Solids from the spacer could have plugged the kill line, or the viscosity or gel strength of the spacer could have been too high to allow pressure to be transmitted through the kill line. BP, Deepwater Horizon *Accident Investigation Report*, app. Q, 3, 9.

[18] Testimony of Ronnie Sepulvado (BP), Hearing before the Deepwater Horizon Joint Investigation Team, July 20, 2010, 129.

[19] Testimony of Leo Lindner, 309-11, 320; Testimony of John Guide, July 22, 2010, 324; Internal BP document (BP-HZN-MBI 129043); Internal BP document (BP-HZN-MBI 129240); Internal BP document (BP-HZN-MBI 129256); Internal BP document (BP-HZN-MBI 129268). The Resource Conservation and Recovery Act (RCRA) identifies materials that are hazardous waste and regulates how hazardous waste is to be managed and disposed of. 42 U.S.C. §§ 6921-6939f. RCRA regulations, however, identify exceptions to material which might otherwise be treated as hazardous waste, including "[d]rilling fluids, produced waters, and other wastes associated with the exploration, development, or production of crude oil, natural gas, or geothermal energy." 40 C.F.R. § 261.4.

[20] Testimony of Leo Lindner, 308-11, 320; Internal BP document (BP-HZN-MBI 129256); Testimony of Greg Meche, 216-18.

[21] Internal BP document (BP-HZN-BLY 47100). It is unclear whether BP or M-I SWACO came up with the original idea to use the lost circulation material as spacer, but BP ultimately approved its use. Testimony of John Guide, July 22, 2010, 323. BP well site leader Ronnie Sepulvado stated that M-I SWACO mud engineer Leo Lindner had presented the idea to him on the rig, but that he assumed he had talked to either BP or M-I SWACO engineers onshore first. Testimony of Ronnie Sepulvado, 126-31. For his part, Lindner testified that he broached the subject with Murry Sepulvado (Lindner may have misidentified the well site leader), but that "it wasn't an idea that I came up with." Testimony of Leo Lindner, 297.

[22] Internal BP document (BP-HZN-MBI 129268); Murry Sepulvado, interview.

[23] Testimony of Leo Lindner, 275-76.

[24] Internal BP document (BP-HZN-MBI 133083); Testimony of Leo Lindner, 276. In contrast to other accounts, the BP *Deepwater Horizon* Accident Investigation Report indicates that 424 barrels of spacer and 30 barrels of freshwater were pumped. BP, Deepwater Horizon *Accident Investigation Report*, 83.

[25] Internal BP document (BP-HZN-MBI 133083); BP, Deepwater Horizon *Accident Investigation Report*, 83; Smith, *Review of Operational Data Preceding Explosion on Deepwater Horizon in MC252*, 9; Testimony of Jimmy Harrell (Transocean), Hearing before the Deepwater Horizon Joint Investigation Team, May 27, 2010, 33; Internal BP document (BP-HZN-CEC 20187, 20204).

[26] BP, Deepwater Horizon *Accident Investigation Report*, 84.

[27] Transocean has indicated that it believes that 100 barrels of spacer remained beneath the BOP, suggesting that two-thirds of the annular volume between the drill pipe and casing was filled with spacer rather than seawater. Ambrose, interview. Generally consistent with Transocean's view, Dr. John Smith, an independent expert, has estimated that there was spacer at least 1,830 feet below the mudline. Smith, *Review of Operational Data Preceding Explosion on Deepwater Horizon in MC252*, 9, 18.

[28] The drill pipe pressure that should have been expected here was 1,610 psi. Instead, the reading was 2,325 psi. Smith, *Review of Operational Data Preceding Explosion on Deepwater Horizon in MC252*, 9; Internal Halliburton document (HAL_48974).

[29] None of the temporary abandonment procedures that the BP shoreside team prepared included expected pressures for the beginning of the negative pressure test. BP depended on the well site leaders to prepare such calculations. Murry Sepulvado, interview. There is no evidence that anyone present at the start of the test had calculated what pressure ought to have been expected on the drill pipe. One rig crew member, Randy Ezell, reported that toolpusher Jason Anderson had a form with expected drill pipe pressures, but there is no evidence in any of the accounts of the negative pressure test that this form was consulted. Nor is there evidence that Anderson, who worked the evening shift, would have been in the drill shack at this point. Randy Ezell (Transocean), interview with Commission staff, September 16, 2010.

[30] Testimony of Ross Skidmore (Swift), Hearing before the Deepwater Horizon Joint Investigation Team, July 20, 2010, 386. The kill line pressure had leveled off at 1,250 psi after the rig crew had completed displacing it with seawater. *Ibid;* Internal Halliburton document (HAL _48974). Skidmore said the drill pipe was bled to 1,200 psi, an insignificant difference.

[31] BP, Deepwater Horizon *Accident Investigation Report*, 24; Internal Halliburton document (HAL_48974).

[32] Testimony of John Smith (Expert witness), Hearing before the National Commission, November 9, 2010, 139-40. Spacer in the annulus between the drill pipe and the casing would cause the drill pipe pressure to increase and the kill line pressure to drop due to a phenomenon called the u-tube effect. A u-shaped tube with two differently weighted fluids on each side will tend to show increased pressure on one end of the tube as the heavier fluid pushes against the lighter fluid. At Macondo, the heavy weight of the spacer that was only on the annular (or kill line) side would push against the lighter seawater below it and exert pressure on the drill pipe. At the same time, the heavy fluid would act as a barrier to pressure being felt on the kill line.

[33] Evidence suggests that the crew may have recognized the pressure readings were abnormal and ascribed it to u-tubing. Internal BP document (BP-HZN-CEC 20188); Testimony of Chris Pleasant (Transocean), Hearing before the Deepwater Horizon Joint Investigation Team, May 28, 2010, 116. However, it appears that the u-tube effect was attributed to supposed residual mud in the kill line rather than spacer beneath the BOP. Testimony of Lee Lambert (BP), Hearing before the Deepwater Horizon Joint Investigation Team, July 20, 2010, 387.

[34] Smith, *Review of Operational Data Preceding Explosion on Deepwater Horizon in MC252*, 18-19; Testimony of John Smith, November 9, 2010, 140-41.

[35] Smith, *Review of Operational Data Preceding Explosion on Deepwater Horizon in MC252*, 9; Internal Halliburton document (HAL _48974); Testimony of John Smith, July 23, 2010, 283.

[36] Internal Halliburton document (HAL _48974)(23 barrels); Internal BP document (BP-HZN-MBI 129629). Although this is the account of the witness at the cement unit, there are other estimates. Two other witnesses described a similar gain as the amount bled by bringing the drill pipe pressure down from 2,325 to 1,250 psi. Testimony of Ross Skidmore, 386 (25 barrels); Internal BP document (BP-HZN-CEC 20226)(25 bbl). Their testimony does not offer how much was bled to bring the drill pipe pressure from 1,250 down to 260 psi. BP has at times suggested that this approximate 23-barrel bleed included the later 15-barrel bleed. BP, Deepwater Horizon *Accident Investigation Report*, 25; Internal BP document (BP-HZN-MBI 129637). There is also general agreement that 60 to 65 barrels were bled "total" to bring the drill pipe pressure to 0 psi. Internal BP document (BP-HZN-CEC 20211); Internal BP document (BP-HZN-CEC 20338, 203347). However, it is unclear whether the "0" refers to the first or last time 0 psi was reached on the drill pipe.

[37] According to one well site leader, volumes expected to be bled should always be calculated ahead of time. Murry Sepulvado, interview. There is no evidence that any of the crew had prepared estimates of how many barrels of seawater would be bled, nor is there any reference in their accounts as to how the volumes bled compared to what they were expecting. Testimony of Darryl Bourgoyne (Expert witness), Hearing before the National Commission, November 9, 2010, 149-50.

[38] Smith, *Review of Operational Data Preceding Explosion on Deepwater Horizon in MC252*, 10; Internal Halliburton document (HAL_48974).

[39] By comparison, only approximately 6.5 barrels were needed during the positive pressure test to increase the pressure from 0 to 2,500 psi. Internal BP document (BP-HZN-MBI 136948). A return of four times as many barrels when reducing the pressure by half as much should have been seen as anomalous.

[40] Smith identified four negative pressure tests that took place, only two of which were recognized by the crew. Testimony of John Smith, July 23, 2010, 272; Smith, *Review of Operational Data Preceding Explosion on Deepwater Horizon in MC252*, 18.

[41] Testimony of Randy Ezell (Transocean), Hearing before the Deepwater Horizon Joint Investigation Team, May 28, 2010, 279-80. The leakage beneath the annular preventer after displacement is not unusual. Murry Sepulvado, interview.

[42] Some have theorized that the fluid level was falling at this time not because the annular was leaking, but because the well was losing returns. The drill pipe pressure was therefore rising because the well was flowing, not because spacer was leaking beneath the BOP. Phil Rae, "The Genesis of the Deepwater Horizon Blowout Full Report," *Energy Tribune*, December 8, 2010. The theory itself suffers from a number of shortcomings. It posits that the well was losing returns and flowing at the same time. And even if the well was losing returns, if the annular preventer was closed it would have had to have been leaking in order for the fluid in the riser to fall. Finally, rig crew accounts state that mud levels in the riser were falling. Kaluza said that "some of the mud had dropped." Internal BP document (BP-HZN-CEC 20187). And Harrell stated that "there was fluid coming out of the riser, dropping down in the riser u-tube." Testimony of Jimmy Harrell, 35.

[43] Internal BP document (BP-HZN-CEC 20187). This observation explains how the crew members could have identified that the fluid levels were falling, though it took place as the riser was being topped off.

[44] Testimony of Steve Bertone (Transocean), Hearing before the Deepwater Horizon Joint Investigation, July 19, 2010, 33.

[45] Testimony of Daun Winslow, 78; Testimony of Randy Ezell, 279-80.

[46] Internal BP document (BP-HZN-CEC 20201); Testimony of Randy Ezell, 279-80; Internal BP document (BP-HZN-CEC 20226)(20 bbl). Other accounts say the riser was filled with more mud. Internal Halliburton document (HAL_48974)(50 bbl); Testimony of Chris Pleasant, 115 (60 bbl).

[47] On the other hand, the crew may not have realized that the dropping fluid levels in the riser meant that fluid was leaking beneath the BOP. Chris Pleasant, a subsea engineer, said that Anderson recognized that mud in the riser had been lost but was "convinced that we didn't lose no mud through the annular" and that as a group, "[w]e never really had a clear understanding of where the fluid went to." Testimony of Chris Pleasant, 115-16, 133. Some testimony suggests that the crew believed that mud, rather than spacer, was leaking beneath the BOP (though this still should have triggered concerns, as heavy mud could confound the test as well as spacer). Testimony of Lee Lambert, 288-89; Internal BP document (BP-HZN-CEC 20174-201).

[48] Internal BP document (BP-HZN-CEC 20187, 20201). Kaluza states that "nothing had been bled off that I know of" at the time he arrived. However, he also states that the drill pipe pressure was 1,260 psi when he arrived. Kaluza had surely missed the bleeding of the drill pipe from 2,325 to 1,250 psi to match the kill line. Given his description of what was occurring on the rig floor when he arrived, he likely also missed the bleed of the drill pipe from 1,250 to 260 psi. Internal Halliburton document (HAL_48974).

[49] According to one BP well site leader, it is common to have such leaks at the annular preventer. The annular preventer is designed to hold pressure from the bottom, not the top. If large amounts of fluid had leaked through, as had happened here, it would be necessary to displace it back to above the BOP. Murry Sepulvado, interview. BP wells team leader also stated that he would have expected the rig crew to flush the spacer above the BOP after learning that it had leaked below the annular preventer. John Guide (BP), interview with Commission staff, January 19, 2011.

[50] Internal BP document (BP-HZN-CEC 20188); BP, Deepwater Horizon *Accident Investigation Report*, 85; Smith, *Review of Operational Data Preceding Explosion on Deepwater Horizon in MC252*, 11; Internal Halliburton document (HAL_48974).

[51] BP, Deepwater Horizon *Accident Investigation Report*, 85; Smith, *Review of Operational Data Preceding Explosion on Deepwater Horizon in MC252*, 19; John Smith (Expert witness), interview with Commission staff, September, 14, 2010.

[52] When the crew on this occasion shut in the drill pipe, it closed the internal blowout preventer (IBOP). The IBOP is a valve in the top drive (a device suspended from the derrick which turns the drill string below it) on the rig. As the drill pipe pressure sensor was downstream of the IBOP, closing the IBOP prevented the drill pipe pressure from being monitored. When the IBOP was opened, the pressure at the cementing unit increased to 773 psi in less than a minute. However, it is likely that the pressure had been gradually building up at the IBOP while it had been closed. Smith, *Review of Operational Data Preceding Explosion on Deepwater Horizon in MC252*, 11.

[53] *Ibid.*; Internal Halliburton document (HAL_48974).

[54] Internal BP document (BP-HZN-CEC 20201); BP, Deepwater Horizon *Accident Investigation Report*, 85-86.

[55] Internal BP document (BP-HZN-CEC 20177, 20189, 20201-02, 20204). According to Kaluza, he wanted to discuss with Vidrine which line Vidrine wanted to monitor the negative pressure test on.

[56] Internal BP document (BP-HZN-CEC 20202-04); Internal BP document (BP-HZN-MBI 129623); Internal BP document (BP-HZN-MBI 129629); Testimony of Chris Haire (Halliburton), Hearing before the Deepwater Horizon Joint Investigation Team, May 28, 2010, 247. However, Halliburton cementer Chris Haire's report of a 15-barrel return is confusing given that he places it after the drill pipe pressure reaches 1,400 psi. *Ibid.*

[57] If witness testimony is accurate, it would appear that at this point there was good communication between the kill line and the drill pipe.

[58] Internal BP document (BP-HZN-MBI 129629); BP, Deepwater Horizon *Accident Investigation Report*, 86; Smith, *Review of Operational Data Preceding Explosion on Deepwater Horizon in MC252*, 11.

[59] Smith, *Review of Operational Data Preceding Explosion on Deepwater Horizon in MC252*, 11; Internal Halliburton document (HAL_48974); BP, Deepwater Horizon *Accident Investigation Report*, 86.

[60] Smith, *Review of Operational Data Preceding Explosion on Deepwater Horizon in MC252*, 19. The BP investigation also focused on this point in the negative pressure test as a moment of critical interpretation, stating that 1,400 psi on the drill pipe was "unexplained unless it was caused by pressure from the reservoir." BP, Deepwater Horizon *Accident Investigation Report*, 89.

[61] The BP investigation found that "[t]he 1,400 psi drill pipe pressure observed during the negative pressure test best matched communication with the M56A sand through the annulus cement barrier and shoe track barriers." BP, Deepwater Horizon *Accident Investigation Report*, 216.

[62] Internal BP document (BP-HZN-CEC 20201-02, 20204); Internal BP document (BP-HZN-CEC 20346).

[63] Witnesses consistently refer only to two negative pressure tests, one conducted on the drill pipe and one conducted on the kill line. Internal BP document (BP-HZN-CEC 20353-54); Internal BP document (BP-HZN-CEC 20190); Testimony of Jimmy Harrell, 88; Testimony of Randy Ezell, 68.

[64] Internal BP document (BP-HZN-CEC 20342, 20348); Internal BP document (BP-HZN-CEC 20205); Internal BP document (BP-HZN-CEC 20339); Internal BP document (BP-HZN-CEC 20213); Internal BP document (BP-HZN-MBI 129621); Testimony of Lee Lambert, 334; Ezell, interview. Some of the above witness accounts include toolpusher Randy Ezell as part of the discussion. Ezell, however, testified that he left the drill shack before the drill pipe pressure reached 1,400 psi. Testimony of Randy Ezell, 38-39. Two M-I SWACO mud engineers, Gordon Jones and Blair Manuel, and a Dril-Quip service technician, Charles Credeur, may have been present on the rig floor but may not have taken part in the discussion. Harrell may have been present during an earlier discussion about the negative pressure test—likely regarding the leaking annular—but not concerning the pressure abnormalities. Testimony of Jimmy Harrell, 89-90.

[65] Internal BP document (BP-HZN-CEC 20334, 20339, 20342, 20346, 20352); Internal BP document (BP-HZN-CEC 20177-78, 20190-20221, 20204-05); Testimony of Lee Lambert, 292.

[66] Testimony of Lee Lambert, 292.

[67] Smith, interview, September 7, 2010.

[68] Internal BP document (BP-HZN-CEC 20204).

[69] Internal BP document (BP-HZN-CEC 20178).

[70] Testimony of Lee Lambert, 395-96.

[71] Testimony of Jimmy Harrell, 117; Internal BP document (BP-HZN-CEC 20204); Internal BP document (BP-HZN-OSC 1438).

[72] Internal BP document (BP-HZN-CEC 20200). *See also* Chapter 4.5 on TA procedures.

[73] According to Ezell, Vidrine "wasn't happy with the results from the first test." Testimony of Randy Ezell, 300.

[74] Smith, *Review of Operational Data Preceding Explosion on Deepwater Horizon in MC252*, 12; Internal Halliburton document (HAL_48974); Internal BP document (BP-HZN-CEC 20191, 20205). When the crew initially bled the kill line, 0.6 barrels were bled off to reach 0 psi. When the kill line was shut in, pressure rose to 30 psi. The crew then bled the pressure down to 0 psi again, bleeding off 0.2 more barrels. Internal BP document (BP-HZN-CEC 20205); Internal BP document (BP-HZN-CEC 20351-52).

[75] Internal BP document (BP-HZN-CEC 20339, 20348, 20352).

[76] Internal BP document (BP-HZN-CEC 20177-78, 20190, 20204-05). According to Kaluza, the bladder effect was first discussed at the end of the negative pressure test on the drill pipe, to explain the rise in drill pipe pressure to 1,400 psi. The bladder effect was also then discussed during the test on the kill line as an explanation for how there could be 1,400 psi on the drill pipe despite no flow on the kill line.

[77] Internal BP document (BP-HZN-MBI 262896-97).

[78] Internal BP document (BP-HZN-CEC 20342, 20359).

[79] Testimony of Darryl Bourgoyne, 174-75; Testimony of John Smith, November 9, 2010, 175-76; Testimony of Steve Lewis (Expert witness), Hearing before the National Commission, November 9, 2010, 176-77; Murry Sepulvado, interview; Ronnie Sepulvado (BP), interview with Commission staff, October 26, 2010.

[80] Testimony of Darryl Bourgoyne, 174-75; Murry Sepulvado, interview; Ronnie Sepulvado, interview; Testimony of Bill Ambrose, 208.

[81] According to Transocean offshore installation manager Jimmy Harrell, both the well site leader and the toolpusher were interpreting the negative pressure test data. Testimony of Jimmy Harrell, 91. Although he was not on the rig floor during the interpretation of results, Harrell understood the negative pressure test to have been successful. *Ibid.*, 117. According to Pat O'Bryan, BP vice president for drilling and completions, Transocean's toolpusher and driller would be able to interpret the results of a negative pressure test. Testimony of Pat O'Bryan (BP), Hearing before the Deepwater Horizon Joint Investigation Team, August 26, 2010, 449-50. And according to John Guide, BP wells team leader, the company man was "one of the people" who were supposed to determine if the negative pressure test was successful or not. Testimony of John Guide, July 22, 2010, 161-62.

[82] Testimony of Randy Ezell, 311-12; Testimony of Paul Johnson (Transocean), Hearing before the Deepwater Horizon Joint Investigation Team, August 23, 2010, 356. Though Transocean has also stated it is unclear whether or not its rig personnel had the training and experience to interpret the negative pressure test. Testimony of Bill Ambrose, 207; Internal BP document (BP-HZN-CEC 20334, 20339, 20342, 20346, 20352); Internal BP document (BP-HZN-CEC 20177-78, 20190-20221, 20204-05); Testimony of Lee Lambert, 292.

[83] Vidrine may have made a call to Mark Hafle onshore during the negative pressure test but not talked about the results of the test. Internal BP document (BP-HZN-CEC 20339, 20352); Internal BP document (BP-HZN-CEC 20245). There is testimony from the rig crew that Kaluza called John Guide after the first negative pressure test. Testimony of Chris Pleasant, 117-18. Guide has denied this, and there is no evidence of this in BP's notes of its interviews with Kaluza. Testimony of John Guide, July 22, 2010, 175. Nor is there any conclusive evidence of this in logs of telephone calls made from the rig. While Guide made several brief calls to the rig during the negative pressure test (all under five minutes) in an attempt to determine how the executives' visit was going, he never spoke with the well site leaders. Guide, interview, January 19, 2011; Benjamin Powell (BP legal team), letter to Commission staff, December 22, 2010, telephone log attachment. Ezell states that the rig crew never asked him about the 1,400 psi during the test, though several witness accounts place him in the drill shack for at least some portion of the discussion. Ezell, interview; Internal BP document (BP-HZN-CEC 20342, 20348); Internal BP document (BP-HZN-CEC 20205); Internal BP document (BP-HZN-CEC 20339); Internal BP document (BP-HZN-CEC 20213).

[84] Testimony of Mark Bly, 286-90; Testimony of Bill Ambrose, 286-90; Murry Sepulvado, interview.

[85] Testimony of John Guide, July 22, 2010, 133-34; Testimony of John Guide (BP), Hearing before the Deepwater Horizon Joint Investigation Team, October 7, 2010 187-92; Testimony of Darryl Bourgoyne, 156-58.

[86] John Guide (BP), interview with Commission staff, September 17, 2010. Morel indicated in his internal BP interview that "if negative test unsuccessful the decision tree says contact John Guide." Internal BP document (BP-HZN-MBI 21306).

[87] Testimony of John Guide, July 22, 2010, 234-37. Transocean OIM Harrell had judged even the first test to be a successful one. Testimony of Jimmy Harrell, 26.

[88] Powell, letter, telephone log attachment. According to notes of BP's internal investigation interview of Vidrine, Vidrine may have described the test as "squirrelly" but assured Hafle that the 1,400 psi did not indicate a problem because "if there had been a kick in the well we would have seen it." Internal BP document (BP-HZN-CEC 20343, 20348). Hafle's reaction to the 1,400 psi, if he was told about it, was apparently to tell Vidrine to "check the line up." Internal BP document (BP-HZN-CEC 20359); Internal BP document (BP-HZN-BLY 61374). Other accounts suggest Vidrine did not specifically tell Hafle about the drill pipe pressure. Internal BP document (BP-HZN-CEC 20234). But Hafle's account states that he had INSITE up on his computer screen during the call, a program that allowed him to access data transmitted from the rig. Internal BP document (BP-HZN-BLY 61374).

[89] Telephone logs show that wells team leader John Guide made several brief phone calls to the rig during the course of the test. (Powell, letter, telephone log attachment). Guide explained that he was calling to inquire about the progress of the VIP tour and did not discuss the course of the negative pressure test. Guide, interview, January 19, 2011.

[90] There are multiple ways that the spacer could have prevented flow from the kill line. Spacer could have clogged the kill line by either a phenomenon known as "bridging," by acting as a gelatin (the stiffening that occurs when the spacer is stationary), or by exerting a higher yield point and yield stress than the seawater above it. Regardless of how it acted, the important point may be that the clogging would not have needed to act as a complete barrier against the 1,400 psi in the drill pipe. Dr. John Smith has indicated that the spacer would only have needed to absorb 100 to 200 psi to prevent flow. Smith, interview, September, 14, 2010. Even if it did not clog the kill line, the heavy spacer beneath the BOP would have handicapped the test. Murry Sepulvado, interview. If the intent in displacing the well with seawater to 8,367 feet was to simulate conditions after abandonment, the presence of heavy spacer beneath the BOP defeated it. The cement shoe at the bottom of the well was expected to hold back formation pressures from below with only a column of water above 8,367 feet on top of it. Like a heavy weight compressing a spring below it, the heavier spacer would have exerted additional pressure from above that would have prevented the shoe cement from bearing the full formation pressure it would face in abandonment. Once displacement of the riser resumed after the negative pressure test, however, this heavy spacer would be removed.

[91] There are problems associated with each of these theories. BP has suggested that a valve connecting the kill line to the BOP may have been accidentally left closed during the negative pressure test. BP, Deepwater Horizon *Accident Investigation Report*, 87. However, that valve would have been opened minutes before during the negative pressure test on the drill pipe. The theory thus assumes that the valve was closed and

then almost immediately forgotten about. Transocean has suggested the kill line may have been clogged with mud, as it was never fully displaced during preparations for the negative pressure test. Ambrose, interview. However, Dr. John Smith has stated that both the volumes pumped and pressures after displacement indicate the kill line was fully displaced with seawater. Smith, email. While well site leader John Guide and drilling engineer Brian Morel have suggested that hydrates from migrating gas may have frozen in the kill line, no evidence has been produced suggesting that this actually took place or that gas had made it to the BOP as early as the time of the negative pressure test. Guide, interview, September 17, 2010; Internal BP document (BP-HZN-CEC 20247).

[92] Deepwater Horizon Study Group, *The Macondo Blowout: 3rd Progress Report* (December 5, 2010), app. B, 10; National Academy of Engineering, *Interim Report on Causes of Deepwater Horizon Oil Rig Blowout and Ways to Prevent Such Events* (November 16, 2010), 9-10.

[93] Smith, *Review of Operational Data Preceding Explosion on Deepwater Horizon in MC252*, 19.

[94] For this reason, Transocean indicated "that spacer placement became ever so important but may have been overlooked. And that added confusion, and in that regards the test became more complicated." Testimony of Bill Ambrose, 207.

[95] Internal BP document (BP-HZN-BLY 47100).

[96] Wells team leader John Guide agreed that personnel on the rig should have done so. Guide, interview, January 19, 2011.

[97] Testimony of John Guide, July 22, 2010, 333; Testimony of Daun Winslow, 194-95.

[98] Testimony of John Guide, July 22, 2010, 333. Well site leader Don Vidrine stated that there is "[n]o standard procedure on how to do these...leave to rig on how to do procedure." Internal BP document (BP-HZN-CEC 20335).

[99] Testimony of Daun Winslow, 194-95. The previous negative pressure tests performed by the *Deepwater Horizon* crew at the Kodiak II and Tiber wells had been devised by the well site leader Murry Sepulvado and toolpusher Jason Anderson. Their method was to displace the choke, kill, and boost lines with seawater and to displace the drill pipe with spacer and seawater until the drilling mud was above the annular preventer. The method's use of the drill pipe to conduct the negative pressure test explains why the test was initially conducted on the drill pipe, despite the fact that the later APM stated that the negative pressure test would be done "with the kill line." According to Ezell, this method was printed out and laminated by Murry Sepulvado and available in the drill shack. However, neither Murry nor Ronnie Sepulvado recalls such a procedure. Moreover, the procedure was "generic" in the sense that it did not include specific volumes or pressures to be expected on an individual well. Testimony of Leo Lindner, 347-48; Guide, interview, September 17, 2010; Ezell, interview; Murry Sepulvado, interview; Ronnie Sepulvado, interview.

[100] Before unlatching from the well in anticipation of Hurricane Ida, Transocean's *Marianas* conducted a negative pressure test. The negative pressure test was different in several ways. It used base oil rather than seawater. The kill line was displaced rather than the drill pipe. There was no displacement beneath the wellhead. The choke and boost lines were not displaced beforehand. Internal BP document (BP-HZN-MBI 172005).

[101] Testimony of Leo Lindner, 271-72.

[102] In September 2010, BOEMRE, the agency formerly known as MMS, proposed to update its regulations. The new regulations require that a negative pressure test be performed on intermediate and production casing strings, that test procedures and criteria be provided on the permit application, and that the results of the test be available for inspection. 30 C.F.R. § 250.423(c); Oil and Gas and Sulphur Operations in the Outer Continental Shelf—Increased Safety Measures for Energy Development on the Outer Continental Shelf, 75 Fed. Reg. 63346, 63373 (October 14, 2010).

[103] Negative pressure tests are done only if the well will experience a similar underbalanced pressure condition during temporary abandonment. In many wells (especially land wells) the well is abandoned in an overbalanced state, so a negative pressure test is not necessary. Testimony of John Smith, November 9, 2010, 153; Darryl Bourgoyne (Expert witness), email to Commission staff, December 24, 2010.

[104] Smith, *Review of Operational Data Preceding Explosion on Deepwater Horizon in MC252*, 17.

[105] *Ibid.*

[106] Oil and Gas and Sulphur Operations in the Outer Continental Shelf—Increased Safety Measures for Energy Development on the Outer Continental Shelf, 75 Fed. Reg. 63346, 63373 (October 14, 2010); John Smith (Expert witness), interview with Commission staff, October 26.

[107] Murry Sepulvado, interview.

[108] Internal BP document (BP-HZN-CEC 20352); Internal BP document (BP-HZN-CEC 20191-92); Internal BP document (BP-HZN-CEC 20264).

[109] Internal BP document (BP-HZN-MBI 262896-97); Internal BP document (BP-HZN CEC 20196).

[110] Testimony of Bill Ambrose, 291; Internal Transocean document (TRN-HCEC 5402).

[111] Testimony of Bill Ambrose, 205-07.

[112] Smith, interview, October 26, 2010; Darryl Bourgoyne (Expert witness), interview with Commission staff, October 26, 2010.

[113] Guide, interview, January 19, 2010; Pat O'Bryan (BP), interview with Commission staff, December 17, 2010.

[114] Murry Sepulvado, interview.

[115] Internal BP document (BP-HZN-MBI 126180, 126189-90). Morel was certainly unfamiliar with whether there were any regulatory requirements for the negative pressure test. He asked in one email, apparently sent after the APM had been submitted, "Anyone know if there is any requirements in the MMS regs for a negative test, can't find any specifics?" and later said "If anyone else has any ideas of where something else might be let me know." Internal BP document (BP-HZN-BLY 47094); Internal BP document (BP-HZN-MBI 128655).

[116] Internal BP document (BP-HZN-MBI 126585-86).

[117] Testimony of Jimmy Harrell, 26-28; Ezell, interview. Harrell's testimony indicates that the omission of the negative pressure test may have occurred during briefing on April 19 or April 20. Testimony of Jimmy Harrell, 76-77, 115-16. It seems difficult to understand that Kaluza would have omitted the test from a briefing on the morning of April 20, as he (1) had discussed the test during the rig call earlier that morning, (2) had asked Lindner how they conducted the negative pressure test on earlier wells after the rig call, and (3) had just received Morel's Ops Note, which included the negative pressure test. Gregory Walz (BP), interview with Commission staff, October 6, 2010; Testimony of Leo Lindner, 271-72; Internal BP document (BP-HZN-MBI 195580).

[118] Independent experts have stated that these calculations should have been part of the temporary abandonment procedure. Testimony of John Smith, November 9, 2010, 147; Testimony of Darryl Bourgoyne, 148.

[119] Internal BP document (BP-HZN-CEC 20206).

[120] Internal BP document (BP-HZN-BLY 61380).

[121] Internal BP document (BP-HZN-MBI 133083).

[122] According to Kaluza's account in his interview with BP investigators, he left the rig floor after Leo Lindner's 3 p.m. safety meeting and did not arrive until Wyman Wheeler was filling the riser with mud after the fluid level had fallen. Internal BP document (BP-HZN-CEC 20200-01). According to BP's own investigators' notes, Kaluza "was in office and did not know volume." Internal BP document (BP-HZN-CEC 20207). However, Randy Ezell's testimony suggests that Kaluza may have been present at the time that he and the executive tour arrived at the drill shack. Testimony of Randy Ezell, 279-80.

[123] If Kaluza was not on the rig floor as the annular leaked, it may explain his statement to investigators that "spacer was above the top annular" even though it had by this point migrated beneath it. Internal BP document (BP-HZN-CEC 20201); Internal BP document (BP-HZN-CEC 20174).

[124] Murry Sepulvado, interview.

[125] Darryl Bourgoyne (Expert witness), email to the Commission staff, December 18, 2010; Steve Lewis (Expert witness), email to Commission staff, December 29, 2010. BP wells team leader John Guide also stated that he would expect well site leaders to be in the drill shack when the negative pressure test is run. Guide, interview, January 19, 2011. BP management appears to encourage well site leader presence on the rig floor. Kaluza's most recent performance evaluation before the blowout criticized him for not spending enough time there, warning that safety could not be assured from sitting in the well site leader's office. It is not clear, however, what occasions the evaluation was referring to, as the Chief Counsel's team was unable to interview Kaluza or his evaluator. Kaluza was criticized for "giving priority to WSL office preparation for meetings;" "can't assure HSE [Health, Safety, and Environment] and rig operation performance or be aware of the details of how the crews are executing their jobs from WSL office;" "he should spend more time on the

rig floor;" and "should spend more time out on deck and other parts of the rig to work with the crews towards safety performance improvement." Internal BP document (BP-HZN-MBI 193097).

[126] Internal BP document (BP-HZN-CEC 20207).

[127] Internal BP document (BP- HZN-CEC 20174, 20185-87, 20201, 20204).

[128] Lewis, email. A former well site leader has stated that while he would not have felt that examining it personally would have been required, he would have examined the spacer if his time allowed. Bourgoyne, email, December 28, 2010.

[129] Testimony of Mark Bly, 286-90; Testimony of Bill Ambrose, 286-90. Morel stated that his Ops Note directed that results of the positive pressure test on the surface cement plug were to be sent to Houston, but there was "no similar requirement for negative test." Internal BP document (BP-HZN-CEC 20235).

[130] Guide, interview, September 17, 2010; Internal BP document (BP-HZN-MBI 21306).

[131] Internal BP document (BP-HZN-MBI 21330); Internal BP document (BP-HZN-MBI 129616); Testimony of John Guide, July 22, 2010, 197.

Chapter 4.7

[1] Transocean asserts that a reduced pump efficiency during the final displacement potentially "skew[ed] the measurement of returns and potentially mask[ed] the entry of hydrocarbons into the well." Transocean legal team, letter to Commission staff, November 5, 2010. Even if true, this assertion does not alter the Chief Counsel's team's findings. The analysis in this section is based on data anomalies that *are* apparent (despite any error in pump efficiency). A correct pump efficiency would only have made *more* anomalies apparent. Furthermore, if the pump efficiency did indeed decrease, rig personnel properly monitoring the data by performing volumetric calculations should have detected the change during the displacement itself and taken actions to resolve the discrepancy. Darryl Bourgoyne (Expert witness), interview with Commission staff, November 23, 2010.

[2] Rig personnel can augment an existing barrier, such as by increasing the weight of the mud in the well, or put in place a separate barrier, such as by closing in the well with the BOP.

[3] Hearing to Review Recent Issues in Offshore Oil and Gas Development, Before the S. Comm. on Energy and Natural Resources, 111th Cong. (May 11, 2010)(statement of F.E. Beck, Texas A&M University).

[4] There are several more parameters that rig personnel use to detect whether a kick is developing, including rate of penetration and changes in the salinity and electrical resistivity of mud. American Petroleum Institute, *Recommended Practice for Well Control Operations 59*, 2nd ed. (May 2006), 33 ("API RP 59").

[5] Internal BP document (BP-HZN-BLY 61693).

[6] API RP 59, 33; Doug Slitor (MMS), "Drilling Operations and Equipment" (presentation to Commission staff, August 3, 2010), 33.

[7] A displacement will also not appear to be closed-loop, even if all fluids come into and out of the pits, if the pits involved in the fluid transfer are not all selected as part of the active pit system. For example, when fluid going into the well is taken from an active pit, but fluid coming out of the well is returned to a reserve pit.

[8] Transocean personnel typically performed this calculation by hand, periodically throughout a displacement. Allen Seraile (Transocean), interview with Commission staff, January 7, 2011. The Sperry-Sun system may calculate volume-in automatically. Testimony of Joseph Keith (Halliburton), Hearing before the Deepwater Horizon Joint Investigation Team, December 7, 2010, part 1, 193. But it does not compare volume-in and volume-out to compute pit gain automatically. Testimony of John Gisclair (Halliburton), Hearing before the Deepwater Horizon Joint Investigation Team, December 7, 2010, part 2, 137.

[9] Testimony of John Gisclair (Halliburton), Hearing before the Deepwater Horizon Joint Investigation Team, October 8, 2010, 100.

[10] API RP 59, 33. Because the two numbers are derived in different ways (one a measurement, the other a calculation), the difference between them need not be zero so much as constant. Testimony of John Gisclair, October 8, 2010, 101.

[11] API RP 59, 33.

[12] *Ibid.*, 34. Rig personnel should carefully investigate each of these other phenomena as well. Rig heave can be accounted for by monitoring for several heave cycles. Thermal expansion would be exceedingly slow and should be watched. And ballooning would have to be fingerprinted and applies only if open hole sections are exposed, which was not the case at the time of the Macondo explosion. Darryl Bourgoyne (Expert witness), email to Commission staff, December 16, 2010.

[13] The amount of residual flow is rig-specific and can be as high as 120 barrels. Commission staff site visit to *Deepwater Nautilus*, September 9, 2010.

[14] Testimony of John Guide (BP), Hearing before the Deepwater Horizon Joint Investigation Team, October 7, 2010, part 2, 199.

[15] Testimony of John Guide, October 7, 2010, 199; Testimony of John Gisclair (Halliburton), Hearing before the National Commission, November 8, 2010, 230. "When you're staring at these traces, you're going to have to wait a significant number of minutes in some cases to notice a certain trend." Testimony of John Gisclair, November 8, 2010, 230.

[16] Drill pipe pressure is actually represented by stand pipe pressure. The stand pipe is a line connecting the pumps to the drill pipe (via the kelly hose and top drive). The pressure sensor is located on that line. Commission staff site visit to *Deepwater Nautilus*, September 9, 2010.

[17] Testimony of John Gisclair, October 8, 2010, 135.

[18] Bourgoyne, email, December 16, 2010.

[19] Darryl Bourgoyne (Expert witness), interview with Commission staff, October 26, 2010; John Smith (Expert witness), interview with Commission staff, October 26, 2010.

[20] Bourgoyne, interview, October 26, 2010; Smith, interview, October 26, 2010.

[21] John Smith (Expert witness), interview with Commission staff, September 7, 2010.

[22] API RP 59, 34.

[23] Darryl Bourgoyne (Expert witness), email to Commission staff, November 23, 2010.

[24] API RP 59, 34; Testimony of John Gisclair, October 8, 2010, 227-29; Bourgoyne, interview, October 26, 2010; Smith, interview, October 26, 2010.

[25] API RP 59, 34. But "Until a confirmation can be made as to whether the cause is a hole or a well kick, a kick should be assumed." *Ibid.*

[26] Bourgoyne, interview, October 26, 2010; Smith, interview, October 26, 2010.

[27] BP's decision tree on monitoring wellbore pressure, included in its Macondo well drilling program, identifies an increase in gas as an indication of increasing pore pressure and a reason to stop drilling and check for flow. Internal BP document (BP-HZN-BLY 39344).

[28] Steve Lewis (Expert witness), interview with Commission staff, September 2, 2010.

[29] Bourgoyne, email, December 16, 2010.

[30] Internal Transocean document (TRN-HCEC 90727); Bill Ambrose (Transocean), interview with Commission staff, September 30, 2010.

[31] Internal Halliburton document (MC252_001_ST00BP01_EOWR).

[32] *Ibid.*

[33] Testimony of John Gisclair, November 8, 2010, 233. "Most of the data that is in that Sperry database was transmitted to us realtime from Transocean." *Ibid.*

[34] Internal BP document (BP-HZN-BLY 38391); Testimony of John Gisclair, November 8, 2010, 234.

[35] Testimony of John Gisclair, November 8, 2010, 233; Testimony of Joseph Keith, 113.

[36] Testimony of John Gisclair, October 8, 2010, 96; Commission staff site visit to *Deepwater Nautilus*, September 9, 2010; Internal Halliburton document (MC252_001_ST00BP01_EOWR).

[37] Commission staff site visit to *Deepwater Nautilus*, September 9, 2010.

[38] *Ibid.*; Bill Ambrose (Transocean), interview with Commission staff, September 21, 2010; Testimony of John Gisclair, October 8, 2010, 109-10. Indeed, setting an alarm for pit volume change is typically standard operating procedure. *Ibid.,* 109-10.

[39] Ronnie Sepulvado (BP), interview with Commission staff, October 26, 2010.

[40] Joseph Keith (Halliburton), interview with Commission staff, October 6, 2010; Darryl Bourgoyne (Expert witness), interview with Commission staff, September 10, 2010; Commission staff site visit to *Deepwater Nautilus*, September 9, 2010.

[41] Testimony of John Gisclair, October 8, 2010, 99; Internal BP document (BP-HZN-MBI 21292).

[42] Internal Halliburton document (MC252_001_ST00BP01_EOWR).

[43] Testimony of John Gisclair, October 8, 2010, 97.

[44] Internal BP document (BP-HZN-MBI 21292).

[45] Testimony of John Gisclair, October 8, 2010, 109; Testimony of Joseph Keith, 113.

[46] Another difference between the sensors was their mode of measurement. The Hitec flow-out sensor measured flow-out in terms of a percentage of total flow; the Sperry-Sun sensor measured it in terms of gallons per minute. Randy Ezell (Transocean), interview with Commission staff, September 16, 2010. In terms of crew reliance, Transocean and BP rig personnel looked at the Hitec flow-out meter. Testimony of John Guide (BP), Hearing before the Deepwater Horizon Joint Investigation Team, July 22, 2010, 302. The Sperry Drilling mudlogger did not understand the readings from the Hitec flow-out sensor and did not rely on it. Keith, interview; Testimony of Joseph Keith, 172-73.

[47] Testimony of Joseph Keith, 152-53.

[48] Commission staff site visit to *Deepwater Nautilus*, September 9, 2010; Murry Sepulvado (BP), interview with Commission staff, December 10, 2010. Returns first go to the gumbo box, then split off to either the shakers or overboard. Commission staff site visit to *Deepwater Nautilus*, September 9, 2010.

[49] Internal Transocean document (TRN-HCEC 90727). "There were no reported discrepancies for pit volumes, [flow-out] indicators, mud pump strokes or weight," and the inspectors' "scan of the pages did not show any anomalies." *Ibid.* All of the gauges were in good condition and had proper calibration labels; and the driller and electronics technician confirmed "the condition of the drilling instruments and that the system operated without any problems." *Ibid.*

[50] Testimony of Stephen Bertone (Transocean), Hearing before the Deepwater Horizon Joint Investigation Team, July 19, 2010, 30; Testimony of Michael Williams, Hearing before the Deepwater Horizon Joint Investigation Team, July 23, 2010, 42; Seraile, interview.

[51] Testimony of Stephen Bertone, 195-96; Testimony of John Guide, July 22, 2010, 63; Testimony of Michael Williams, July 23, 2010, 104.

[52] Testimony of Stephen Bertone, 197-200; Testimony of Michael Williams, 42, 99-101. On a prior well, a crash of the *Horizon* A-chair coincided with a well control incident. *Ibid.*

[53] Testimony of John Guide, July 22, 2010, 63; Testimony of Stephen Bertone, 199; Testimony of Michael Williams, 102-03. They may have also replaced some of the chairs' software. Testimony of Paul Johnson (Transocean), Hearing before the Deepwater Horizon Joint Investigation Team, August 23, 2010, part 2, 164. Transocean planned to entirely upgrade the hardware and software at a later date, after the *Horizon* left the Macondo well. Internal Transocean document (TRN-HCEC 90727); Testimony of Paul Johnson, 164. According to one witness, Transocean was waiting to see how the upgraded package worked on a sister rig before installing it on the *Horizon*. Testimony of Michael Williams, 99.

[54] Testimony of John Guide, July 22, 2010, 63; Testimony of Stephen Bertone, 200; Testimony of Paul Johnson, 164-65; Seraile, interview.

[55] Internal Transocean document (TRN-HCEC 90727).

[56] Testimony of Stephen Bertone, 199. Even if the chairs had crashed on April 20, the Sperry-Sun system would have remained operational. Seraile, interview.

[57] Testimony of Bill Ambrose (Transocean), Hearing before the National Commission, November 8, 2010, 239; Testimony of John Gisclair, November 8, 2010, 240-41.

[58] Internal BP document (BP-HZN-MBI 21442).

[59] Testimony of Micah Burgess (Transocean), Hearing before the Deepwater Horizon Joint Investigation Team, May 29, 2010, 94-95.

[60] Internal BP document (BP-HZN-BLY 38370, 38378).

[61] Testimony of Mark Bly (BP), Hearing before the National Commission, November 8, 2010, 247-48; Testimony of Micah Burgess, 84.

[62] Commission staff site visit to *Deepwater Nautilus*, September 9, 2010.

[63] Testimony of Darryl Bourgoyne, 179.

[64] *Ibid.*; Testimony of Micah Burgess, 84. "Roles as the driller would be just to – I mean, to maintain, watching the well, know drilling, and make sure, you know, I had the authority to shut the well in at any time I had any doubt about anything." Testimony of Micah Burgess, 84.

[65] Internal BP document (BP-HZN-MBI 21434).

[66] *Ibid.*; Testimony of Micah Burgess, 97.

[67] Testimony of Allen Seraile (Transocean), Hearing before the Deepwater Horizon Joint Investigation Team, May 29, 2010, 129-30.

[68] Internal BP document (BP-HZN-MBI 139541).

[69] Commission staff site visit to *Deepwater Nautilus*, September 9, 2010.

[70] Seraile, interview.

[71] Testimony of Mark Bly, November 8, 2010, 249.

[72] Internal Halliburton document (MC252_001_ST00BP01_EOWR); Testimony of Mark Bly, November 8, 2010, 249; Testimony of Joseph Keith, 18. "Sperry Drilling Services (Unit # 82418) was contracted to perform surface data logging and pore pressure prediction services by BP Exploration and Production for the Macondo Prospect 001 ST00BP00 in Mississippi Canyon Block 252." Internal Halliburton document (MC252_001_ST00BP01_EOWR).

[73] Internal Halliburton document (MC252_001_ST00BP01_EOWR); Testimony of John Gisclair, October 8, 2010, 140; Testimony of Joseph Keith, 27.

[74] Testimony of Micah Burgess, 100.

[75] Cathleenia Willis (Halliburton), interview with Commission staff, October 21, 2010.

[76] Testimony of Joseph Keith, 22-23.

[77] Internal BP document (BP-HZN-BLY 38334).

[78] Internal Halliburton document (MC252_001_ST00BP01_EOWR); Willis, interview.

[79] Testimony of Mark Bly, November 8, 2010, 247; Testimony of Ronnie Sepulvado (BP), Hearing before the Deepwater Horizon Joint Investigation Team, July 20, 2010, 78.

[80] Bourgoyne, interview, September 10, 2010; Steve Lewis (Expert witness), interview with Commission staff, September 21, 2010; Ezell, interview. Transocean argues that the BP company man should monitor all critical operations and that the final displacement was a critical operation. Transocean legal team, interview with Commission staff, August 17, 2010. BP disagrees. Mark Bly (BP), interview with Commission staff, September 8, 2010.

[81] Lee Lambert (BP), interview with Commission staff, September 17, 2010; Ezell, interview; Testimony of Joseph Keith, 110-11; Ronnie Sepulvado (BP), interview with Commission staff, August 20, 2010.

[82] Internal BP document (BP-HZN-MBI 21458).

[83] *Ibid.*

[84] Ezell, interview.

[85] Internal BP document (BP-HZN-MBI 21456).

[86] Ezell, interview.

[87] Internal BP document (BP-HZN-MBI 21448).

[88] Commission staff site visit to *Deepwater Nautilus*, September 9, 2010; Ezell, interview.

[89] This feature was called *INSITE Anywhere*. Testimony of Michael Beirne (BP), Hearing before the Deepwater Horizon Joint Investigation Team, October 6, 2010, part 2, 18-19, 34.

[90] Testimony of Bill Ambrose, 240; Testimony of John Guide, July 22, 2010, 418-419.

[91] Testimony of Brett Cocales, Hearing before the Deepwater Horizon Joint Investigation Team, August 27, 2010, 233-35.

[92] Testimony of John Guide, October 7, 2010, 152; Internal Halliburton document (MC252_001_ST00BP01_EOWR); Internal Halliburton document (HAL_50546)(log showing BP, Anadarko, MOEX, and Halliburton personnel accessing INSITE).

[93] Testimony of Michael Beirne, 18-19. Co-owners Anadarko and MOEX had a contractual right to the data, and their personnel accessed it. *Ibid.*, 18-19, 96-97; Internal BP document (BP-HZN-MBI 173481-83, 174919, 175868).

[94] Testimony of Mark Bly, November 8, 2010, 234; Testimony of Mark Bly (BP), Hearing before the National Academy of Engineering, September 26, 2010; Mike Zanghi (BP), interview with Commission staff, December 15, 2010; Internal BP document (BP-HZN-MBI 175868).

[95] Testimony of John Gisclair, October 8, 2010, 103; Internal Halliburton document (MC252_001_ST00BP01_EOWR).

[96] Testimony of John Gisclair, October 8, 2010, 107-08; Testimony of John Gisclair, December 7, 2010, 25-26.

[97] Testimony of Brett Cocales, 234; Testimony of Joseph Keith, 242.

[98] Testimony of Paul Johnson, 332; Testimony of Bill Ambrose, 240.

[99] Testimony of Bill Ambrose, 238.

[100] Commission staff site visit to *Deepwater Nautilus*, September 9, 2010; Testimony of Bill Ambrose, 223. Some data display systems record metadata of user settings including what screen the user has up at any given time, what alarm thresholds the user has set, when an alarm activates, and when the user acknowledges an alarm. Dan Jenkins (Oilfield Instrumentation), interview with Commission staff, November 12, 2010. This does not appear to have been the case with the Sperry-Sun and Hitec systems on the *Deepwater Horizon*.

[101] Testimony of Bill Ambrose, 223.

[102] Testimony of John Gisclair, November 8, 2010, 228.

[103] Keith, interview; Seraile, interview; Ronnie Sepulvado, interview, October 26, 2010.

[104] Ronnie Sepulvado, interview, October 26, 2010.

[105] Testimony of Bill Ambrose, 224; Testimony of John Gisclair, October 8, 2010, 107, 122; Testimony of John Gisclair, December 7, 2010, 70; Willis, interview. The driller would have had standpipe or drill pipe pressure on his screen. Testimony of Bill Ambrose, 224.

[106] Testimony of John Gisclair, November 8, 2010, 230.

[107] Internal BP document (BP-HZN-MBI 133083).

[108] Internal BP document (BP-HZN-MBI 139541).

[109] Internal BP document (BP-HZN-MBI 21387).

[110] Testimony of Charles Credeur (Dril-Quip), Hearing before the Deepwater Horizon Joint Investigation Team, May 29, 2010, 62-63; Testimony of Heber Morales (Transocean), Hearing before the Deepwater Horizon Joint Investigation Team, May 29, 2010, 143-44; Internal BP document (BP-HZN-MBI 129622); Brandon Bouillion (Weatherford), interview with Commission staff, October 22, 2010. One of an assistant driller's responsibilities is to make up connections and equipment. Internal BP document (BP-HZN-MBI 21434).

[111] Internal BP document (BP-HZN-MBI 139543); Testimony of John Gisclair, November 8, 2010, 249.

[112] Internal BP document (BP-HZN-MBI 21406); Sperry-Sun data, April 20, 2010, 20:02. "Sperry-Sun data" references an internal Halliburton document (HAL_48973-74).

It appears that the crew had a short meeting in the drill shack prior to displacing the riser. The details of the meeting are unclear. The drill crew typically uses such meetings to review the remaining steps in the operation and discuss expected volumes and pump strokes, but not pressures. Caleb Holloway (Transocean), interview with Commission staff, December 20, 2010; Seraile, interview.

[113] Testimony of John Gisclair, October 8, 2010, 129.

[114] Testimony of Joseph Keith, 98, 190.

[115] Testimony of Joseph Keith, 99; Testimony of Ronnie Sepulvado, July 20, 2010, 137.

[116] Testimony of Joseph Keith, 146; Sperry-Sun data, April 20, 2010.

[117] John Smith, *Review of Operational Data Preceding Explosion on Deepwater Horizon in MC252* (July 1, 2010), 13.

[118] Testimony of Leo Lindner (M-I SWACO), Hearing before the Deepwater Horizon Joint Investigation Team, July 19, 2010, 275.

[119] John Gisclair (Halliburton), interview with Commission staff, September 20, 2010; Seraile, interview.

[120] This behavior was expected. Testimony of Joseph Keith, 134.

[121] Transocean legal team, "April 20 End of Well Activities" (presentation to Commission staff, November 5, 2010); Sperry-Sun data, April 20, 2010.

[122] Smith, *Review of Operational Data Preceding Explosion on Deepwater Horizon in MC252*, 13.

[123] Gisclair, interview.

[124] Testimony of Joseph Keith, 178-81.

[125] Testimony of Bill Ambrose, 220.

[126] BP, Deepwater Horizon *Accident Investigation Report* (September 8, 2010), 25.

[127] Internal BP document (BP-HZN-BLY 61369-74); Internal BP document (BP-HZN-MBI 21406).

[128] Testimony of Joseph Keith, 50.

[129] *Ibid.*, 51, 102, 114-15.

[130] *Ibid.*, 102, 183-84.

[131] *Ibid.*, 102 ("Five minutes, eight minutes. No longer than that."), 184 ("About 10 minutes. That's what it seemed like.").

[132] Nonpublic Transocean document (presentation to Commission staff, August 5, 2010), 14; BP, Deepwater Horizon *Accident Investigation Report*, 25-26.

[133] BP, Deepwater Horizon *Accident Investigation Report*, 92. "At that point in time it masked what was the biggest inflow at that point." Testimony of Bill Ambrose, 221. "It would have been impossible to tell in realtime what increase might have been due to the trip tank drain or what increases might have been due to a well influx." Testimony of John Gisclair, December 7, 2010, 12-14, 55.

[134] Testimony of John Gisclair, November 8, 2010, 229. Emptying the trip tanks does not affect drill pipe pressure, so that action does not explain the increase. Bill Ambrose (Transocean), interview with Commission staff, August 17, 2010.

[135] Testimony of John Gisclair, November 8, 2010, 228; Bly, interview; Willis, interview.

[136] Testimony of John Gisclair, November 8, 2010, 228; BP, Deepwater Horizon *Accident Investigation Report*, 92; Seraile, interview; Willis, interview.

[137] Testimony of Joseph Keith, 238.

[138] Internal BP document (BP-HZN-MBI 170828).

[139] Testimony of Joseph Keith, 120-21, 196.

[140] Internal BP document (BP-HZN-MBI 139549).

[141] *Ibid.*

[142] The *Deepwater Horizon's* residual flow signature was a straight line evenly going down. Seraile, interview.

[143] Ambrose, interview, September 21, 2010.

[144] Sperry-Sun data, April 20, 2010; Internal BP document (BP-HZN-MBI 139549).

[145] Steve Lewis (Expert witness), interview with Commission staff, September 7, 2010; Darryl Bourgoyne (Expert witness), interview with Commission staff, November 23, 2010; Keith, interview (watched flow line for 30 to 45 seconds); Seraile, interview (*Horizon* residual flow was typically three to five minutes). According to assistant driller Seraile, a credible visual confirmation of no flow required five to seven minutes after the pumps stopped. Here, the crew diverted returns within two minutes. Seraile, interview.

[146] Gisclair, interview; Bourgoyne, interview, November 23, 2010.

[147] Internal BP document (BP-HZN-MBI 13950); Gisclair, interview.

[148] Seraile, interview; Murry Sepulvado, interview; Ronnie Sepulvado, interview, August 20, 2010.

[149] Bourgoyne, email, December 16, 2010.

[150] The non-zero reading on the Sperry-Sun flow-out sensor, from 9:09 to 9:21 p.m., likely reflects fluid left over in the flow line after the gate to route returns overboard closed. Gisclair, interview; Bly, interview; Smith, interview, October 26, 2010.

[151] Testimony of Jimmy Harrell (Transocean), Hearing before the Deepwater Horizon Joint Investigation Team, May 27, 2010, 28; Internal BP document (BP-HZN-MBI 21305).

[152] Transocean legal team, interview, August 17, 2010 (sustained drill pipe pressure increase was anomalous); Testimony of John Gisclair, October 8, 2010, 135 ("a curiosity"); BP, Deepwater Horizon *Accident Investigation Report*, 93 (anomalous); Smith, interview, September 7, 2010 (fishy).

[153] Ambrose, interview, September 21, 2010; Testimony of John Gisclair, October 8, 2010, 228-29 ("So it's not your typical indicator of a kick but it, again, is something to give pause").

[154] Testimony of Bill Ambrose, 225-26. This underscores the notion that current kick detection methodology depends on the right person being in the right place at the right time. *Ibid.*, 246.

[155] Keith testified that, if he had seen the increase, he would have notified the drill crew. Testimony of Joseph Keith, 215.

[156] *Ibid.*, 216, 230-31.

[157] The PRV exists "for the purpose of protecting the pump from excessive high-pressure overloads." Internal BP document (BP-HZN-BLY 49091). The PRV is generally "piped directly into the mud tanks" so that, when it blows, fluid goes to the pits. *Ibid.* On the *Horizon*, the PRVs were probably routed to the slugging pit. Internal Transocean document (TRN-HCEC 68478). When the PRV blew, the data showed a gain in the slugging pit (pit 11).

[158] Seraile, interview; BP, Deepwater Horizon *Accident Investigation Report*, 97; Keith, interview. It is unlikely that the PRV blew because of pressure from the well. John Smith (Expert witness), interview with Commission staff, October 14, 2010.

[159] Holloway, interview. That group also included Adam Weise, Shane Roshto, and Roy Wyatt Kemp.

[160] Testimony of Bill Ambrose, 380-82. That valve is on the choke manifold, on the rig floor. Commission staff site visit to *Deepwater Nautilus*, September 9, 2010; Holloway, interview; Seraile, interview.

[161] Testimony of Randy Ezell (Transocean), Hearing before the Deepwater Horizon Joint Investigation Team, May 28, 2010, 282.

[162] Testimony of Bill Ambrose, 384; Testimony of John Gisclair, October 8, 2010, 138.

[163] Testimony of Bill Ambrose, 80-82. "When that happened, it had a very strange trend, and over a period of about seven minutes it started to build pressure." *Ibid.*

[164] Experts have suggested at least two explanations for this anomaly. One explanation is that the kill line pressure was anomalous because it did not sufficiently match the drill pipe pressure. Testimony of Bill Ambrose, 381. Another explanation is that the kill line pressure was anomalous because it was higher than it should have been, at a time when the crew was not pumping fluid down the kill line. In this latter scenario,

the crew closed the kill line valve once they saw the pressure rising. Bourgoyne, interview, November 23, 2010.

[165] Testimony of Bill Ambrose, 384. "The anomaly was that now that the kill line had been opened, the differential was causing some concern." *Ibid.*

[166] *Ibid.*

[167] Testimony of David Young (Transocean), Hearing before the Deepwater Horizon Joint Investigation Team, May 27, 2010, 258-59. Tabler entered shortly afterward. *Ibid.*, 276.

[168] David Young (Transocean), interview with Commission staff, November 19, 2010.

[169] Testimony of David Young, 276.

[170] Young, interview.

[171] Testimony of David Young, 276.

[172] Seraile, interview; Smith, interview, September 7, 2010.

[173] "With months of work we've determined that it appears as the kick was coming in, the influx was coming in, it was changing heights of fluid columns in the well. And the geometry of the well was such that as the 14-pound mud that was – When this all started, there was about 500 barrels of 14-pound mud below the drill pipe, and that had been pushed up into the BOP, and as it hit the BOP it was – it kept a constant pressure, a sign that fluids are moving." Testimony of Bill Ambrose, 380-82. "The drill pipe pressure increased until 21:35 hours as 14.17 ppg mud rose above the stinger and then declined after that as hydrocarbons rose above the stinger. Modeling indicated that this fluctuation was consistent with the change in fluid densities (spacer, seawater, mud and hydrocarbons) moving through the various cross-sectional areas in the well." BP, Deepwater Horizon *Accident Investigation Report*, 101. Fluid flow is complex. The two-phase flow pattern, in which gas migrates up through the mud instead of completely displacing the mud, can effect unexpected pressure fluctuations. Adam T. Bourgoyne, Jr., "University uses on-campus abandoned well to simulate deepwater well-control operations," *Oil and Gas Journal* (May 31, 1982), 141; Peter G. McFadden, "The Pressure Behavior of a Shut-In Well Due to the Upward Migration of a Gas Kick" (thesis, December 1984), 8-9.

[174] Holloway, interview.

[175] *Ibid.*; Smith, *Review of Operational Data Preceding Explosion on Deepwater Horizon in MC252*, 14; BP, Deepwater Horizon *Accident Investigation Report*, 100.

[176] Holloway, interview.

[177] *Ibid.*

[178] Testimony of Bill Ambrose, 381.

[179] Testimony of John Gisclair, November 8, 2010, 30.

[180] Seraile, interview; Internal Transocean document (TRN-HCEC 5415). It is the responsibility of the drill crew "to shut-in the well as quickly as possible if a kick is indicated or suspected." *Ibid.* Shutting in the well also would have involved notifying senior Transocean and BP personnel. *Ibid.*

[181] Testimony of David Young, 259-60.

[182] *Ibid.*, 297.

[183] Young, interview.

[184] *Ibid.*

[185] *Ibid.*

[186] *Ibid.*

[187] Holloway, interview.

[188] Testimony of Joseph Keith, 20-21.

[189] Testimony of Charles Credeur, 62-63; Bouillion, interview. BP estimates mud overflow at 9:40 p.m. BP, Deepwater Horizon *Accident Investigation Report*, 103; Transocean estimates it at 9:43 p.m. Testimony of Bill Ambrose, 383.

[190] Smith, *Review of Operational Data Preceding Explosion on Deepwater Horizon in MC252*, 14. It appears that it was the *Horizon*'s practice to "monitor the flow from the riser via the flowline to the trip tank." Internal Transocean document (TRN-HCEC 25883); Seraile, interview; Ezell, interview.

[191] Testimony of Chris Pleasant, 166 (BOP panel showed lower annular closed); Testimony of Bill Ambrose, 252; BP, Deepwater Horizon *Accident Investigation Report*, 104. Though testimony indicates that Anderson activated the lower annular, Transocean has contended that the rig crew in fact activated the upper annular, not the lower annular.

[192] Internal BP document (BP-HZN-MBI 139551); Internal BP document (BP-HZN-MBI 21421); Internal BP document (BP-HZN-MBI 142485); Internal BP document (BP-HZN-CEC 20334); Internal BP document (BP-HZN-CEC 20357); BP, Deepwater Horizon *Accident Investigation Report*, 103.

[193] Testimony of Randy Ezell, 283.

[194] Testimony of Yancy Keplinger (Transocean), Hearing before the Deepwater Horizon Joint Investigation Team, October 5, 2010, part 1, 150, 219, 254-55; Testimony of Andrea Fleytas (Transocean), Hearing before the Deepwater Horizon Joint Investigation Team, October 5, 2010, part 2, 13, 39. Floorhand Caleb Holloway called Revette via radio during this time but received no response. David Barstow, David Rohde, and Stephanie Saul, "Deepwater Horizon's Final Hours," *New York Times*, December 25, 2010.

[195] Internal BP document (BP-HZN-MBI 139551).

[196] Testimony of Randy Ezell, 283.

[197] Testimony of Yancy Keplinger, 149, 219.

[198] *Ibid.*, 252.

[199] Testimony of Paul Meinhart (Transocean), Hearing before the Deepwater Horizon Joint Investigation Team, May 29, 2010, 29. Meanwhile, Young had gone by the subsea office and mentioned to Chris Pleasant and Allen Seraile that the cement job was delayed because they were "having some sort of differential pressure problem" up on the rig floor. Pleasant and Seraile (who was getting ready to come on tour) turned to live video feed of the rig floor and saw the mud. Pleasant instantly picked up the phone and called the rig floor—all three lines—but got no answer. Young, interview; Testimony of Chris Pleasant, 103-04, 121.

[200] Bly, interview.

[201] BP, Deepwater Horizon *Accident Investigation Report*, 103-04; Testimony of Bill Ambrose, 252-53. It is worth noting that Pleasant and Vidrine, once on the bridge, looked at the BOP panel and saw only an indication that an annular had been activated, not the variable bore ram. Internal BP document (BP-HZN-MBI 21420); Internal BP document (BP-HZN-MBI 142485).

[202] Smith, *Review of Operational Data Preceding Explosion on Deepwater Horizon in MC252*, 23-24; BP, Deepwater Horizon *Accident Investigation Report*, 146.

[203] Smith, *Review of Operational Data Preceding Explosion on Deepwater Horizon in MC252*, 15; Bourgoyne, email, November 23, 2010; Testimony of Joseph Keith, 124.

[204] Internal BP document (BP-HZN-MBI 21424); Internal BP document (BP-HZN-MBI 21427).

[205] Internal BP document (BP-HZN-MBI 21424); Internal BP document (BP-HZN-MBI 139551).

[206] Testimony of Chad Murray (Transocean), Hearing before the Deepwater Horizon Joint Investigation Team, May 27, 2010, 335-36.

[207] Internal BP document (BP-HZN-MBI 21267). Keith did not notice any anomalies in the data during the displacement, with the exception of one instance of pit gain that he learned reflected emptying of the sand traps. Testimony of Joseph Keith, 178-81, 193. The first time Keith became aware that the well was flowing was, "When it sounded like it was raining outside my unit and I started smelling gas coming through my purge system." *Ibid.*, 32.

[208] Testimony of Jimmy Harrell, 64, 128-129118.

[209] BP, Transocean, and Halliburton representatives all agree that if skilled personnel had observed the Sperry-Sun data, they would have noticed indicators sufficiently clear to cause alarm. Testimony of Mark Bly, November 8, 2010, 241-42; Testimony of Bill Ambrose, 241-42; Testimony of John Gisclair, November 8, 2010, 241-42.

[210] Some have suggested that the rig pumping mud to the *Damon Bankston* interfered with well monitoring during the final displacement. Donald Winter et al., *Interim Report on Causes of the Deepwater Horizon Oil*

Rig Blowout and Ways to Prevent Such Events (November 17, 2010), 10-11. This is incorrect. Pumping to the Bankston had ceased by 5:17 p.m., well before the start of the displacement of the riser. Internal BP document (BP-HZN-MBI 139482).

[211] Murry Sepulvado, interview.

[212] Internal BP document (BP-HZN-BLY 61692); Ronnie Sepulvado, interview, October 26, 2010; Seraile, interview.

[213] Bourgoyne, email, November 23, 2010. "The rig crew could decide to fill pits in the active system from the sea chest system, then line up the pump to those pits." *Ibid.*

[214] It is worth noting that, even if the gas meter were used for monitoring, it is likely that rig personnel would not have seen gas until after most of the mud was ejected from the well. Bourgoyne, email, December 16, 2010.

[215] Earlier in the evening, the crew had been using the cranes to off-load equipment onto a supply ship. The starboard crane had been down for repair. Testimony of Heber Morales, 143. By the time of the displacement, it was apparently working and in use. Young, interview. The crew may have been testing it. Internal BP document (BP-HZN-MBI 139588); Testimony of Joseph Keith, 35. Or they may have resumed off-loading equipment. There was also a crane operator in the gantry crane on the port aft deck helping the rig personnel at the bucking unit put together the tools necessary for running the lockdown sleeve. Testimony of Micah Sandell (Transocean), Hearing before the Deepwater Horizon Joint Investigation Team, May 29, 2010, 7-8; Testimony of Heber Morales, 143-44; Testimony of Ross Skidmore (Swift), Hearing before the Deepwater Horizon Joint Investigation Team, July 20, 2010, 217-19; Testimony of Joseph Keith, 35; Internal BP document (BP-HZN-MBI 129622); Internal BP document (BP-HZN-MBI 139583); Internal BP document (BP-HZN-MBI 139588); Internal BP document (BP-HZN-MBI 21277).

[216] Internal BP document (BP-HZN-MBI 21441).

[217] Testimony of John Gisclair, October 8, 2010, 102; Testimony of Joseph Keith, 35, 198; Willis, interview.

[218] API RP 59, 46. Persistent rig movement can also make it difficult to calibrate the flow-out sensor. Testimony of Joseph Keith, 24-25.

[219] Bourgoyne, email, December 16, 2010; Testimony of Joseph Keith, 25.

[220] Testimony of Jimmy Harrell, 28; Testimony of Leo Lindner, 279; BP, Deepwater Horizon *Accident Investigation Report*, 91; Internal BP document (BP-HZN-MBI 21305).

[221] Testimony of Leo Lindner, 272; Testimony of Joseph Keith, 191; Internal BP document (BP-HZN-MBI 21250).

[222] Testimony of Bill Ambrose, 221; Ambrose, interview, August 17, 2010.

[223] Ambrose, interview, August 17, 2010; Testimony of John Smith (Expert witness), Hearing before the Deepwater Horizon Joint Investigation Team, July 23, 2010, 297.

[224] Gisclair, interview; Ronnie Sepulvado, interview, August 20, 2010. Another example is the rig's flow-out meters. According to BP well site leader Ronnie Sepulvado, the rig's flow-out meters "had inherent variability and were not completely accurate." Internal BP document (BP-HZN-BLY 61694). Measuring flow-out can be challenging because there are few flow-out meters able to handle such a wide variety of fluids and pump rates. Bourgoyne, interview, September 10, 2010; Testimony of John Smith, 295.

[225] Testimony of John Gisclair, October 8, 2010, 102.

[226] Testimony of Mark Bly, November 8, 2010, 246-47; Testimony of Bill Ambrose, 246-47.

[227] Commission staff site visit to *Deepwater Nautilus*, September 9, 2010; Ambrose, interview, September 21, 2010; Testimony of John Gisclair, October 8, 2010, 109-10.

[228] Testimony of John Gisclair, October 8, 2010, 109-10.

[229] BP, Deepwater Horizon *Accident Investigation Report*, app. W.

[230] Testimony of Darryl Bourgoyne, 170-72.

[231] Testimony of John Gisclair, November 8, 2010, 228. "The standpipe pressure, especially that first increase, it basically draws a straight line, and it's very difficult to spot that 100-pound increase over that extended period using that particular presentation." *Ibid.*

[232] Representatives from BP, Transocean, and Sperry-Sun all agree that kick indications were clear enough that, if observed by skilled personnel, they would have allowed the rig crew to have responded earlier. Testimony of Bill Ambrose; Testimony of John Gisclair, November 8, 2010, 242; Testimony of Mark Bly, November 8, 2010, 242.

[233] Of the kicks that occurred during drilling operations, a "majority of them were observed when drilling new hole." Per Holand and Pal Skalle, *Deepwater Kicks and BOP Performance* (SINTEF, July 2001), 11, 45. It is worth noting that Transocean specifically drilled its crew for this situation. Internal Transocean document (TRN-USCG_MMS 30540).

[234] Per Holand and Pal Skalle, *Deepwater Kicks and BOP Performance*, 11, 45-46.

[235] John Guide (BP), interview with Commission staff, January 19, 2011 (if they had already passed the negative test, it maybe would not be surprising if he did not check for flow all the time); Testimony of John Smith, 301 ("They think they've already proven that the well is safe...if it's sealed up and you've proved it's sealed up and it's not going to leak, well, that's a reason to [reduce] your rigor."); Bourgoyne, interview, September 10, 2010; Lewis, interview, September 21, 2010.

[236] Internal Transocean document (TRN-PC 3227); David Borthwick, *Report of the Montara Commission of Inquiry* (The Montara Commission of Inquiry, Australia, June 2010), 116.

[237] Internal Transocean document (TRN-PC 3227); David Borthwick, *Report of the Montara Commission of Inquiry* (The Montara Commission of Inquiry, Australia, June 2010), 116.

[238] Anadarko legal team, interview with Commission staff, September 29, 2010; Testimony of Steve Lewis (Expert witness), Hearing before the National Commission, November 9, 2010, 65, 79-80.

[239] Local Impact of the Deepwater Horizon Oil Spill, Before Subcomm. on Oversight and Investigations of the H. Comm. on Energy and Commerce, 111th Cong. (2010)(statement of Courtney Kemp), 62 (deceased husband characterized well as having "so many problems and so many things were happening...it was just kind of out of hand"); Testimony of Micah Burgess, 117 ("It was a difficult well."); Internal BP document (MC 252-1_DDR)(the Macondo well sustained numerous lost circulation events, two previous kicks, a ballooning event, and trouble with LOTs); E.C. Thomas (Expert witness), interview with Commission staff, October 27, 2010 (the Horizon crew encountered more problems than usual).

[240] Internal BP document (BP-HZN-CEC 20200); Ross Skidmore (Swift), interview with Commission staff, December 21, 2010 (after the last casing string is run, floorhands are often doing preventative maintenance duties and thinking about the next job); Testimony of Greg Meche (M-I SWACO), Hearing before the Deepwater Horizon Joint Investigation Team, May 28, 2010, 216-217.

[241] Testimony of Ross Skidmore, 263-64.

[242] Murry Sepulvado, interview (always monitors final displacements from the rig floor); Guide, interview, January 19, 2011 (would have expected the well site leader to be watching the data during the displacement and performing flow checks himself); Bourgoyne, email, November 23, 2010. After the explosion, well site leader Don Vidrine apparently felt "like he should have been on the [rig] floor when this happened." Internal BP document (BP-HZN-BLY 72265).

[243] This critique would apply equally to the Sperry Drilling mudlogger, if he had noticed the pressure anomalies. There is no evidence that he did.

[244] In addition to the anomalies discussed in this chapter, the negative pressure test repeatedly showed that the bottomhole cement job did not isolate hydrocarbons (Chapter 4.6). Testimony of Mark Bly, November 8, 2010, 203-04.

[245] Testimony of Bill Ambrose, 381; Seraile, interview.

[246] Testimony of John Gisclair, November 8, 2010, 30.

[247] Testimony of Bill Ambrose, 380-86.

[248] *Ibid.*

[249] *Ibid.*, 205, 207, 277; Transocean legal team, interview with Commission staff, December 10, 2010.

[250] Testimony of Bill Ambrose, 205 ("when the approval came back [from BP] that it was a good negative test, our people proceeded ahead on good faith that it was a good test"); Transocean legal team, interview, December 10, 2010.

[251] Transocean legal team, interview, December 10, 2010.

[252] BP's Engineering Technical Practice on Simultaneous Operations (GP 10-75) directs its personnel to conduct a risk assessment of simultaneous operations "in order to identify the risks across the complete range of well activities" and to "ensure all well activities...are carried out in a safe and controlled manner, when these activities are performed in the same space and time as another operation." Internal BP document (BP-HZN-OSC 8019). It is not clear whether BP personnel associated with the *Deepwater Horizon* conducted any such risk assessment with respect to rig activities going on during the final displacement operation. At least one witness has suggested that other well site leaders would not have allowed so many rig operations. Keith, interview.

[253] Testimony of John Gisclair, November 8, 2010, 218; Smith, *Review of Operational Data Preceding Explosion on Deepwater Horizon in MC252*, 22.

[254] For example, Transocean suggests that the driller may have watched the screens for 60 seconds and then turned his back to line up pump 2. Testimony of Bill Ambrose, 225-26.

[255] Ezell, interview; Testimony of Paul Johnson, 329-30.

[256] Transocean legal team, interview with Commission staff, September 21, 2010; Testimony of Paul Johnson, 389 (design decisions affect the crew's ability to control the well).

[257] "Prior to commencing operations, it is advisable that all involved parties understand the objectives, procedures, and hazards. All personnel should be encouraged to report abnormal conditions. Alertness and speed of communication are critical factors in well control." API RP 59, 40.

[258] The pre-tour meetings that day did not involve any special instructions about the negative pressure test or kick detection. Seraile, interview. "I guess they figured it was going to go the same way as having all these other barriers in place." *Ibid.*

[259] Guide, interview, January 19, 2011 (would expect rig crew and well site leader to perform direct visual flow checks—not just rely on telemetry—whenever they shut down the pumps).

[260] Industry experts recommend a well control plan be "worked out beforehand.... To be successful, subsurface conditions must be predicted, detected, and controlled." API RP 59, 35. It is unlikely that the *Horizon*'s driller knew what drill pipe pressures to expect during the displacement. Ambrose, interview, September 21, 2010.

[261] Rig personnel could communicate with each other through radio, telephone, overhead pages, and messenger. Willis, interview; Seraile, interview.

[262] Both Ronnie and Murry Sepulvado, alternative BP well site leaders for the *Horizon*, took actives role in ensuring proper communication between the drill crew and mudlogger. They did so by periodically speaking with the drill crew and mudlogger when there was a change in rig operations and confirming that the two entities had kept each other informed of the change. Murry Sepulvado, interview; Willis, interview.

[263] Willis, interview.

[264] *Ibid.*

[265] Testimony of Joseph Keith 31-32.

[266] This included that the drill crew was switching returns from the active pits to the reserve pits, dumping the sand traps into the active pits, emptying the trip tanks, transferring fluid from the active pits (pits 9 and 10) to reserve pit 6 during the sheen test, and fixing a blown PRV. *Ibid.*, 40, 178, 193; Keith, interview.

[267] Testimony of Joseph Keith, 177-78; Transocean legal team, interview, December 10, 2010.

[268] Testimony of John Gisclair (Halliburton), Hearing before the National Academy of Engineering, September 26, 2010, 18, 63.

[269] *Ibid.*; Willis, interview.

[270] Internal BP document (BP-HZN-MBI 193469).

[271] Internal BP document (BP-HZN-CEC 20094).

[272] Internal BP document (BP-HZN-OSC 5641-44).

[273] Internal BP document (BP-HZN-MBI 193469); Internal BP document (BP-HZN-CEC 20094).

[274] Internal BP document (BP-HZN-CEC 20094).

[275] Internal BP document (BP-HZN-MBI 128017).

[276] Internal BP document (BP-HZN-MBI 128018).

[277] David Sims (BP), interview with Commission staff, December 14, 2010.

[278] Internal BP document (BP-HZN-OSC 5522).

[279] Internal BP document (BP-HZN-MBI 127999, 128002, 128018-20); Zanghi, interview.

[280] Internal BP document (BP-HZN-MBI 128002, 128018).

[281] Zanghi, interview; Internal BP document (128017)(ERA would focus "primarily on the execute stage of well construction"); Internal BP document (128017, 128002, 128018)(preset configurations for drilling, tripping, running casing, and cementing; no mention of abandonment or completion).

[282] Internal BP document (BP-HZN-MBI 127996, 128009).

[283] Testimony of Pat O'Bryan (BP), Hearing before the Deepwater Horizon Joint Investigation Team, August 26, 2010, 412; Testimony of Brett Cocales, 233.

[284] Testimony of Brett Cocales, 233-35; Testimony of Gregory Walz (BP), Hearing before the Deepwater Horizon Joint Investigation Team, October 7, 2010, part 2, 21. It is worth noting that, after the blowout, BP has been re-evaluating its planned ERA program and is in the process of reshaping it to include more attention to well control issues. Zanghi, interview.

[285] Internal BP document (BP-HZN-OSC 4095).

[286] *Ibid.*

[287] Internal BP document (BP-HZN-MBI 213677).

[288] Internal BP document (BP-HZN-OSC 4095); Internal Halliburton documents (MC252_001_ST00BP00 Sperry Rpt 037); Internal Halliburton document (MC252_001_ST00BP01_EOWR); Internal BP document (BP-HZN-MBI 213677)(there was a 35- to 40-barrel influx).

[289] Internal BP document (BP-HZN-OSC 4095); Seraile, interview.

[290] Internal BP document (BP-HZN-MBI 197360).

[291] Guide, interview, January 19, 2011.

[292] Internal BP document (BP-HZN-OSC 1490); Internal BP document (BP-HZN-OSC 1493).

[293] BP, Deepwater Horizon *Accident Investigation Report*, 107; Internal BP document (BP-HZN-MBI 113015); Internal BP document (BP-HZN-MBI 198127). One TIGER team member went to the *Deepwater Horizon* for several days to implement the suggested improvements.

[294] Internal BP document (BP-HZN-MBI 198126, 113018-19)(emphasis added)("The entire breadth of pore-pressure indicators need to be evaluated under higher scrutiny.... Thus far on this well, it has been shown that one or more pore-pressure indicators have provided ambiguous or even contradictory data. A more robust analysis of all indicators would allow us to better discern systemic pore-pressure changes from localized anomalies.").

[295] Internal BP document (BP-HZN-MBI 113018).

[296] BP, Deepwater Horizon *Accident Investigation Report*, 107; Internal BP document (BP-HZN-BLY 61465). Although BP personnel deemed the drill crew's response inadequate, Transocean personnel did not. Transocean legal team, interview, September 21, 2010.

[297] John Guide (BP), interview with Commission staff, September 17, 2010.

[298] Internal BP document (BP-HZN-MBI 113018).

[299] Internal BP document (BP-HZN-BLY 61465-66).

[300] Guide, interview, January 19, 2011.

[301] Internal BP document (BP-HZN-MBI 113018-19)("drilling practices," "drilling techniques," "drilling parameters").

[302] Internal Transocean document (TRN-HCEC 16041).

[303] Internal Transocean document (TRN-HCEC 15880).

[304] Internal Transocean document (TRN-HCEC 16041).

[305] *Ibid.*

[306] Internal BP document (BP-HZN-BLY 38354-55, 38357-58). The desired completion fluid was base oil, so the plan was to clean up the well by swapping out mud first to seawater and then to base oil. Internal BP document (BP-HZN-BLY 38357-58).

[307] "The seawater was pumped via one pit with a constant top up (no volume control)." Internal BP document (BP-HZN-BLY 38358).

[308] Internal BP document (BP-HZN-BLY 38354-55, 38357-58).

[309] Testimony of Daun Winslow (Transocean), Hearing before the Deepwater Horizon Joint Investigation Team, August 24, 2010, 139.

[310] Rig personnel noticed an increase in flow show, increase in pits, and problems at the shakers, but made different interpretations due to "changes in clean up parameters on the one pit fill, perceived rig trimming issues; and potential blockage in lines from shakers to pits." Internal BP document (BP-HZN-BLY 38358, 38361).

Chapter 4.8

[1] Also known as a "degasser" or "po' boy degasser."

[2] Internal BP document (BP-HZN-BLY 56833).

[3] Internal Transocean document (TRN-HCEC 5708)(Transocean Well Control Handbook). The crew conducted an EDS drill on January 31, 2010. Internal BP document (BP-HZN-BLY 55870). The *Horizon's* EDS also successfully activated and sheared drill pipe on June 30, 2003. Internal BP document (BP-HZN-MBI 136652).

[4] Internal BP document (BP-HZN-BLY 56833).

[5] *Ibid.*

[6] Testimony of Christopher Pleasant (Transocean), Hearing before the Deepwater Horizon Joint Investigation Team, May 28, 2010, 165.

[7] Internal BP Document (BP-HZN-CEC 45983).

[8] There is an inversely proportional relationship between pressure and volume for an ideal gas. Though actual conditions in the wellbore might lead to different behavior than what is predicated for an ideal gas, the gas almost certainly expanded exponentially—or very near to exponentially.

[9] BP's analysis indicates that gas traveled up the 5,000-foot riser in about two minutes. BP, Deepwater Horizon *Accident Investigation Report* (September 8, 2010), 27.

[10] Testimony of Bill Ambrose (Transocean), Hearing before the National Commission, November 8, 2010, 386; Testimony of Mark Bly (BP), Hearing before the National Commission, November 8, 2010, 245.

[11] Internal Transocean document (TRN-HCEC 5607).

[12] Internal BP document (BP-HZN-MBI 213402).

[13] Internal Transocean document (TRN-HCEC 5607).

[14] Testimony of Bill Ambrose, 257.

[15] Detailed information on diverters is available in API, *Recommended Practice for Diverter Systems Equipment and Operations*, 2nd ed. (November 2001, reaffirmed March 1, 2007)("API RP 64").

[16] Internal Transocean document (TRN-HCEC 5608).

[17] Internal BP document (BP-HZN-BLY 49092).

[18] Confidential industry expert, interview with Commission staff.

[19] Internal BP document (BP-HZN-BLY 49092); Internal BP document (BP-HZN-BLY 49094).

[20] Internal BP document (BP-HZN-BLY 49094).

[21] Confidential industry expert, interview.

[22] Confidential industry expert, letter to Commission staff.

[23] *Ibid.*

[24] *Ibid.*

[25] Internal Transocean document (TRN-HCEC 5606); Confidential industry expert, letter.

[26] Confidential industry expert, letter.

[27] Gas rises very quickly in deepwater. Internal Transocean document (TRN-HCEC 5607).

[28] BP's analysis estimates the time at 9:40 p.m. BP, Deepwater Horizon *Accident Investigation Report*, 28. Transocean contends that the time was 9:43 p.m. Testimony of Bill Ambrose, 383. These estimates are generally consistent with the testimony of witnesses.

[29] Internal Transocean document (TRN-HCJ 120914).

[30] Testimony of Micah Sandell (Transocean), Hearing before the Deepwater Horizon Joint Investigation Team, May 29, 2010, 10.

[31] Testimony of David Young (Transocean), Hearing before the Deepwater Horizon Joint Investigation Team, May 27, 2010, 264.

[32] Testimony of Bill Ambrose, 244.

[33] *Ibid.*

[34] Testimony of Micah Sandell, 9-10.

[35] Internal BP document (BP-HZN-CEC 20357).

[36] Testimony of Micah Sandell, 9-10.

[37] *Ibid.*

[38] Darryl Bourgoyne (Expert witness), interview with Commission staff, September 9, 2010.

[39] Confidential industry expert, interview; Internal Transocean Document (TRN-HCEC 5606).

[40] Deepwater Horizon Study Group, *The Macondo Blowout: 3rd Progress Report* (December 5, 2010), app. B, 12.

[41] Testimony of Micah Sandell, 10.

[42] Internal BP document (BP-HZN-BLY 61525).

[43] Testimony of Jimmy Harrell (Transocean), Hearing before the Deepwater Horizon Joint Investigation Team, May 27, 2010, 98-99.

[44] BP's modeling of the flow of mud and gas onto the rig indicates that the mud gas separator equipment may have failed. BP, Deepwater Horizon *Accident Investigation Report*, 115. The Chief Counsel's team is not aware of a sophisticated model that has been completed by an independent party at this time.

[45] Internal BP document (BP-HZN-MBI 129187).

[46] Confidential industry expert, interview.

[47] John Smith, *Review of Operational Data Preceding Explosion on Deepwater Horizon in MC252* (July 1, 2010), 14; Testimony of Bill Ambrose, 252-53; BP, Deepwater Horizon *Accident Investigation Report*, 28. According to subsea supervisor Chris Pleasant, he reset the lower annular regulator pressure to 1,500 psi from 1,900 psi soon after 9 p.m. Testimony of Christopher Pleasant, 120. According to BP, 1,500 psi was the normal regulator pressure setting for both annular preventers. BP, Deepwater Horizon *Accident Investigation Report*, 145. According to an independent expert, this would not have affected the ability to close the annular but it may have affected the equipment's ability to seal fully. Confidential industry expert, interview with Commission staff; Internal BP document (BP-HZN-BLY 61257).

[48] Testimony of Christopher Pleasant, 123. BP post-explosion models also suggest an annular was activated at 9:41 p.m. BP, Deepwater Horizon *Accident Investigation Report*, 104.

[49] Internal Transocean document (TRN-HCJ 121107); Internal BP document (BP-HZN-BLY 61563).

[50] Data on well pressures are consistent with the closing of the annular or the variable bore ram. BP, Deepwater Horizon *Accident Investigation Report*, 103-04. Transocean has suggested the rig crew closed both variable bore rams. Transocean legal team, interview with Commission staff, December 10, 2010. Post-explosion information indicates that the rig crew may have activated the upper variable bore ram and perhaps the middle pipe ram. Response teams closed the upper variable bore ram but pumped only 1.5 gallons of hydraulic fluid, instead of the 28 gallons typically required to close the ram. This indicates the rig crew had likely already closed the upper variable bore ram. BP, Deepwater Horizon *Accident Investigation Report*, 162. Post-explosion pressure measurements also suggest the rig crew may have activated the middle pipe ram. *Ibid.*

[51] Testimony of Bill Ambrose, 256; Internal Transocean document (TRN-HCEC 5487). Because of Macondo, some industry experts now question the appropriateness of typical kick response procedures.

[52] BP, Deepwater Horizon *Accident Investigation Report*, 44.

[53] Darryl Bourgoyne (Expert witness), interview with Commission staff, September 10, 2010; Testimony of Bill Ambrose, 256.

[54] MMS regulation 30 CFR § 250.416(e) requires that the BOP description include information that shows the blind shear rams installed as capable of shearing the drill pipe in the hole under maximum anticipated surface pressures.

[55] BP, Deepwater Horizon *Accident Investigation Report*, 175.

[56] *Ibid.*

[57] *Ibid.*, 103-04; Internal Transocean document (TRN-HCEC 5487).

[58] Internal Transocean document (TRN-HCEC 5487).

[59] Internal Transocean document (TRN-HCEC 5543).

[60] BP, Deepwater Horizon *Accident Investigation Report*, 29.

[61] Internal Transocean document (TRN-HCJ 121107).

[62] *Ibid.*

[63] Testimony of Daun Winslow (Transocean), Hearing before the Deepwater Horizon Joint Investigation Team, August 23, 2010, 447-448.

[64] Internal Transocean document (TRN-HCJ 121107).

[65] Confidential industry expert, interview.

[66] Testimony of William Stoner (Transocean), Hearing before the Deepwater Horizon Joint Investigation Team, May 28, 2010, 380. The engines are equipped with shutdown devices. During an audit of the rig in April, Engine 3 was out of service and its shutdown function was likely not tested. Internal Transocean document (TRN-USCG_MMS 38670).

[67] Testimony of Michael Williams (Transocean), Hearing before the Deepwater Horizon Joint Investigation Team, July 23, 2010, 11-17.

[68] Testimony of Paul Meinhart (Transocean), Hearing before the Deepwater Horizon Joint Investigation Team, May 29, 2010, 30; Testimony of Douglas Brown (Transocean), Hearing before the Deepwater Horizon Joint Investigation Team, May 26, 2010, 100-01.

[69] Testimony of Douglas Brown, 130.

[70] Testimony of Michael Williams, 16.

[71] Testimony of Paul Meinhart, 30; Testimony of Jimmy Harrell, 65; Testimony of Stephen Bertone (Transocean), Hearing before the Deepwater Horizon Joint Investigation Team, July 19, 2010, 116.

[72] Internal BP document (BP-HZN-CEC 20285).

[73] Testimony of Chris Pleasant, 123.

[74] BP, Deepwater Horizon *Accident Investigation Report*, 150; Internal BP document (BP-HZN-MBI 143364).

[75] Internal Transocean document (TRN-HCJ 121114); Testimony of Jimmy Harrell, 21, 66-67, 70-71; Internal BP document (BP-HZN-CEC 20285).

[76] BP, Deepwater Horizon *Accident Investigation Report*, 152.

[77] The MUX cable reels were located in the moon pool , in the center of the rig underneath the rig floor. BP, Deepwater Horizon *Accident Investigation Report*, 151.

[78] Testimony of Curt Kuchta (Transocean), Hearing before the Deepwater Horizon Joint Investigation Team, May 27, 2010, 190.

[79] Internal BP document (BP-HZN-BLY 39346); Testimony of Joseph Keith (Halliburton), Hearing before the Deepwater Horizon Joint Investigation Team, December 7, 2010, 237; Allen Seraile (Transocean), interview with Commission staff, January 7, 2011; Testimony of Darryl Bourgoyne (Expert witness), Hearing before the National Commission, November 9, 2010, 179-80.

[80] BP's modeling indicates that "the flowing conditions could have prevented an annular preventer from fully closing and sealing around the drill pipe." BP, Deepwater Horizon *Accident Investigation Report*, 146.

[81] Internal Halliburton document (HAL_48973); BP representatives, interview with Commission staff, September 8, 2010.

[82] Smith, *Review of Operational Data Preceding Explosion on Deepwater Horizon in MC252*, 24.

[83] Internal Halliburton document (HAL_48973); Smith, *Review of Operational Data Preceding Explosion on Deepwater Horizon in MC252*, 23.

[84] Transocean asserts that the variable bore ram was closed. Testimony of Bill Ambrose, 252. Witnesses only saw the annular activated. Internal BP document (BP-HZN-MBI 21420).

[85] Internal Halliburton document (HAL_48973); Smith, *Review of Operational Data Preceding Explosion on Deepwater Horizon in MC252*, 24.

[86] BP, Deepwater Horizon *Accident Investigation Report*, 172. According to Transocean, the proposal to convert the lower annular to a stripping annular was approved on July 29, 2006. Internal BP document (BP-HZN-MBI 136646).

[87] Internal BP document (BP-HZN-MBI 136646).

[88] Darryl Bourgoyne (Expert witness), interview with Commission staff, December 18, 2010.

[89] Some industry experts—including Darryl Bourgoyne at Louisiana State University—suggest that the diverter should be set automatically to go overboard. If the event is not an emergency, then the rig crew will have time to send the influx to the mud gas separator.

[90] Internal Transocean document (TRN-HCEC 5607).

[91] *Ibid.*

[92] Confidential industry expert, interview.

[93] BP, Deepwater Horizon *Accident Investigation Report*, 117.

[94] *Ibid.*, app. V, 52.

[95] *Ibid.*

[96] Steve Lewis (Expert witness), email to Commission staff, January 11, 2011.

[97] The Chief Counsel's team thus disagrees with Transocean's assertion that flow rate and volume would have been so extreme that it necessarily would have overwhelmed the diverter system had the crew sent flow overboard.

[98] BP, Deepwater Horizon *Accident Investigation Report*, 117. The Chief Counsel's team also requested information on the slip joint from Transocean, but it was unavailable at the publication of this report.

[99] *Ibid.*, 114.

[100] *Ibid.*, 117.

[101] NOAA, "Meteorological data from NOAA's Station PSTL1," http://www.ndbc.noaa.gov/view_text_file.php?filename=pstl142010.txt.gz&dir=data/stdmet/Apr/; Internal BP document (BP-HZN-MBI 129187).

[102] NOAA, "Meteorological data from NOAA's Station PSTL1," http://www.ndbc.noaa.gov/view_text_file.php?filename=pstl142010.txt.gz&dir=data/stdmet/Apr/; Internal BP document (BP-HZN-MBI 129187).

[103] Wind often changes, and BP's report has the wind earlier in the day blowing the opposite direction. BP, Deepwater Horizon *Accident Investigation Report*, app. V, 22.

[104] The port overboard line was unavailable because the *Damon Bankston* was on that side of the rig.

[105] Deepwater Horizon Study Group, *The Macondo Blowout: 3rd Progress Report,* app. B, 13.

[106] David Izon, E.P. Danenberger and Melinda Mayes, "Absence of fatalities in blowouts encouraging in MMS study of OCS incidents," *Drilling Contractor 90* (July/August 2007). The study's authors also noted that a previous study found that "22 of the 41 diverter uses were considered successful." *Ibid.*

[107] Confidential industry expert, interview; Deepwater Horizon Study Group, *The Macondo Blowout: 3rd Progress Report,* app. B, 13; BP, Deepwater Horizon *Accident Investigation Report*, 44.

[108] BP, Deepwater Horizon *Accident Investigation Report*, 44.

[109] Rachel Clingman (Transocean legal team), letter to Commission staff, November 16, 2010.

[110] BP agrees: "Transocean's shut-in protocols did not fully address how to respond in high flow emergency situations after well control has been lost. Well control actions taken prior to the explosion suggest the rig crew was not sufficiently prepared to manage an escalating well control situation." BP, Deepwater Horizon *Accident Investigation Report*, 44.

[111] Rachel Clingman (Transocean legal team), letter to Commission staff, November 1, 2010, att. *The Deepwater Horizon—Crew and Safety*, 7-8.

[112] Testimony of Bill Ambrose, 252.

[113] Clingman, letter.

[114] Internal Transocean document (TRN-HCEC 5609-5610).

[115] Even at the ninth step, the procedures are ambiguous. They say merely to "be prepared...to send the mud overboard," not directly to send the mud overboard. Moreover, the 10th step prescribes continued circulation through the mud gas separator. Internal Transocean document. *Ibid.*

[116] In an interview after the explosion with BP investigators, Mark Hafle, the senior drilling engineer for Macondo, agreed. He observed that rig crews conduct frequent drills but not for worst-case scenarios. Internal BP document (BP-HZN-BLY 61371).

Chapter 4.9

[1] 30 C.F.R. § 250.442; 30 C.F.R. § 250.515(b); 30 C.F.R. § 250.1624(b)(1).

[2] Internal BP document (BP-HZN-CEC 46007).

[3] Internal Cameron document (CAM-GR 101); Internal Transocean document (TRN-HCEC 5543); Testimony of Daun Winslow (Transocean), Hearing before the Deepwater Horizon Joint Investigation Team, August 24, 2010, 137.

[4] To avoid having a tool joint lay across the blind shear rams, rig crews engage in a procedure whereby they close an annular preventer, strip the pipe upward (through the closed annular) until it hits a tool joint, and then lower the pipe a predetermined amount so that the tool joint is out of the way and there is only straight pipe across the rams. This procedure requires time and a sequence of steps, which an emergency situation may not permit. Commission staff site visit to the *Deepwater Nautilus*, September 9, 2010.

[5] Internal BP document (BP-HZN-BLY 59392-93).

[6] Darryl Bourgoyne (Expert witness), interview with Commission staff, January 22, 2011; BP legal team, interview with Commission staff, January 25, 2011. Most rigs use range 2 drill pipe, which averages 31 feet in

length per interval. However, deepwater rigs may use range 3 drill pipe, which comes in 38- to 45-foot lengths. Bourgoyne, interview, January 22, 2011.

[7] Adding an additional ram may require a reconfiguration of drilling rigs in order to accommodate a taller BOP stack.

[8] West Engineering Services, *Mini Shear Study for U.S. Minerals Management Service* (December 2002), 3.

[9] *Ibid.*

[10] Michael Montgomery (West Engineering Services), letter to Commission staff, November 12, 2010, 4.

[11] West Engineering Services, *Mini Shear Study for U.S. Minerals Management Service*, 3.

[12] *Ibid.*, 12.

[13] *Ibid.*

[14] BP, Deepwater Horizon *Accident Investigation Report* (September 8, 2010), app. H, 234; Internal BP document (BP-HZN-BLY 52594). The test took five seconds to shear the pipe and 12 seconds to completely close the shear rams.

[15] BP, Deepwater Horizon *Accident Investigation Report*, 156.

[16] Internal BP document (BP-HZN-MBI 136652).

[17] 30 C.F.R. § 250.449(b), (d).

[18] 30 C.F.R. § 250.516; Internal BP document (BP-HZN-MBI 133302).

[19] 30 C.F.R. § 250.446(b) requires visual inspection of subsea BOPs at least once every three days. According to witness testimony, ROVs inspected the Deepwater Horizon BOP daily to monitor for equipment irregularities including leaks. Testimony of Tyrone Benton (Oceaneering), Hearing before the Deepwater Horizon Joint Investigation Team, July 23, 2010, 244. However, according to daily operations reports there were some days, including February 18, when ROVs did not dive to the BOP. Internal BP document (BP-HZN-MBI 135192).

[20] The blind shear ram was tested to 914 psi. Internal Transocean document (TRN-USCG_MMS 26242). According to daily drilling reports, the rig crew pressure tested the blind shear ram on February 6, 2010; February 9, 2010; March 21, 2010; March 26, 2010; and April 1, 2010. Internal BP document (BP-HZN-OSC 4074); Internal BP document (BP-HZN-OSC 4099); Internal BP document (BP-HZN-OSC 4191); Internal BP document (BP-HZN-OSC 4222); Internal Transocean document (TRN-USCG_MMS 26242).

[21] Internal BP document (BP-HZN-OSC 4116)(February 11, 2010); Internal BP document (BP-HZN-OSC 4139)(February 14, 2010); Internal BP document (BP-HZN-OSC 4207)(February 24, 2010); Internal BP document (BP-HZN-OSC 4037)(March 1, 2010); Internal BP document (BP-HZN-OSC 4070)(March 5, 2010); Internal BP document (BP-HZN-OSC 4158)(March 16, 2010); Internal BP document (BP-HZN-OSC 4222)(March 26, 2010); Internal Transocean document (TRN-USCG_MMS 26242)(April 1, 2010); Internal Transocean document (TRN-USCG_MMS 26266)(April 8, 2010); Testimony of Mark Hay (Transocean), Hearing before the Deepwater Horizon Joint Investigation Team, August 25, 2010, 249.

[22] Darryl Bourgoyne (Expert witness), interview with Commission staff, January 24, 2011.

[23] 30 C.F.R. § 250.448(b).

[24] Internal BP document (BP-HZN-OSC 619).

[25] BP, Deepwater Horizon *Accident Investigation Report*, app. H, 230. According to daily drilling reports, the blind shear ram was tested at various pressures during drilling, including 15,000 psi on the surface (February 6, 2010), 6,500 psi (February 9, 2010), 2,400 psi (March 21, 2010), 1,800 psi (March 26, 2010), and 914 psi (April 1, 2010). Internal BP document (BP-HZN-OSC 4074); Internal BP document (BP-HZN-OSC 4099); Internal BP document (BP-HZN-OSC 4191); Internal BP document (BP-HZN-OSC 4222); Internal Transocean document (TRN-USCG_MMS 26242).

[26] 30 C.F.R. § 250.448(c). The upper annular was rated to withstand 10,000 psi closed on pipe or 5,000 psi closed on an open hole. BP, Deepwater Horizon *Accident Investigation Report*, app. H, 227. After conversion to a stripping annular, the lower annular was rated to withstand 5,000 psi. BP, Deepwater Horizon *Accident Investigation Report*, 172. Post-explosion examination of the blue pod found regulated pressure on the lower annular preventer was set to approximately 1,700 psi. *Ibid*, 145. Transocean has recently contended the rig crew activated the upper annular, not the lower annular.

[27] Internal BP document (BP-HZN-MBI 130883).

[28] Internal BP document (BP-HZN-OSC 619).

[29] Internal BP document (BP-HZN-OSC 925, 931-32).

[30] Steve Lewis (Expert witness), interview with Commission staff, September 28, 2010.

[31] Darryl Bourgoyne (Expert witness), interview with Commission staff, December 18, 2010.

[32] *Ibid.*

[33] Internal Transocean document (TRN-USCG_MMS 11646).

[34] BP, Deepwater Horizon *Accident Investigation Report*, 175-76. ROVs can also activate the rams electronically through subsea controls.

[35] Nonpublic BP document (presentation to Commission staff, September 8, 2010), 13.

[36] Internal BP document (BP-HZN-CEC 46012-13).

[37] Internal BP document (BP-HZN-CEC 46012).

[38] Internal Transocean document (TRN-USCG_MMS 38822).

[39] BP, Deepwater Horizon *Accident Investigation Report*, 29.

[40] Internal BP document (BP-HZN-CEC 18896); BP, Deepwater Horizon *Accident Investigation Report*, 162.

[41] BP, Deepwater Horizon *Accident Investigation Report*, 29.

[42] Internal BP document (BP-HZN-CEC 18896).

[43] Internal Department of the Interior document (OSC-DWH BOEM-WDC-B06-00001-0005), 374-79.

[44] Internal BP document (BP-HZN-MBI 136647).

[45] BP, Deepwater Horizon *Accident Investigation Report*, 172.

[46] Internal BP document (BP-HZN-MBI 136648).

[47] Internal BP document (BP-HZN-CEC 18896).

[48] *Ibid.*

[49] Testimony of Harry Thierens (BP), Hearing before the Deepwater Horizon Joint Investigation Team, August 25, 2010, 104–05.

[50] The modification was made in 2004, and according to BP notes, it was "overlooked at the time" to change the ROV hot stab connections. Internal BP document (BP-HZN-MBI 136648); Internal BP document (BP-HZN-CEC 18896).

[51] Internal BP document (BP-HZN-CEC 18896).

[52] Internal Transocean document (TRN-HCEC 5708); Testimony of Daun Winslow (Transocean), Hearing before the Deepwater Horizon Joint Investigation Team, August 23, 2010, 515.

[53] BP, Deepwater Horizon *Accident Investigation Report*, 155.

[54] *Ibid.*; Darryl Bourgoyne (Expert witness), interview with Commission staff, January 21, 2011.

[55] Transocean legal team, interview with Commission staff, December 10, 2010.

[56] BP, Deepwater Horizon *Accident Investigation Report*, 156.

[57] Internal Cameron document (CAM-GR 209). Each SEM has an AMF card that recognizes when power, communication, and hydraulics are lost. BP, Deepwater Horizon *Accident Investigation Report*, 154 and app. X, 2.

[58] Internal Cameron document (CAM-GR 209). According to manufacturer specifications, communication between the control pods must also be lost in order for the AMF/deadman to activate. It is not certain whether the control pods communicate locally or through the MUX cables running to the rig.

[59] *Ibid.* The rigid conduit line running from the rig continually charged BOP stack accumulators. BP, Deepwater Horizon *Accident Investigation Report*, 160. Pressure sensors compare ambient pressure at the seabed with pressure in the conduit line. If conduit pressure is equal or less than ambient hydrostatic pressure, the AMF sequence continues. Internal Cameron document (CAM-GR 8041).

[60] BP, Deepwater Horizon *Accident Investigation Report*, 175-76.

[61] Internal Cameron document (CAM-GR 209); Internal BP document (BP-HZN-BLY 56840).

[62] Internal Transocean document (TRN-USCG_MMS 38826); BP, Deepwater Horizon *Accident Investigation Report*, 150.

[63] BP, Deepwater Horizon *Accident Investigation Report*, 151.

[64] Transocean legal team, interview with Commission staff, September 30, 2010.

[65] BP, Deepwater Horizon *Accident Investigation Report*, 151. Transocean questions whether hydraulic power was severed between the rig and BOP during the explosion. Transocean legal team, interview, September 30, 2010.

[66] Transocean legal team, interview, September 30, 2010.

[67] Internal BP document (BP-HZN-CEC 18896); BP, Deepwater Horizon *Accident Investigation Report*, 29.

[68] Transocean legal team, interview, December 10, 2010.

[69] Internal BP document (BP-HZN-BLY 56143).

[70] Transocean legal team, interview, September 30, 2010.

[71] Internal Transocean document (TRN-HCEC 5708).

[72] Rather, Transocean representatives indicated that neither the AMF system nor batteries necessary to power the system are regularly tested. Transocean legal team, interview, September 30, 2010. A deadman system surface test typically cuts the power and hydraulics to the BOP stack and confirms the rams are closed. Internal BP document (BP-HZN-CEC 18893). Testing the deadman subsea is a risk because cutting power and hydraulics to the BOP stack could drain the system batteries. The system could also fail to restart as designed. However, according to an internal BP memo, these are "manageable risks." Internal BP document (BP-HZN-CEC 18894).

[73] Records indicate "working on yellow pod" on February 1 and 2, 2010. Internal BP document (BP-HZN-BLY 55870). Both pods were function tested and powered on February 5, 2010. BP Internal document (BP-HZN-OSC 4065).

[74] BP, Deepwater Horizon *Accident Investigation Report*, 153. The 27-volt battery was comprised of three connected 9-volt battery packs. *Ibid.*, app. X, 2.

[75] *Ibid.*, app. X, 2. The sequence may have stopped prior to energizing the conduit and ambient pressure sensors, which according to BP testing requires at least 14.4 volts. *Ibid.*, app. X, 4.

[76] *Ibid.*, app. X, 4.

[77] Testimony of William Stringfellow (Transocean), Hearing before the Deepwater Horizon Joint Investigation Team, August 25, 2010, 421.

[78] BP, Deepwater Horizon *Accident Investigation Report*, app. X, 3. The Chief Counsel's team received protocols used during this initial blue pod testing but awaits more comprehensive results from the ongoing government forensic testing of the blowout preventer.

[79] *Ibid.*

[80] After the AMF sequence, the SEM, AMF card, and AMF controller are powered down to preserve battery life. *Ibid.*, app. X, 2.

[81] Transocean legal team, interview, September 30, 2010.

[82] Important records relating to the blowout preventer were lost with the sinking of the rig. This includes the event logger, which continuously records data from both pods and records all BOP functions activated from the control panels. BP, Deepwater Horizon *Accident Investigation Report*, app. H, 232. This lost information may have helped to determine what the battery charges were at the time of the incident.

[83] Internal Cameron document (CAM-GR 252). Cameron recommends replacing pod batteries at the following times, whichever is earliest: after one year of on-time operation, when a battery is actuated 33 times, or five years after date of purchase.

[84] Testimony of William Stringfellow, 348-49.

[85] Internal Transocean document (TRN-USCG_MMS 38660).

[86] Transocean subsea supervisor Chris Pleasant testified that pod batteries were not replaced immediately before the crew lowered the BOP in February 2010. Testimony of Chris Pleasant (Transocean), Hearing before the Deepwater Horizon Joint Investigation Team, May 28, 2010, 199. Senior subsea engineer Mark Hay testified that the blue pod batteries are "replaced within a year." Testimony of Mark Hay, 262-63.

[87] BP, Deepwater Horizon *Accident Investigation Report*, 167.

[88] Internal Transocean document (TRN-HCJ 98117)(Pod #2 December 29, 2005 replacement, Cost: $17,786.13); Internal Transocean document (TRN-PC 3787).

[89] BP, Deepwater Horizon *Accident Investigation Report*, app. H, 233.

[90] *Ibid.*; Earl Shanks et al., "Deepwater BOP Control Systems—A Look at Reliability Issues" (paper, Offshore Technology Conference, Houston, TX, May 2003), 1.

[91] BP, Deepwater Horizon *Accident Investigation Report*, app. H, 233; Cameron representative, interview with Commission staff, August 10, 2010.

[92] Internal BP document (BP-HZN-BLY 55870).

[93] BP, Deepwater Horizon *Accident Investigation Report*, 154. These results were witnessed and accepted on May 8, 2010. Internal Cameron document (CAMCG 3011).

[94] Internal Cameron document (CAMCG 2999).

[95] Commission staff site visit to *Deepwater Nautilus*, September 9, 2010; BP, Deepwater Horizon *Accident Investigation Report*, 173.

[96] Testimony of Mark Hay, 262. An April 2010 Transocean rig condition assessment identified an error message on the yellow pod indicating a valve was mismatched. Internal Transocean document (TRN-USCG_MMS 38657). However, the audit did not clarify which solenoid valve and left the issue unresolved. It is not certain whether this message could be an indication of a faulty solenoid. Internal Transocean document (TRN-USCG_MMS 38658).

[97] Darryl Bourgoyne (Expert witness), interview with Commission staff, December 14, 2010.

[98] Internal BP document (BP-HZN-BLY 56842).

[99] BP, Deepwater Horizon *Accident Investigation Report*, 156.

[100] Internal BP document (BP-HZN-BLY 55870); Internal Transocean document (TRN-HCEC 5708). The autoshear is tested by sending a signal that simulates LMRP disconnection, then verifying the shuttle valves for the selected rams worked as intended. This can be done through "dry firing," meaning with no hydraulic pressure. Internal BP document (BP-HZN-CEC 18894).

[101] Internal BP document (BP-HZN-CEC 18896); BP, Deepwater Horizon *Accident Investigation Report*, 48.

[102] Internal BP document (BP-HZN-CEC 18896); BP, Deepwater Horizon *Accident Investigation Report*, 156.

[103] Nonpublic BP document (presentation to Commission staff on BOP), 15.

[104] Preliminary evidence shows the upper annular preventer closed around the drill pipe. Therefore, even if pipe severed above the annular preventer, the annular would have prevented that pipe from falling below the annular and interfering with the closure of the blind shear ram.

[105] BP, Deepwater Horizon *Accident Investigation Report*, app. H, 228.

[106] *Ibid.*, app. H, 231. The BOP stack accumulators also provided hydraulic fluid should the rig crew directly activated the blind shear ram from one of the rig's control panels. *Ibid.*

[107] *Ibid.*, 160.

[108] Internal BP document (BP-HZN-BLY 55870); Internal BP document (BP-HZN-OSC 4049). Accumulator levels are generally checked prior to or during an accumulator drill. Bourgoyne, interview, January 22, 2011.

[109] According to BP's internal investigation, if only three accumulators were charged there would not be sufficient pressure to shear the pipe and seal the well. BP, Deepwater Horizon *Accident Investigation Report*, 160.

[110] *Ibid.*, 170-71.

[111] *Ibid.*, 170.

[112] *Ibid.*, 171.

[113] Internal BP document (BP-HZN-MBI 135553).

[114] Darryl Bourgoyne (Expert witness), email to Commission staff, December 15, 2010.

[115] *Ibid.*

[116] Bourgoyne, interview, January 22, 2011. According to a 2010 West Engineering study on blowout prevention equipment reliability, "a very large percentage" of control system failures can be identified by function tests. Internal Cameron document (CAM-GR 15911). According to this study, lower pressures may not allow leaks to be indentified. Internal Cameron document (CAM-GR 15918).

[117] Bourgoyne, interview, January 22, 2011.

[118] *Ibid.*

[119] Testimony of Mark Hay, 242.

[120] The rig crew identified leaks 1, 2, and 5 before the incident. Rachel Clingman (Transocean legal team), letter to Commission staff, November 1, 2010.

[121] Bourgoyne, interview, December 18, 2010. Post-explosion log books identified more leaks, some of which may have developed during the response effort and some of which were identified prior to the incident. Internal BP document (BP-HZN-MBI 135553).

[122] BP, Deepwater Horizon *Accident Investigation Report*, 170.

[123] *Ibid.*, 156; Clingman, letter.

[124] Testimony of John Guide (BP), Hearing before the Deepwater Horizon Joint Investigation Team, July 22, 2010, 193-95; BP, Deepwater Horizon *Accident Investigation Report*, 169.

[125] Internal BP document (BP-HZN-MBI 135226); Internal BP document (BP-HZN-MBI 192017). The leak may have been recognized even earlier than February 23. While inspecting leak 2 on February 19, the rig crew isolated the pressure supply to the yellow pod and, using the ROV, observed a pilot system leak on the yellow pod. This may have been referring to leak 1. Clingman, letter.

[126] Clingman, letter; BP, Deepwater Horizon *Accident Investigation Report*, 169.

[127] Clingman, letter. According to witness testimony, this leak was discovered approximately two weeks after landing the BOP on the wellhead. Testimony of Mark Hay, 244. According to BP's internal investigation, this leak was not identified until post-explosion ROV intervention. BP, Deepwater Horizon *Accident Investigation Report*, 169.

[128] BP, Deepwater Horizon *Accident Investigation Report*, 170.

[129] Internal Transocean document (TRN-USCG_MMS 38843).

[130] BP, Deepwater Horizon *Accident Investigation Report*, 170.

[131] Internal Transocean document (TRN-USCG_MMS 38843).

[132] BP, Deepwater Horizon *Accident Investigation Report*, 171.

[133] Clingman, letter.

[134] *Ibid.*

[135] 30 C.F.R. § 250.466(f).

[136] Elmer Danenberger (Expert witness), email to Commission staff, January 5, 2011.

[137] Internal BP document (BP-HZN-OSC 4199-4202).

[138] Internal BP document (BP-HZN-MBI 135226); Internal BP document (BP-HZN-MBI 135558).

[139] John Guide, interview with Commission staff, September 17, 2010.

[140] Ronnie Sepulvado, interview with Commission staff, August 20, 2010.

[141] Clingman, letter.

[142] Internal BP document (BP-HZN-OSC 4175-77).

[143] Clingman, letter.

[144] Danenberger, email, January 5, 2011.

[145] Bourgoyne, interview, January 22, 2011.

[146] According to senior subsea supervisor Owen McWhorter, this leak was brought to everyone's attention. Clingman, letter. BP's John Guide called Transocean's Paul Johnson and confirmed the leak did not affect the ram's functionality. Guide, interview.

[147] Internal BP document (BP-HZN-MBI 225124).

[148] Clingman, letter.

[149] *Ibid.*

[150] *Ibid.*

[151] *Ibid.*; Internal BP document (BP-HZN-MBI 135553).

[152] Testimony of Mark Hay, 195.

[153] American Petroleum Institute, *Recommended Practices for Blowout Prevention Equipment Systems for Drilling Wells*, 3rd ed. (March 1997, reaffirmed September 2004), 56 ("API RP 53 § 18.10.3").

[154] *Ibid.*

[155] Testimony of William Stringfellow, 359.

[156] Internal Cameron document (CAM-GR 261); Bourgoyne, interview, January 22, 2011.

[157] Earl Shanks et al., *Deepwater BOP Control Systems—A Look at Reliability Issues* (May 2003), 2.

[158] Internal Transocean document (TRN-HCEC 90585).

[159] 30 C.F.R. § 250.446(a).

[160] API RP 53 § 18.10.3.

[161] Internal Cameron document (CAM-GR 261-62).

[162] Testimony of Mark Hay, 205; Testimony of John Sprague (BP), Hearing before the Deepwater Horizon Joint Investigation Team, December 8, 2010, part 2, 79-80.

[163] Internal BP document (BP-HZN-MBI 136213); Internal BP document (BP-HZN-MBI 136230).

[164] Internal Transocean document (TRN-USCG_MMS 38652).

[165] Internal BP document (BP-HZN-MBI 136230); Testimony of John Sprague, 79-80.

[166] Internal Transocean document (TRN-USCG_MMS 38652).

[167] Internal Transocean document (TRN-USCG_MMS 38662). The BOP's fail-safe valves, designed to shut off choke and kill lines remotely and automatically, had also not been certified since December 13, 2000. Internal Transocean document (TRN-USCG_MMS 38656).

[168] 30 C.F.R. § 250.446(a); Danenberger, email, January 5, 2011. To use a stack that was out of compliance, BP may have needed to request a regulatory departure under 30 C.F.R. § 250.142. Elmer Danenberger (Expert witness), email to Commission staff, January 16, 2011.

[169] Internal Transocean document (TRN-HCEC 66722); Testimony of Eric Neal (MMS), Hearing before the Deepwater Horizon Joint Investigation Team, May 11, 2010, 318-21.

[170] Internal MMS document.

[171] Testimony of Mark Hay, 255-56.

[172] Internal Transocean document (TRN-HCEC 11560).

[173] Testimony of William Stringfellow, 414-15.

[174] *Ibid.*, 415.

[175] *Ibid.*, 413-14 (emphasis added)(condition of BOPs is tracked in the RMS).

[176] *Ibid.*

[177] *Ibid.*, 426-27, 439.

[178] BP, Deepwater Horizon *Accident Investigation Report*, app. AA, 2.

[179] Internal BP document (BP-HZN-MBI 136649).

[180] Internal BP document (BP-HZN-MBI 136646).

[181] Internal Transocean document (TRN-USCG_MMS 38855).

[182] *Ibid.*

[183] BP, Deepwater Horizon *Accident Investigation Report*, app. AA, 1. According to Transocean, conditions after the explosion permitted the control pods to be retrieved despite this modification. Transocean legal team, interview with Commission staff, November 2, 2010.

[184] BP, Deepwater Horizon *Accident Investigation Report*, app. AA, 1.

[185] *Ibid.*, app. AA, 2.

[186] *Ibid.*

[187] Internal Transocean document (TRN-USCG_MMS 38855).

[188] Cameron representative, interview.

[189] BP, Deepwater Horizon *Accident Investigation Report*, 172.

[190] *Ibid.*, app. AA, 1.

[191] *Ibid.*, 172.

[192] Internal BP document (BP-HZN-MBI 136649).

[193] Internal BP document (BP-HZN-MBI 136648).

[194] *Ibid.*

[195] BP, Deepwater Horizon *Accident Investigation Report*, app. AA, 2.

[196] Internal BP document (BP-HZN-MBI 136647).

[197] Internal BP document (BP-HZN-MBI 136648).

[198] Internal BP document (BP-HZN-MBI 136649).

[199] *Ibid.*

[200] *Ibid.*

[201] Internal Transocean document (TRN-HCJ 96972).

[202] BP, Deepwater Horizon *Accident Investigation Report*, 172; Internal BP document (BP-HZN-MBI 136646).

[203] Internal BP document (BP-HZN-BLY 56004).

[204] Internal Transocean document (TRN-USCG_MMS 38855).

[205] *Ibid.*

[206] *Ibid.*

[207] Internal BP document (BP-HZN-MBI 136648).

Chapter 4.10

[1] Testimony of Daun Winslow (Transocean), Hearing before the Deepwater Horizon Joint Investigation Team, August 24, 2010, 87.

[2] Internal Transocean document (Transocean, "Deepwater Horizon Safety and Maintenance Overview," presentation to U.S. House of Representatives Committee on Energy and Commerce, August 2010), 13.

[3] *Ibid,* 8, 10.

[4] Testimony of Pat O'Bryan (BP), Hearing before the Deepwater Horizon Joint Investigation Team, August 26, 2010, 411.

[5] John Guide (BP), interview with Commission staff, September 17, 2010.

[6] Testimony of Pat O'Bryan, 365.

[7] Internal BP document (BP-HZN-MBI 198309).

[8] Internal Transocean document (TRN-HCEC 5322). Transocean also used a system called "FOCUS" for tracking audit items and rig crew training. David Young (Transocean), interview with Commission staff, November 19, 2010.

[9] Testimony of Paul Johnson (Transocean), Hearing before Deepwater Horizon Joint Investigation Team, August 23, 2010, 325. Previously, Transocean used an Empac system. Internal BP document (BP-HZN-MBI 136212).

[10] Internal Transocean document (TRN-HCEC 5340); Testimony of Stephen Bertone (Transocean), Hearing before the Deepwater Horizon Joint Investigation Team, July 19, 2010, 231. The Rig Management System was the approved Computerized Maintenance Management System (CMMS) for all Transocean installations. Witness testimony commonly refers to the RMS as the rig maintenance system. Testimony of Stephen Bertone (Transocean), July 19, 2010, 70.

[11] Internal Transocean document (TRN-HCEC 5340).

[12] Internal Transocean document (TRN-HCEC 5343).

[13] Preventative maintenance refers generally to normally scheduled maintenance, including planned maintenance, scheduled overhauls, condition monitoring, and daily and weekly equipment checks and measurements. Internal Transocean document (TRN-HCEC 5358).

[14] Testimony of Stephen Bertone, 109.

[15] *Ibid.*

[16] Internal Transocean document (TRN-HCEC 5340).

[17] Young, interview. Young suggested that a lack of knowledge among the crew may not have been problematic because supervisors were familiar with the system. Young handled entries into the system for his team such that, according to Young, a lack of system familiarity did not inhibit operations. David Young, interview.

[18] Internal Transocean document (Transocean, "Deepwater Horizon Safety and Maintenance Overview," presentation to U.S. House of Representatives Committee on Energy and Commerce, August 2010), 16.

[19] Testimony of Stephen Bertone, 233.

[20] *Ibid.*

[21] *Ibid.* According to chief engineer technician Mike Williams, there would be up to four listings for the same job. Testimony of Mike Williams (Transocean), Hearing before the Deepwater Horizon Joint Investigation Team, July 23, 2010, 76.

[22] Allen Seraile (Transocean), interview with Commission staff, January 7, 2011.

[23] Testimony of Stephen Bertone, 232-33.

[24] Seraile, interview.

[25] Testimony of Mike Williams, 75. Transocean has suggested that the rig crew described the implementation of the RMS as a "speed bump." Internal Transocean document (Transocean, "Deepwater Horizon Safety and Maintenance Overview," presentation to U.S. House of Representatives Committee on Energy and Commerce, August 2010), 24.

[26] Internal Transocean document (TRN-HCEC90590).

[27] *Ibid.*

[28] *Ibid.*

[29] Internal Transocean document (TRN-HCEC 90910-11).

[30] Testimony of Paul Johnson, 352; Internal BP document (BP-HZN-OSC 4254). However, some forms of maintenance, including BOP maintenance, were exempt from this provision. *Ibid.*

[31] Internal Transocean document (TRN-HCEC 90584).

[32] Internal Transocean document (TRN-HCEC 90585).

[33] Guide, interview.

[34] Testimony of Daun Winslow (Transocean), Hearing before the Deepwater Horizon Joint Investigation Team, August 23, 2010, 508. Winslow's full title is Operations Manager, Performance, North American Division. *Ibid, 437.*

[35] Testimony of Mike Williams, 161.

[36] Internal Transocean document (Transocean, "Deepwater Horizon Safety and Maintenance Overview," presentation to U.S. House of Representatives Committee on Energy and Commerce, August 2010, slide 14). Internal Transocean document (TRN-HCEC 90585).

[37] Internal Transocean document (TRN-HCEC 90585).

[38] Internal Transocean document (TRN-HCEC 90603).

[39] Transocean legal team, interview with Commission staff, August 6, 2010.

[40] Internal Transocean document (TRN-HCEC 5384).

[41] Testimony of Paul Johnson, 290-91.

[42] For example, the September 2009 audit and April 2010 rig condition assessment both identified excessive silicon on mud pump covers that could cause pump failure. Internal BP document (BP-HZN-MBI 136221); Internal Transocean document (TRN-USCG_MMS 38637). The September 2009 audit identified the pipe racking system (PRS) as requiring maintenance before drilling operations resumed; the April 2010 rig condition assessment found the PRS in only "fair" condition. Internal BP document (BP-HZN-MBI 136240); Internal Transocean document (TRN-USCG_MMS 38666). According to assistant driller Allen Seraile, the rig crew was aware of major problems with the PRS. Seraile, interview.

[43] Internal Transocean document (TRN-HCEC 90955).

[44] Internal Transocean document (TRN-HCEC 90968-69).

[45] Testimony of Mike Williams, 115.

[46] Guide, interview.

[47] Internal Transocean document (TRN-HCEC 90901).

[48] Testimony of Michael Saucier (MMS), Hearing before the Deepwater Horizon Joint Investigation Team, May 12, 2010, 12-15.

[49] Testimony of Paul Johnson, 120.

[50] *Ibid.*

[51] *Ibid.*

[52] The rig audit surveys the condition of the drilling machinery while the marine audit surveys the ability of the vessel itself. Testimony of John Guide (BP), Hearing before the Deepwater Horizon Joint Investigation Team, July 22, 2010, 96. BP conducts assessments on all vessels owned by third parties that BP employs in the Gulf of Mexico. Testimony of Neil Cramond (BP), Hearing before the Deepwater Horizon Joint Investigation Team, August 23, 2010, 21-30. The September 2009 audit followed up a previous BP rig audit and included a standard marine assessment as prescribed by the International Maritime Contractors Association (IMCA). The audit also reviewed the condition of the dynamic positioning system. *Ibid,* 94-95.

[53] Internal BP document (BP-HZN-MBI 136222).

[54] *Ibid.*; Testimony of Neil Cramond, 94.

[55] The new system showed duplicates and listed jobs with no work history as immediately due because there was no history in the new system. Internal Transocean document. (Transocean, "Deepwater Horizon Safety and Maintenance Overview," presentation to U.S. House of Representatives Committee on Energy and Commerce, August 2010, slides 16, 24.) Duplicate maintenance orders and orders from other rigs may have also inflated the number of overdue items. *Ibid,* 233.

[56] Internal BP document (BP-HZN-MBI 136222); Testimony of Paul Johnson, 314-15, 321-22; Testimony of Neil Cramond, 35-36. The *Horizon* was already in an out-of-service period at the time of the audit so rather than stopping ongoing operations, BP requested the rig remain shut down until some repairs were made. Testimony of Paul Johnson, 248. This included repairs to the rig's pipe racking system, as one of the racking arms was bent. *Ibid*, 315.

[57] Testimony of Daun Winslow, August 24, 2010, 187.

[58] BP, Deepwater Horizon *Accident Investigation Report*, app. Y, 1 (September 8, 2010), 1.

[59] Angel Rodriguez of BP verified a Transocean spreadsheet of maintenance items. Testimony of Neil Cramond, 100-01. Tracking of other audits may not have been as focused. An April 2010 email from John Guide to Paul Johnson indicated 36 of 37 action items from an August 2009 environmental audit may have remained outstanding as of April 2010. Internal BP document (BP-HZN-MBI 251718).

[60] Testimony of Neil Cramond, 74. Four days after the audit ended on September 17, Transocean's Paul Johnson emailed BP's John Guide to inform him Transocean was satisfied the rig could resume operations because audit items were either closed out or had "robust mitigation measures in place." Internal BP document (BP-HZN-BLY 55679). The rig audit took place September 13-17. Internal BP document (BP-HZN-MBI 136212).

[61] Brett Cocales, Angel Rodriguez, and sometimes John Guide would attend these meetings. Testimony of Paul Johnson, 325, 354-55. While meetings were initially weekly, the companies later began meeting biweekly or monthly to discuss audit progress. *Ibid*, 325.

[62] *Ibid*, 355. BP well site leaders stayed informed by attending 8:30 a.m. rig meetings or had rig crew members email them regarding broken parts. Ronnie Sepulvado (BP), interview with Commission staff, August 20, 2010.

[63] Internal Transocean document (TRN-HCEC 115553).

[64] Testimony of Paul Johnson, 247.

[65] *Ibid*, 316; Testimony of Neil Cramond, 115.

[66] Sepulvado, interview.

[67] Testimony of Neil Cramond, 116-18.

[68] Testimony of John Guide, 201.

[69] Internal Transocean document (TRN-HCEC 66722); Testimony of Eric Neal (MMS), Hearing before the Deepwater Horizon Joint Investigation Team, May 11, 2010, 318.

[70] Internal MMS document.

Chapter 5

[1] BP, Deepwater Horizon *Accident Investigation Report* (September 8, 2010), 32 (identifying eight "physical or operational barriers" that failed and concluding that "[i]f any of these critical factors had been eliminated, the outcome of *Deepwater Horizon* events on April 20, 2010, could have been either prevented or reduced in severity").

[2] BP's operating management system (OMS) sets out minimum expectations for "leadership," "people and competence," "procedures," "working with contractors," "technology," and "risk" (among others). Internal BP document (BP-HZN-MBI 208576-77). Transocean's management system has principles on "leadership," "policies and procedures," "organization," "risk management," "training and competence," and "communications." Internal Transocean document (TRN-USCG_MMS 32700). Exxon's operations integrity management system (OIMS) is also quite similar. ExxonMobil, *Operations Integrity Management System*.

[3] Internal BP document (BP-HZN-MBI 208603).

[4] Internal BP document (BP-HZN-MBI 208611).

[5] Internal BP document (BP-HZN-MBI 222521).

[6] *Ibid.*

[7] David Sims (BP), interview with Commission staff, February 1, 2011. At about the same time, Sims had drafted a longer version of the email to Guide that he never sent. Internal BP document (BP-HZN-MBI 222540). He wrote, "Everything else is someone else's fault. You criticize nearly everything we do on the rig but don't seem to realize that you are responsible for everything we do on the rig." *Ibid.* He also wrote, "You seem to not want to make a decision so that you can criticize it later." *Ibid.* Sims also pointed out a number of times when episodes on the rig—including a kick, a crane incident, and an injury—were not reported in a timely manner to onshore members of the Macondo team. *Ibid.* Sims later explained that he was frustrated with Guide but later realized he probably should have cut Guide a bit more slack because a member of Guide's family had recently passed away. Sims, interview, February 1, 2011.

[8] Internal BP document (BP-HZN-MBI 212781). The next day, BP senior drilling engineer Mark Hafle asked a colleague, "Have you been within earshot of any of the Sims / Guide conversations lately?" Internal BP document (BP-HZN-MBI 212781).

[9] David Sims (BP), interview with Commission staff, December 14, 2010; Testimony of John Sprague (BP), Hearing before the Deepwater Horizon Joint Investigation Team, December 8, 2010, part 2, 10-11.

[10] Sims, interview, December 14, 2010; Testimony of Gregory Walz (BP), Hearing before the Deepwater Horizon Joint Investigation Team, October 7, 2010, part 2, 119.

[11] Internal BP document (BP-HZN-CEC 21533).

[12] Internal BP document (BP-HZN-BLY 125441).

[13] Sims, interview, February 1, 2011.

[14] Internal BP document (BP-HZN-BLY 61356).

[15] Internal BP document (BP-HZN-BLY 61470).

[16] Internal BP document (BP-HZN-MBI 255906).

[17] John Guide (BP), interview with Commission staff, January 19, 2011.

[18] Internal BP document (BP-HZN-BLY 61372).

[19] Sims, interview, February 1, 2011; Internal BP document (BP-HZN-BLY 125441).

[20] BP, Deepwater Horizon *Accident Investigation Report*, 36.

[21] Internal BP document (BP-HZN-BLY 125439).

[22] Internal BP document (BP-HZN-BLY 125436).

[23] Testimony of Brett Cocales (BP), Hearing before the Deepwater Horizon Joint Investigation Team, August 27, 2010, 113; Randy Ezell (Transocean), interview with Commission staff, September 16, 2010; Testimony of Paul Johnson (Transocean), August 23, 2010, 136-38. Halliburton's Vincent Tabler admitted that he read the OptiCem report but did not review the portion indicating "SEVERE gas flow" potential. Testimony of

Vincent Tabler (Halliburton), Hearing before the Deepwater Horizon Joint Investigation Team, August 25, 2010, 17.

[24] Internal BP document (BP-HZN-BLY 61350).

[25] Internal BP document (BP-HZN-MBI 254886).

[26] Internal BP document (BP-HZN-MBI 193521).

[27] The plan required personnel to call shore for "Any HSSE incident" or "Anytime operations deviate from agreed plan." Internal BP document (BP-HZN-MBI 193528).

[28] Internal BP document (BP-HZN-BLY 61374).

[29] Internal BP document (BP-HZN-MBI 308059).

[30] Internal BP documents (BP-HZN-MBI 262896-7).

[31] Gregory Walz (BP), interview with Commission staff, October 6, 2010; Brett Cocales (BP), interview with Commission staff, September 16, 2010; John Guide (BP), interview with Commission staff, September 17, 2010; Sims, interview, December 14, 2010; Pat O'Bryan (BP), interview with Commission staff, December 17, 2010.

[32] It is perhaps not surprising that neither Kaluza nor Vidrine called back to shore. Kaluza was new to the *Deepwater Horizon*, whereas Vidrine had been there for months and Revette and Anderson were experienced veterans on the rig. In Kaluza's 2009 performance review, BP management observed, "It sometimes appears that Bob is trying too much to impress the Houston office by attempting to have all the answers to any questions that may arise." Internal BP document (BP-HZN-MBI 193095-98). As for Vidrine, BP management praised him in his 2009 performance review because he only called to shore when needed. Internal BP document (BP-HZN-OSC 9089).

[33] Guide, interview, September 17, 2010; Murry Sepulvado (BP), interview with Commission staff, December 10, 2010.

[34] Testimony of Mark Bly (BP), Hearing before the National Commission, November 8, 2010, 286. Even that "expectation" would have been insufficient, as it would have required the well site leaders to be confused. Thus, Kaluza and others twice called back to shore for guidance when they encountered difficulties attempting to convert the float equipment. See Chapter 4.3. Whether well site leaders are confused or not, they should be required to call back to shore for a second opinion whenever there is anything unusual or unexpected with critical tests, such as the negative pressure test.

[35] Testimony of John Sprague, 177 (agreeing that it does not take long to call back to shore).

[36] Internal BP document (BP-HZN-BLY 38354).

[37] Internal BP document (BP-HZN-BLY 38355).

[38] *Ibid.*

[39] Internal BP document (BP-HZN-BLY 38361).

[40] *Ibid.*

[41] Internal BP document (BP-HZN-BLY 38362).

[42] Internal Transocean document (HQS-OPS-ADV-09).

[43] *Ibid.*

[44] *Ibid.*

[45] *Ibid.*

[46] *Ibid.*

[47] Ezell, interview; Testimony of Paul Johnson, 340-342, 344-347; Testimony of Daun Winslow (Transocean), Hearing before the Deepwater Horizon Joint Investigation Team, August 24, 2010, 132-36;Testimony of Jerry Canducci, Hearing before the Deepwater Horizon Joint Investigation Team, December 9, 2010, part 1, 171-172.

[48] Transocean representatives, interview with Commission staff, December 10, 2010.

[49] Internal Transocean document (TRN-PC 3227-30).

[50] Internal Transocean document (TRN-PC 3227).

[51] Internal Transocean document (TRN-PC 3229).

[52] Testimony of Daun Winslow, 122-23.

[53] Internal Transocean document (TRN-PC 3229); Transocean legal team, interview, December 10, 2010. As an aside, it appears that BP sometimes shared lessons from well control events on certain wells with rig crews operating other wells. Internal BP document (BP-HZN-MBI 221988). It is unclear whether the company did this on a systematic basis. In general, it appears that the industry as a whole does not systematically share incident reports and "lessons learned" broadly across companies.

[54] Internal Transocean document (TRN-PC 3227). Even after the blowout, Transocean continues to insist that drilling and completions are so different—like "apples and oranges"—as to render the North Sea incident irrelevant. Transocean representatives, interview, December 10, 2010.

[55] Testimony of Daun Winslow, 129.

[56] Transocean legal team, interview with Commission staff, December 10, 2010.

[57] Testimony of Paul Johnson, 385. Transocean sent out another, more detailed advisory on April 14, 2010. Internal Transocean document (TRN-PC 3227-230). However, it sent the advisory only to its North Sea fleet; it did not send it to the Gulf of Mexico. And again, it unduly limited it to completion operations—the title of the advisory was "Loss of Well Control During Completions Operations." Internal Transocean document (TRN-PC 3227-30).

[58] Transocean legal team, interview, December 14, 2010.

[59] Internal BP document (BP-HZN-MBI 199122).

[60] Morel was relatively new to drilling engineering and BP. He was assigned to the exploration team in 2008, where he helped to plan two wells before being transferred to Macondo to help Hafle. Hafle, by contrast, was the senior drilling engineer of the two and has been involved with deepwater drilling since 1993 and has personally been involved in between 20 and 50 wells. Testimony of Mark Hafle (BP), Hearing before the Deepwater Horizon Joint Investigation Team, May 28, 2010, 8-9. Hafle was on vacation during much of the week leading up to the blowout.

[61] Internal BP document (BP-HZN-MBI 195579).

[62] Testimony of John Sprague, 175-76.

[63] *Ibid.*, 193-94. Sprague further acknowledged that drill plans should be provided to the rig as soon as possible, admitting that "You can plan and understand things a lot better the sooner they get there, isn't that true?" *Ibid.*, 192.

[64] Internal BP document (BP-HZN-MBI 199122).

[65] Internal BP document (BP-HZN-MBI 126145).

[66] Internal BP document (BP-HZN-MBI 126333).

[67] Internal BP document (BP-HZN-MBI 126585-586).

[68] Internal BP document (BP-HZN-CEC 21260-279).

[69] Internal BP document (BP-HZN-MBI 126585-86).

[70] *Ibid.*

[71] *Ibid.*

[72] Internal BP document (BP-HZN-MBI 126982).

[73] Internal BP document (BP-HZN-MBI 127489).

[74] *Ibid.*

[75] Internal BP document (BP-HZN-CEC 21281-301).

[76] Internal BP document (BP-HZN-CEC 21288-91).

[77] *Ibid.*

[78] Internal BP document (BP-HZN-MBI 127909). The April 16 APM provides so little detail on the negative pressure test that members of BP's Macondo team after the fact disagree as to what it means. Testimony of Gregg Walz, Hearing before the Deepwater Horizon Joint Investigation Team, October 8, 2010, 147-48, 150-51 (Acknowledging it could be subject to different interpretations but that he interpreted it to mean that the negative pressure test should be performed after displacing mud from 8,367 feet to above the BOP); Guide, interview, September 16, 2010 (Interpreting steps 2 and 3 to be subparts of step 1); Ronnie Sepulvado (BP), interview with Commission staff, August 20, 2010 (interpreting step 1 to require performing the test at the mudline with base oil and step 3 as a separate flow check after fully displacing the riser).

[79] Internal BP document (BP-HZN-MBI 127906).

[80] Internal BP document (BP-HZN-BLY 47094).

[81] Internal BP document (BP-HZN-CEC 8574).

[82] Internal BP document (BP-HZN-CEC 22433-34). Other examples include the last-minute decisions regarding the combined spacer and long string versus a liner, the latter prompting Transocean's rig manager for the *Deepwater Horizon*, Paul Johnson, to complain to Guide that the crew needed the procedures to plan accordingly. Testimony of Paul Johnson, 360-61.

[83] Internal BP document (BP-HZN-MBI 213550).

[84] Internal BP document (BP-HZN-CEC 21533).

[85] Internal BP document (BP-HZN-MBI 255906).

[86] *Ibid.*

[87] Testimony of Mark Hafle, 8-9.

[88] Both Guide and Morel also may have been overworked, as suggested by some of the hasty, last-minute decisions. In his 2009 performance evaluation of Morel, David Sims wrote that Morel "[w]ants and needs a high work load" and went on to describe all of the different projects he was involved with. Internal BP document (BP-HZN-OSC 9029). As for Guide, when asked by the MBI panel for any recommendations going forward, BP operations engineer Brett Cocales testified, "I would say with the kind of load that a wells team leader is undertaking, that additional resources would be of benefit to that person," including "[a]dditional people to handle the multitasking areas that that person has to undertake." Testimony of Brett Cocales, 268-69.

[89] Testimony of John Sprague, 33-34.

[90] *Ibid.*, 34-37; Internal BP document (BP-HZN-MBI 208674). Section 3.1.13 of BP's Drilling and Well Operations Practice (DWOP) manual provides: "Temporary or permanent changes to personnel, systems, procedures, programmes, safety and critical data, equipment, facilities, materials or substances should be made using the management of change process." Internal BP document (BP-HZN-OSC 7222).

[91] Testimony of John Guide, Hearing before the Deepwater Horizon Joint Investigation Team, July 22, 2010, 110-11.

[92] Testimony of John Sprague, 33-34.

[93] Internal Transocean document (TRN-HCEC 5402-797). Although the near miss in the North Sea on December 23, 2009 did not occur during end-of-well activities, it took place during similar circumstances— the single barrier to flow had been tested. The incident prompted Transocean to remind its North Sea fleet that "tested barriers can fail" and that "high vigilance" is necessary "when reduced to one barrier underbalanced." Internal BP document (BP-HZN-BLY 38362).

[94] David Borthwick, *Report of the Montara Commission of Inquiry* (The Montara Commission of Inquiry, Australia, June 2010), 5, 26. According to the *Report of the Montara Commission of Inquiry*, released on November 24, 2010, many of the technical and managerial causes of the Montara blowout track those at Macondo. For instance, the Commission of Inquiry concluded that the cement job in the 9⅝-inch casing shoe failed, that there were numerous risk factors surrounding the cement job that went unheeded, and that the cement job was not properly pressure tested. *Ibid.*, 7. According to the Commission of Inquiry:

The multiple problems in undertaking the cement job—such as the failure of the top and bottom plugs to create a seal after "bumping," the failure of the float valves and an unexpected rush of fluid—should have raised alarm bells. Those problems necessitated a careful evaluation of what happened, the instigation of pressure testing and, most likely, remedial action. No such careful evaluation was undertaken. The problems were not complicated or unsolvable, and the potential remedies were well known and not costly. This was a failure of "sensible oilfield practice 101."

Ibid. The Commission of Inquiry went on to conclude that while the "absence of tested barriers was a proximate cause of the Blowout," the deeper failure was a systemic failure of management on the part of the operator, PTTEP Australasia. *Ibid.*, 9.

[95] *Ibid.*

[96] Testimony of Steve Lewis (Expert witness), Hearing before the National Commission, November 9, 2010, 64-65.

[97] Testimony of Ross Skidmore (BP), Hearing before the Deepwater Horizon Joint Investigation Team, July 20, 2010, 264.

[98] Ezell, interview.

[99] Internal Transocean document (TRN-HCEC 5609-10).

[100] There were a number of different alarm systems on the *Deepwater Horizon*, including a general alarm audible to the entire rig and localized combustible gas, fire, and toxic gas alarms that were positioned throughout the rig. When triggered, the localized alarms would sound automatically in the affected area and send a signal back to the DPO's panel on the bridge. According to most witness accounts, the general alarm was set to "manual mode," meaning that a person on the bridge had to activate the general alarm in order for it to sound to the entire rig. See, e.g., Testimony of Yancy Keplinger (Transocean), Hearing before the Deepwater Horizon Joint Investigation Team, October 5, 2010, part 1, 188, 222 (standard practice to have general alarm in manual mode in order "to permit some human judgment to make a determination as to whether or not a general alarm should be sounded as opposed to having it on some automated system"); Testimony of Jerry Canducci, 124 (purpose of manual mode is to "alert somebody so that they can pass judgment on the efficacy of the system, and when it is deemed to be a proper alarm, then the alarm is sounded to all"). The Chief Counsel's team does not express a view on the wisdom of having the general alarms in manual rather than automatic mode except to note that if the alarm is in manual mode, Transocean must ensure that its DPOs are trained to deal with emergency situations. See, e.g., U.S. Coast Guard, Navigation and Vessel Inspection Circular No. 2089 (August 4, 1999) § 5.2 (stating that a member of the crew should decide whether to sound the general alarm, that the alarm must be initiated manually and is intended to be sounded by a person on watch or other responsible member of the crew, and that the general alarm may be sounded automatically by a safety monitoring system if a fire alarm is not acknowledged within a reasonable amount of time).

[101] Testimony of Andrea Fleytas, Hearing before the Deepwater Horizon Joint Investigation Team, October 5, 2010, part 2, 13-14.

[102] *Ibid.* Fleytas did tell the engine control room that the rig was experiencing a well control situation, as she had just received a phone call from the drill floor indicating as much. *Ibid.*

[103] *Ibid.*, 65.

[104] *Ibid.*, 53-55.

[105] *Ibid.*, 40. Asked whether any of the alarms on the panel were from the engine control room, she responded, "There were so many alarms. There were hundreds of them on that page, so I don't remember if those were some of them." *Ibid.*

[106] Section 15.2.7 of BP's Drilling and Well Operations Practice (DWOP) manual provides: "Kick detection, diverter, circulating, stripping, and shut-in drills shall be held regularly until the designated company representative is satisfied that each crew demonstrates suitable BP standards." Internal BP document (BP HZN-OSC 7267). It is not clear what, if any drills, BP required on emergency kick detection or diverter situations.

[107] Ezell, interview.

[108] Nonpublic BP document (presentation to Commission staff, August 9, 2010), 6.

[109] Testimony of Mark Bly, 335.

[110] For example, see Rick Godfrey's questioning of Gagliano. Testimony of Jesse Gagliano, Hearing before the Joint Investigation Team, August 24, 2010, 291.

[111] Internal BP document (BP-HZN-OSC 9338).

[112] Internal BP document (BP-HZN-MBI 255509).

[113] *Ibid.*

[114] Internal BP document (BP-HZN-MBI 128542); Internal BP document (BP-HZN-MBI 110151).

[115] Internal BP document (BP-HZN-MBI 128542).

[116] *Ibid.*

[117] Sims, interview, February 1, 2011.

[118] Internal BP document (BP-HZN-BLY 125446).

[119] Internal BP document (BP-HZN-MBI 212826).

[120] Internal BP document (BP-HZN-MBI 61225).

[121] Testimony of Daniel Oldfather (Weatherford), Hearing before the Deepwater Horizon Joint Investigation Team, October 7, 2010, 14.

[122] Jesse Gagliano (Halliburton), interview with the U.S. House of Representatives Committee on Energy and Commerce, June 11, 2010; Testimony of Gregg Walz, 53-54.

[123] Testimony of Nathaniel Chaisson (Halliburton), Hearing before the Deepwater Horizon Joint Investigation Team, August 24, 2010, 437-38.

[124] Gagliano, interview.

[125] Testimony of Jesse Gagliano, August 24, 2010, 335, 360.

[126] Internal BP document (BP-HZN-MBI 137370).

[127] Guide, interview, January 19, 2011.

[128] BP and Transocean agree on this. Testimony of Mark Bly, 246-47; Bill Ambrose (Transocean), Hearing before the National Commission, November 8, 2010, 246-247.

[129] See Chapter 4.7.

[130] Joseph Keith (Halliburton), interview with Commission staff, October 6, 2010; Cathleenia Willis (Halliburton), interview with Commission staff, October 21, 2010.

[131] See Chapter 4.7.

[132] See Chapter 4.7.

[133] Allen Seraile (Transocean), interview with Commission staff, January 7, 2010.

[134] See Chapter 4.7.

[135] Commission staff site visit to *Deepwater Nautilus*, September 9, 2010.

[136] Internal BP document (BP-HZN-MBI 127997-128022).

[137] *Ibid.*

[138] Internal BP document (BP-HZN-MBI 128018).

[139] *Ibid.*

[140] Internal BP document (BP-HZN-MBI 128002).

[141] Internal BP document (BP-HZN-MBI 128009).

[142] During a visit to BP's Houston headquarters to see the Macondo room, BP personnel told the Chief Counsel's team that BP did not constantly monitor data and other information from onshore because doing so tended to disempower personnel on the rig. The Chief Counsel's team does not fully understand that

explanation. In any event, it is inconsistent with BP's pre-blowout plan to implement the ERA advisory system.

[143] It should be noted that Transocean does not send its data in real time back to shore.

[144] The notable exception is the decision to use a long string production casing, which had been the plan all along. However, it was not until the lost circulation event and declaration of early total depth that BP's Macondo team identified many of the risks associated with using a long string at Macondo.

[145] Internal BP document (BP-HZN-OSC 5420). According to the manual, "'Beyond the Best' is the strategic brand for Drilling & Completions focused on improving performance in drilling and completions efficiency. It targets a prize characteri[z]ed by improved safety performance, improved capital efficiency and a bigger contribution to production from new wells." Internal BP document (BP-HZN-OSC 5426).

[146] Internal BP document (BP-HZN-OSC5428); Internal BP document (BP-HZN-OSC 5434).

[147] Internal BP document (BP-HZN-OSC 5463).

[148] Guide, interview, September 17, 2010.

[149] Sims interview, December 14, 2010. Sims told the Chief Counsel's team that except for changes to the well plan, the wells team leader had discretion whether to subject a particular decision to the MOC process. *Ibid.*

[150] The drilling and completions group's Beyond the Best manual requires that "[a] clear process must be in place for management of change throughout any project. This management of change process must be auditable and in place during planning as well as operations." Internal BP document (BP-HZN-OSC 5458). Among its requirements for a management of change process are: "A clear project statement saying when the Management of Change process will be utilized," "[i]ncorporate a risk assessment," and "[h]ave a clear approval structure linked to this change process." *Ibid.*

[151] BP's exploration and production unit's DWOP provides: "Any significant changes to a well programme shall be documented and approved via a formal management of change (MOC) process which includes those on the original approval list." Internal BP document (BP-HZN-OSC 7243).

[152] See Internal BP document (BP-HZN-MBI 143120-22); Internal BP document (BP-HZN-MBI 143242-44); Internal BP document (BP-HZN-MBI00143247-49)(January 27, 2010 MOC for change from *Marianas* to *Deepwater Horizon*); Internal BP document (BP-HZN-MBI 143251-53); Internal BP document (BP-HZN-MBI 143255-57); Internal BP document (BP-HZN-MBI 143259-61); Internal BP document (BP-HZN-MBI 143292-94).

[153] Internal BP document (BP-HZN-MBI 143251-53); Internal BP document (BP-HZN-MBI 143255-57); Internal BP document (BP-HZN-MBI 143259-61); Internal BP document (BP-HZN-MBI 143292-94). The Macondo team twice submitted the long string decision to the MOC process because of an error in the first MOC.

[154] Guide, interview, September 17, 2010.

[155] Internal BP document (BP-HZN-BLY 61195).

[156] Internal BP document (BP-HZN-BLY 61205).

[157] Internal BP document (BP-HZN-OSC 7105).

[158] Internal BP document (BP-HZN-OSC 7128).

[159] O'Bryan, interview.

[160] Internal BP document (BP-HZN-BLY 61475).

[161] See Chapter 4.7. BP's Engineering Technical Practice on Simultaneous Operations (GP 10-75) also directs its personnel to conduct a risk assessment of simultaneous operations "in order to identify the risks across the complete range of well activities" and to "ensure all well activities...are carried out in a safe and controlled manner, when these activities are performed in the same space and time as another operation." Internal BP document. Internal BP document (BP-HZN-OSC 8019). It is not clear whether BP personnel associated the ETP with activities on the *Deepwater Horizon* generally, let alone conducted any such risk assessment with respect to rig activities going on during the final displacement operation. At least one witness has suggested that other well site leaders would not have allowed so many rig operations. Joseph Keith, interview.

[162] See Chapter 4.7.

[163] Internal Transocean document (TRN-HCEC 90501).

[164] Internal BP document (BP-HZN-MBI 225048).

[165] Discussing the shut-down of a drilling rig in March, BP executive Harry Thierens wrote Gregg Walz and several other managers: "time is money after all." Internal BP document (BP-HZN-MBI 225981).

[166] Transocean-BP Drilling Contract No. 980249; Amendment 41 to Drilling Contract No. 980249.

[167] Internal BP document (BP-HZN-MBI 126763).

[168] Internal BP document (BP-HZN-MBI 269181); see also Internal BP document (BP-HZN-BLY 125444).

[169] Sims, interview, February 1, 2011.

[170] Internal Halliburton document (HAL_10648).

[171] Guide, interview, January 19, 2011; Internal BP document (BP-HZN-CEC 22433).

[172] Internal BP document (BP-HZN-OSC 5561).

[173] Internal BP document (BP-HZN-OSC 5558).

[174] Internal BP document (BP-HZN-OSC 5573).

[175] *Ibid.*

[176] Internal BP document (BP-HZN-MBI 117622).

[177] *Ibid.*

[178] *Ibid.*

[179] *Ibid.*

[180] Internal BP document (BP-HZN-MBI 126763).

[181] *Ibid.*

[182] Internal BP document (BP-HZN-OSC 5558).

[183] As of April 13, BP had paid Transocean about $68 million for the rig's day rate. The total cost of the well up to that point was $137 million. Internal BP document (BP-HZN-MBI 126763).

[184] *Ibid.*

[185] *Ibid.*

[186] *Ibid.*

[187] Internal BP document (BP-HZN-MBI 258390).

[188] Internal BP document (BP-HZN-MBI 192546).

[189] Internal BP document (BP-HZN-MBI 19552); Internal BP document (BP-HZN-MBI 192558); Internal BP document (BP-HZN-MBI 192559).

[190] The first AFE was $96.1 million. Internal BP document (BP-HZN-MBI 19552). The total authorizations added up to $154.5 million. Internal BP document (BP-HZN-MBI 192559).

[191] Internal BP document (BP-HZN-MBI 125958).

[192] Testimony of David Sims (BP), Hearing before the Deepwater Horizon Joint Investigation Team, August 25, 2010, 200.

[193] Internal BP document (BP-HZN-MBI 128953).

[194] BP estimated bottom 10% performance at 58.7 days per 10,000 feet. Internal BP document (BP-HZN-BLY 47298).

[195] BP estimated bottom 10% performance at $142.16 million. *Ibid.*

[196] Hafle estimated nonproductive time at 48.3%. Internal BP document (BP-HZN-MBI 128953). BP estimated bottom 10% performance at 39.8%. Internal BP document (BP-HZN-BLY 47298).

[197] Ronnie Sepulvado, interview.

[198] *Ibid.*; Guide, interview, January 26, 2011; Sims, interview; O'Bryan interview.

[199] Testimony of Steve Lewis (Expert Witness), hearing before the National Commission, November 9, 2010, 79.

[200] Internal BP document (BP-HZN-OSC 5420).

[201] Internal BP document (BP-HZN-OSC 5437).

[202] Internal BP document (BP-HZN-OSC 5557).

[203] *Ibid.*

[204] Internal BP document (BP-HZN-OSC 5558).

[205] Internal BP document (BP-HZN-OSC 5437).

[206] Internal BP document (BP-HZN-OSC 5557).

[207] Internal BP document (BP-HZN-OSC 5561).

[208] Sims, interview, December 14, 2010.

[209] *Ibid.*

[210] Internal BP document (BP-HZN-OSC 8982).

[211] For instance, Ross Skidmore, a BP contractor tasked with setting the lockdown sleeve at Macondo, and Merrick Kelley, BP subsea wells team leader and the person at BP responsible for lockdown sleeves in the Gulf of Mexico, had planned per their normal practice to perform a separate "wash trip" to clean out any debris before running the lead impression tool and lockdown sleeve. Guide rejected their plan, telling Skidmore: "We will never know if your million dollar flush run was needed. How does this get us to sector leadership." Internal BP document (BP-HZN-MBI 258507). Guide has subsequently explained that he did not believe a separate wash run was necessary because the crew would already have washed out any debris when tripping out of the hole with the drill pipe. Guide, interview, January 19, 2011; Internal BP document (BP-HZN-MBI 258233). This explanation may be credible, and the Chief Counsel's team cannot say that Guide's decision increased risk at Macondo. Nevertheless, BP experts still questioned Guide's views after the incident.

[212] Confidential Commission staff review of personnel information, December 8, 2010.

[213] Internal BP document (BP-HZN-OSC 7163).

[214] Internal BP document (BP-HZN-MBI 98517).

[215] Internal BP document (BP-HZN-MBI 261533).

[216] Internal BP document (BP-HZN-OSC8980).

[217] *Ibid.*

[218] Internal BP document (BP-HZN-OSC 8982).

[219] Internal BP document (BP-HZN-OSC 8980).

[220] *Ibid.*

[221] Graham Ruddick, "BP says 'every dollar counts' as profits tumble," *The Telegraph*, February 3, 2009.

[222] Internal BP document (BP-HZN-OSC 8981).

[223] Internal BP document (BP-HZN-OSC 9060).

[224] Internal BP document (BP-HZN-OSC 9067).

[225] Internal BP document (BP-HZN-OSC 9068).

[226] Internal BP document (BP-HZN-OSC 9074-75).

[227] BP, Letter to the National Commission, November 22, 2010.

[228] *Ibid.*

[229] BP, "Golden Rules of Safety," http://www.bp.com/liveassets/bp_internet/globalbp/STAGING/global_assets/downloads/B/BPs_Golden_rules_of_safety.pdf.

[230] Internal BP document (BP-HZN-OSC 9025).

[231] *Ibid.*

[232] Internal BP document (BP-HZN-OSC 9054).

[233] Internal BP document (BP-HZN-OSC 9025).

[234] Sims, interview, December 14, 2010.

Chapter 6

[1] Troy Trosclair (MMS), interview with Commission staff, October 1, 2010.

[2] Testimony of Walter Cruickshank (MMS), Hearing before the National Commission, November 9, 2010, 186-87.

[3] U.S. Department of the Interior, Outer Continental Shelf Safety Oversight Board, *Report to the Secretary of the Interior* (September 1, 2010), 6. "APMs have increased by 71% from 1,246 in 2005 to 2,136 in 2009 in the New Orleans District." *Ibid.*

[4] Frank Patton (MMS), interview with Commission staff, October 1, 2010; David Trocquet (MMS), interview with Commission staff, October 1, 2010.

[5] Patton, interview.

[6] 30 C.F.R. § 250.413.

[7] Patton, interview.

[8] 30 C.F.R § 250.428.

[9] Internal BP document (BP-HZN-MBI 133874).

[10] Internal BP document (BP-HZN-MBI 23715).

[11] 30 C.F.R. § 250. 415, 420. Test pressures are determined based on operator calculation of maximum anticipated surface pressure, or the amount of pressure an operator expects to be exerted on casing and subsea equipment. Steve Lewis (Expert witness), email to Commission staff, October 27, 2010. Regulations expressly leave this calculation to operators. 30 C.F.R. § 250.413(f). In calculating this figure, some operators, including BP at the Macondo well, currently assume a well column is 50% gas and 50% drilling fluid. Steve Lewis (Expert witness), interview with Commission staff, September 28, 2010. However, during a blowout the well column can be entirely empty of mud and instead contain 100% gas. *Ibid.* The current industry norm assuming 50% mud in the well column thus underestimates MASP and results in correspondingly low test pressures that do not reflect the worst-case blowout scenario. *Ibid.*

[12] 30 C.F.R. § 250. 415, 420.

[13] Patton, interview.

[14] *Ibid.*

[15] 30 C.F.R § 250.422(a).

[16] Testimony of Walter Cruickshank, 198-99.

[17] MMS regulations do state that "[b]efore removing the marine riser, you must displace the riser with seawater. You must maintain sufficient hydrostatic pressure or take other suitable precautions to compensate for the reduction in pressure and to maintain a safe and controlled well condition." 30 C.F.R § 250.442(e). A negative pressure test is a way to prove that the well will withstand that reduction of pressure as a means of satisfying this requirement. Testimony of John Smith (Expert witness), Hearing before the National Commission, November 9, 2010, 152.

[18] Internal BP document (BP-HZN-MBI 23711). Some have questioned whether the temporary abandonment procedures in the approved permit would have required a negative pressure test to be conducted at the wellhead before displacement, or whether it could be done at 8,367 feet in the middle of displacement (as was actually done at Macondo). Regardless of this argument, it is unlikely that the MMS would have rejected an APM that said the negative pressure test was to be conducted at a depth of 8,367 feet. Patton, interview. Discussed further in Chapter 4.6 (Negative Pressure Test).

[19] Oil and Gas and Sulphur Operations in the Outer Continental Shelf—Increased Safety Measures for Energy Development on the Outer Continental Shelf, 75 Fed. Reg. 63346, 63373 (October 14, 2010).

[20] John Smith (Expert witness), interview with Commission staff, October 26, 2010.

[21] Internal BP document (BP-HZN-OSC 1436).

[22] Internal BP document (BP-HZN-MBI 127906).

[23] During the Commission's public hearing, Dr. Walter Cruickshank stated that the official had also drawn reassurance from the fact that BP planned to conduct a negative pressure test on the surface cement plug prior to abandoning the well and that this test would help ensure safety. Testimony of Walter Cruikshank, 206. This makes little sense: The procedure that BP submitted specifies that the negative pressure test will be done *before* setting the surface cement plug. Internal BP document (BP-HZN-MBI 23711).

[24] Internal BP document (BP-HZN-CEC 20942).

[25] 30 C.F.R § 250.448(c). The upper annular was rated to withstand 10,000 psi closed on pipe or 5,000 psi closed on an open hole. BP, Deepwater Horizon *Accident Investigation Report* (September 8, 2010), app. H, 227. After conversion to a stripping annular, the lower annular was rated to withstand 5,000 psi. *Ibid.*, 172. Post-explosion examination of the blue pod found regulated pressure on the lower annular preventer was set to approximately 1,700 psi. *Ibid.*, 145. According to subsea supervisor Chris Pleasant, he reset the lower annular regulator pressure to 1,500 psi soon after 9 p.m. Testimony of Chris Pleasant (Transocean), Hearing before the Deepwater Horizon Investigation Team, May 28, 2010, 120. According to BP, 1,500 psi was the normal regulator pressure setting for both annular preventers. BP, Deepwater Horizon *Accident Investigation Report*, 145.

[26] Internal BP document (BP-HZN-OSC 867).

[27] Internal BP document (BP-HZN-OSC 925).

[28] Internal BP document (BP-HZN-MBI 107364-65).

[29] 30 C.F.R § 250.446(a).

[30] American Petroleum Institute, *Recommended Practices for Blowout Prevention Equipment Systems for Drilling Wells,* 3rd ed. (March 1997, reaffirmed September 2004), 56 ("API RP 53 § 18.10.3").

[31] Internal Cameron document (CAM-GR 261-62).

[32] Testimony of Mark Hay (Transocean), Hearing before the Deepwater Horizon Joint Investigation Team, August 25, 2010, 32-33; Testimony of John Sprague (BP), Hearing before the Deepwater Horizon Joint Investigation Team, December 8, 2010, 79-80.

[33] Internal BP document (BP-HZN-MBI 136213).

[34] Internal Transocean document (TRN-USCG_MMS 38652).

[35] Internal Transocean document (TRN-USCG_MMS 38662). The BOP's fail-safe valves, designed to shut off choke and kill lines remotely and automatically, had not been certified since December 13, 2000. Internal Transocean document (TRN-USCG-MMS 38656).

[36] 30 C.F.R § 250.446(a); Elmer Danenberger (Expert witness), interview with Commission staff, January 5, 2011.

[37] Internal MMS document.

[38] Testimony of Eric Neal (MMS), Hearing before the Deepwater Horizon Joint Investigation Team, May 11, 2010, 325.

[39] MMS, *National Potential Incident of Noncompliance (PINC) and Guideline List* (May 2008).

[40] Testimony of Eric Neal, 326.

[41] MMS, *National Potential Incident of Noncompliance (PINC) and Guideline List – Drilling* (May 2008).

Appendix A | Blowout Investigation Team

FRED H. BARTLIT, JR.
Chief Counsel

RICHARD SEARS
Senior Science & Engineering Advisor

SEAN C. GRIMSLEY
Deputy Chief Counsel

SAMBHAV N. SANKAR
Deputy Chief Counsel

J. JACKSON EATON
Counsel

BRENT C. HARRIS
Counsel

JON IZAK MONGER
Counsel

SARITHA KOMATIREDDY TICE
Counsel

JOSEPH B. HERNANDEZ
Paralegal

MICHELLE FARMER
Executive Legal Assistant

GRAPHICS AND LAYOUT BY TRIALGRAPHIX

Appendix B | Commission Staff

Richard Lazarus, *Executive Director*

Tracy Terry, *Deputy Director*

Fred H. Bartlit, Jr., *Chief Counsel*   Jay Hakes, *Director of Research & Policy*

Priya Aiyar
Deputy Chief Counsel

Felicia Barnes
Analyst

Adam Benthem
Analyst

Gordon Binder
Senior Policy Advisor

Paul Bledsoe
Senior Policy Advisor

Jed J. Borghei
Counsel

C. Hobson Bryan
Analyst

Edwin H. Clark, II
Director of Operations

Kate Clark
Senior Analyst

Dave Cohen
Press Secretary

Cindy Drucker
Director of Public Engagement

Katherine Duncan
Analyst

J. Jackson Eaton
Counsel

Michelle Farmer
Executive Legal Assistant

Sean C. Grimsley
Deputy Chief Counsel

David Greenberg
Senior Policy Advisor

Brent C. Harris
Counsel

Lisa K. Hemmer
Senior Legal Advisor

Joseph B. Hernandez
Paralegal

Joel Hewett
Analyst

Christiana James
Staff Assistant

Jill Jonnes
Senior Researcher

Nancy Kete
Senior Analyst

Caitlin Klevorick
Policy Advisor

Emily Lindow
Senior Analyst

Claire Luby
Assistant to the Executive Director

Bethany Mabee
Communications Coordinator

Scott McKee
Analyst

Claudia A. McMurray
Senior Counsel for Congressional and State Relations

Louise Milkman
Chief of Staff to the Executive Director

Jon Izak Monger
Counsel

Shirley Neff
Senior Analyst

Pete Nelson
Director of Communications

Elena Nikolova
Analyst

Jessica O'Neill
Counsel

Paul Ortiz
Senior Legal Advisor

Tony Padilla
Analyst

Tyler Priest
Senior Analyst

Sarah Randle
Analyst

Irwin Redlener
Senior Public Health Advisor

John S. Rosenberg
Chief Editor

Eric Roston
Senior Analyst

Sara Rubin
Analyst

Sambhav N. Sankar
Deputy Chief Counsel

Nicole A. Sarrine
Staff Assistant

Richard Sears
Senior Science & Engineering Advisor

Steven Siger
Counsel

Robert Spies
Senior Environmental Science Advisor

Danielle Stewart
Staff Assistant

Marika Tatsutani
Editor

Johnny Tenorio
Information Technology Officer

Saritha Komatireddy Tice
Counsel

Lloyd Timberlake
Senior Researcher

Clara Vondrich
Counsel

Jason Weil
Analyst

David Weiss
Counsel

Stephen M. Willie
Administrative Officer

Andrea J. Yank
Assistant to Co-Chair William Reilly

Appendix C | Acronyms

AFE	Approval for Expenditure
AMF	Automatic mode function
APB	Annular pressure buildup
APD	Application for permit to drill
API	American Petroleum Institute
APM	Application for permit to modify
bbl	Barrels
BOEMRE	Bureau of Ocean Energy Management, Regulation, and Enforcement
BOP	Blowout preventer
bpm	Barrels per minute
BSR	Blind shear ram
CMMS	Computerized Maintenance Management System
DP	Dynamically positioned
DPO	Dynamic positioning officers
ECD	Equivalent circulating density
EDS	Emergency disconnect system
ERA	Efficient Reservoir Access
ESD	Equivalent static density
ETP	Engineering Technical Practice
FIT	Formation integrity test
gal/sack	Gallons per sack
gpm	Gallons per minute
HSSE	Health, safety, security, and the environment
LDS	Lockdown sleeve
LMRP	Lower marine riser package
LOT	Leak off test
MC 252	Mississippi Canyon Block 252
MD	Measured depth
MMS	Minerals Management Service

MOC	Management of change
MODU	Mobile offshore drilling unit
MUX	Multiplex
OIM	Offshore installation manager
OMS	Operating management system
PINC	Potential incidents and noncompliance
ppg	Pounds per gallon
PRV	Pressure relief valve
psi	Pounds per square inch
RCRA	Resource Conservation and Recovery Act
RMS	Rig Management System
ROV	Remotely operated vehicle
SG	Specific gravity
TD	Total depth
TIGER	Totally Integrated Geological and Engineering Resource
TOC	Top of cement
TVD	Total vertical depth
UWILD	Underwater Inspection in Lieu of Dry-docking

Appendix D | Chevron Laboratory Report Cover Letter

Chevron

Craig Gardner
Team Leader - Cementing

Energy Technology Company
Drilling and Completions Department
3901 Briarpark
Houston, TX 77042
Tel 713.954.6154
Fax 713.954.6177
craig.gardner@chevron.com

Sambhav N. "Sam" Sankar
Deputy Chief Counsel
National Commission on the BP Deepwater Horizon
Oil Spill and Offshore Drilling
One Thomas Circle, 4th Floor
Washington DC 20005

October 26, 2010

NATIONAL COMMISSION ON THE
BP DEEPWATER HORIZON OIL SPILL AND OFFSHORE DRILLING
CEMENT TESTING RESULTS

MR. SAMBHAV N. "SAM" SANKAR

This report summarizes the results of the testing conducted in the cementing laboratory at Chevron's Briarpark facility at the request of the National Commission on the BP Deepwater Horizon Oil Spill and Offshore Drilling.

We conducted these tests using samples of cement and additives supplied by Halliburton and sent to the Chevron laboratory at the request of the Commission. To our knowledge, these materials were supplied by Halliburton as representative of materials used on the Deepwater Horizon but are neither bulk plant samples nor rig samples from the actual job.

The mud sample used in the contamination testing described in this report was supplied by MI Swaco at the Commission's request. It is a sample of drilling fluid from an actual drilling operation (i.e. not laboratory-prepared nor taken from a freshly-built mud in a liquid mud plant). MI Swaco supplied an analysis (mud check) with the sample, and a similar suite of tests were run in the Chevron drilling fluids laboratory to confirm the fluid characteristics. Both the MI Swaco results and the Chevron results compare reasonably well with the field mud check #79 dated April 19, 2010. Copies of the mud reports are contained in the Appendix.

October 26, 2010
Page 2

The testing was based on the Halliburton laboratory report dated April 12, 2010 and contained in Appendix J of the BP report *Deepwater Horizon Accident Investigation Report, September 8, 2010*, Appendix J. Most of the tests were conducted using multiple protocols. API and ISO cementing standards are, for the most part, technically identical standards which allow latitude in test procedures. The Halliburton report does not contain sufficient information to determine the exact test protocol used in the Halliburton lab in all cases. Halliburton elected not to provide additional information clarifying its testing protocols that was requested through the Commission. Therefore, a range of test procedures was selected to encompass a variety of test conditions.

Many of the test results were in reasonable agreement with those reported by Halliburton. However, we were unable to generate stable foam with any of the tests described in Section 9 of this report.

Craig Gardner

Appendix E | Nile and Kaskida

BP faced MMS deadlines on the two projects planned for the *Deepwater Horizon* after Macondo—permanent abandonment of a Nile well and spudding of a Kaskida well. The Chief Counsel's team found that these regulatory deadlines did not significantly compound the already existing time pressure at Macondo.[1]

Schedule When the *Deepwater Horizon* Arrived at Macondo

The high daily cost of employing the *Deepwater Horizon* put pressure not just on the immediate task of drilling, but also on how BP scheduled future projects for the rig. The schedule for a drilling rig should be seamless. Empty days on the calendar waste dollars. BP had to pay Transocean a daily lease fee regardless of whether the *Deepwater Horizon* was drilling or not.[2] Throughout the drilling of the Macondo well, BP focused on how it would keep the rig active after Macondo. Delays at Macondo, equipment delays at another well, and regulatory commitments to MMS complicated the task.

Long before the *Deepwater Horizon* arrived at Macondo, BP began planning work for the rig at future locations.[3] BP's schedule for the *Deepwater Horizon* stretched years into the future, up to 2013.[4] When the *Deepwater Horizon* arrived at Macondo, BP planned to have the rig on location for about 45 days.[5]

BP planned to then send the rig to Nile for 30 days.[6] Nile was in another tract in the Gulf of Mexico, located about a day's voyage from Macondo. BP faced a July 2, 2010 deadline to permanently abandon its well at Nile.[7] Federal regulations require a lease holder to "permanently plug all wells on a lease within 1 year after the lease terminates."[8] Nile had been a productive well for BP, and it would be BP's first permanent abandonment of a subsea producing well in the Gulf of Mexico.[9] The task would be complex, and the rig crew worried about its challenges.[10]

After Nile, the *Deepwater Horizon* would go to Kaskida, located in yet another tract in the Gulf of Mexico leased by BP.[11] Kaskida is about 250 miles southwest of New Orleans and about a four-day voyage from Macondo.[12] In 2006, the *Deepwater Horizon* drilled an exploration well at Kaskida that proved to be a large discovery.[13] MMS required BP to conduct further activities at Kaskida by May 16, 2010 to keep its lease.[14] Federal regulations require activity on an exploration lease every 180 days.[15] MMS regulation 30 C.F.R. § 250.180 specifies that a lease ends after a certain period "unless you are conducting operations on your lease."[16] Drilling counts as operations, so long as the "objective of the drilling" is "to establish production in paying quantities on the lease."[17] Without activity or production, MMS could cancel the lease.[18] BP's original schedule allowed the *Deepwater Horizon* to carry out the abandonment of Nile first and still meet the deadline at Kaskida.[19]

Request to Suspend Operations at Kaskida

While the *Deepwater Horizon* drilled the Macondo well, BP worried that delays for the Kaskida wellhead would leave the rig with too much time after it completed its current well.[20] BP required a first-of-its-kind wellhead at Kaskida.[21] Delivery of that wellhead proved a headache for BP.[22] The emergency seal for the wellhead failed tests.[23] These failures led to an ever-changing set of delivery dates. In February, BP engineering team leader David Sims expressed his concerns to several managers and executives: "Even with the delays we are experiencing on Macondo, I still feel that there is a significant risk that the Horizon will finish the Nile P&A before the DrilQuip 20K wellhead is delivered."[24]

Fearing that the rig might be left idle because of the wellhead delays, BP considered several options. The company contemplated extending work at Macondo itself and having the rig stay longer.[25] It explored alternative projects for the *Deepwater Horizon* after the rig completed both Macondo and Nile.[26] And it thought about having the rig undergo maintenance to fill gaps in the schedule.[27]

Toward the end of March, the *Deepwater Horizon* fell far enough behind schedule at Macondo that BP stopped brainstorming additional projects to occupy the rig and determined that the Nile project would likely no longer fit in before the 180-day clock ran out at Kaskida. If the *Deepwater Horizon* were going to spud Kaskida despite the delay, that left BP two primary options. One option was to go to Nile first and ask MMS for an extension at Kaskida. Another option was to go to Kaskida directly and make alternative arrangements for Nile.

BP weighed going to Kaskida directly.[28] Reasons to go to Kaskida included avoiding the hurricane season in the Gulf of Mexico and maintaining the schedule for work on the well after the *Deepwater Horizon*'s spud.[29] Ultimately, BP concluded that it preferred to have the *Deepwater Horizon* do the Nile project first. Reasons to go to Nile included continuing concern about the wellhead: "[g]oing to Kaskida post Macondo assumes wellhead ready to utilize, currently planned ready ca. 23 April."[30] BP also wanted to complete Nile in time to fit in a previously scheduled crane replacement operation.[31] On April 8, BP vice president of drilling and completions Pat O'Bryan concluded, "Sounds like we should leave [Nile] on the Horizon as originally planned."[32]

Fitting in Nile before going to Kaskida became impossible from a scheduling perspective. BP anticipated that Nile would take about 30 days.[33] Because BP kept the Nile project on the *Deepwater Horizon*'s schedule, BP had no choice but to ask MMS for an extension of the deadline at Kaskida in order to avoid losing the lease. By April 16, BP had only 30 days until the May 16 deadline at Kaskida, not counting transit time to get from one well to the next.[34] Consequently, BP would need a "suspension of operations" at Kaskida. A suspension of operations "extend[s] the term of a lease."[35]

On April 9, Sims began to draft BP's request to MMS for a suspension of operations at Kaskida.[36] On one level, the request to suspend operations was straightforward. A suspension of operations may be granted "when necessary to allow you time to begin drilling or operations when you are prevented by reasons beyond your control, such as unexpected weather, unavoidable accidents, or drilling rig delays."[37] The primary test on "whether you are 'prevented beyond your control' is whether the particular drilling rig was scheduled to conduct operations at your location *before* the lease expiration date."[38] The *Deepwater Horizon* had been scheduled to conduct operations at the location before the expiration date, and it had faced delays at Macondo.

Nonetheless, a suspension of operations is granted only "on a case-by-case basis" and typically for "a short duration."[39] Moreover, the delay at Macondo prevented the *Deepwater Horizon*'s timely arrival at Kaskida only because BP had kept Nile first on the *Horizon*'s schedule. Without Nile, there would be no need for a suspension. BP's situation fit the criteria for a suspension, but not definitively. A member of BP's offshore land negotiation team commented, "While the Nile P&A timing is critical path to us, the MMS unit group may not see it that way and suggest that operation be delayed to avoid the issuance of an SOO."[40] He then remarked that whether MMS would grant the suspension was "anyone's guess."[41] On April 20, BP sent the request for a suspension of operations to MMS.[42]

While waiting to hear from MMS, BP planned to send the *Deepwater Horizon* to Nile.[43] BP sent a team out to the rig to prepare for the move to Nile.[44] Some members of the BP team may have perceived pressure to complete the Macondo well quickly. Before the MMS request went out, BP subsea wells team leader Merrick Kelley emailed BP drilling engineer Brian Morel: "I know you all are under pressure to finish Macondo so we can get Nile P&A moving and not jeopardize the Kaskida well and IFT."[45] Uncertainty about internal BP plans, or uncertainty about MMS's decision, may have prompted concern about time pressure.

Nonetheless, if there was concern, the Chief Counsel's team has found no evidence that it was widespread. BP drilling engineer team leader Gregg Walz, BP wells team leader John Guide, BP well site leader Murry Sepulvado, and Sims said that Nile put no pressure on the temporary abandonment of Macondo.[46] Similarly, Transocean offshore installation manager (OIM) Jimmy Harrell testified that he faced no pressure from BP or Transocean to move on to Nile.[47] Moreover, BP planned to send the rig directly to Kaskida if MMS denied the request to suspend operations and then to ask for an extension at Nile.[48] If that happened, the *Deepwater Horizon* would experience downtime, not pressure.[49] BP planned maintenance to "fill any gaps" if the wellhead arrived late.[50]

Though BP's decisions at Macondo appear to have been biased in favor of saving time and money, the rig's next wells do not appear to have been an important contributing factor. BP followed the rig's schedule closely and, when necessary, took action to relieve the pressure of regulatory deadlines.

---

[1] There has been some suggestion that these deadlines increased the time pressure to finish Macondo. See, e.g., Joel Achenbach, "BP Cost-cutting Measures are Focus of U.S. Inquiry into Gulf Spill," *Washington Post*, October 8, 2010.

[2] Drilling Contract No. 980249.

[3] Internal BP document (BP-HZN-MBI 123225).

[4] Internal BP document (BP HZN MBI 98347).

[5] Internal BP document (BP-HZN-MBI 125958).

[6] Internal BP document (BP-HZN-MBI 123225).

[7] *Ibid.*

[8] 30 CFR § 250.1710.

[9] Internal BP document (BP-HZN-MBI 123225).

[10] Internal BP document (BP-HZN-MBI 21305).

[11] In its internal plans, BP described Kaskida as "one of the largest Paleogene discoveries to date." The discovery had the potential to support the Gulf of Mexico division's goal to "sustain production over

450mboed" (million barrels of oil or equivalent daily). BP expected that the project would break even at oil prices over $55/barrel and that returns would accrue rapidly above those prices. Internal BP document (BP-HZN-MBI 98022).

[12] Press Release, Anadarko Petroleum, BP & Partners Make Discovery at Kaskida Prospect in Gulf of Mexico, August 31, 2006, http://www.rigzone.com/news/article.asp?a_id=35730.

[13] *Ibid.*

[14] Internal BP document (BP-HZN-MBI 123225). Drilling at Kaskida would be cutting edge and challenging. The test that BP planned there in preparation for production would be a "very, very complicated" completion. Testimony of John Guide (BP), Hearing before the Deepwater Horizon Joint Investigation Team, October 7, 2010, part 2, 213. BP described the work as "the deepest and highest pressure completions ever attempted by the industry globally." The reservoir was "tight" and the fluids "relatively viscous." There were also "[s]alt sutures/inclusions" and "salt exit uncertainty." On top of the difficult geology, the size of the casing limited the "number of contingency strings...imparting drilling complexities normally avoided by early appraisal wells." And BP worried about the risks of a "poor quality cement job." To limit the chance of encountering unexpected challenges, BP placed the planned well fewer than five football fields—only 1,400 feet—from the discovery well. BP would also use a relatively large 8½-inch production casing. Internal BP document (BP-HZN-MBI 98022).

[15] 30 CFR § 250.180.

[16] *Ibid.*

[17] *Ibid.*

[18] 30 CFR § 256.77.

[19] Internal BP document (BP-HZN-MBI 123225). In early January, BP contemplated a start at Kaskida as early as mid-March if the *Deepwater Horizon* performed well on the earlier projects. Internal BP document (BP-HZN-MBI 98069).

[20] In February, Richard Harland, the drilling engineer for Kaskida, speculated that the timing looked "tight" if the *Deepwater Horizon* delivered the Macondo well early and came in on schedule for Nile. If Nile also came in early, there was a "real schedule issue." Internal BP document (BP-HZN-MBI 100909).

[21] Internal BP document (BP-HZN-MBI 98027).

[22] Testimony of Gregory Walz (BP), Hearing before the Deepwater Horizon Joint Investigation Team, 105; Pat O'Bryan (BP), interview with Commission staff, December 17, 2010; David Sims (BP), interview with Commission staff, December 14, 2010.

[23] Internal BP document (BP-HZN-MBI 100909).

[24] Internal BP document (BP-HZN-MBI 107569).

[25] Sims wrote Pat O'Bryan, BP vice president of drilling and completions, "Extend work on Macondo (bypass or deepen)...while waiting on 20K wellhead delivery." Internal BP document (BP-HZN-MBI 108874). He then reiterated the original schedule for the *Deepwater Horizon*: "Do Macondo, Nile, then Kaskida IFT."*Ibid.* At about the same time, on March 7, Sims explored engineering possibilities to deepen the Macondo well. He wrote BP senior drilling engineer Mark Hafle that he had been exploring whether there might be interest in "deepening or bypass core to buy us some time for Kaskida wellhead to arrive." Internal BP document (BP-HZN-MBI 109036). He then asked if a liner or a tieback would be better fit if the well went deeper than planned. *Ibid.* A few days later, BP also considered having the rig undergo maintenance to fill gaps in the schedule. Internal BP document (BP-HZN-MBI 113535).

[26] Sims inquired if there was "any work of relatively short duration (+/-1 month) that we could plan and execute if necessary, to allow the wellhead time to be delivered." Internal BP document (BP-HZN-MBI 107569). In response to Sims' request, BP engineering manager John Sprague asked another team to evaluate if there was anything in the division's "[h]opper" that might be available for the *Deepwater Horizon* after Kaskida. Internal BP document (BP-HZN-MBI 108133). On March 1, Sprague heard back that a colleague was "look[ing] through the hopper for any possibilities." *Ibid.* Without good options forthcoming, Sims asked whether another well might be approved by MMS. He wrote on March 5, "For instance, if we needed to accelerate Tiber to follow Nile P&A because the wellhead equipment is not ready could we have regulatory approval to spud by mid-April?" Internal BP document (BP-HZN-MBI 108827).

[27] Sims asked John Guide if Transocean had "put together a wishlist of work for the rig." Guide replied that "we do have a high level wish list." Internal BP document (BP-HZN-MBI 113536).

[28] Sims, interview.

[29] Internal BP document (BP-HZN-MBI 118963).

[30] *Ibid.*

[31] Internal BP document (BP-HZN-MBI 123225).

[32] Internal BP document (BP-HZN-MBI 119027).

[33] Internal BP document (BP-HZN-MBI 123225).

[34] It would take about a day to get to Nile from Macondo.

[35] 30 CFR § 250.169.

[36] Internal BP document (BP-HZN-MBI 123225).

[37] 30 CFR § 250.175.

[38] NTL No. 2000-G17.

[39] *Ibid.*

[40] Internal BP document (BP-HZN-MBI 127785).

[41] *Ibid.*

[42] Internal BP document (BP-HZN-MBI 127785).

[43] Testimony of Brett Cocales (BP), Hearing before the Deepwater Horizon Joint Investigation Team, August 27, 2010, 11; Testimony of Gregory Walz, 104.

[44] Testimony of Ross Skidmore, 210.

[45] Internal BP document (BP-HZN-MBI 126333).

[46] Walz, interview; Sims, interview; Murry Sepulvado, interview; Guide, interview, January 19, 2011.

[47] Testimony of Jimmy Harrell (Transocean), Hearing before the Deepwater Horizon Joint Investigation Team, May 27, 2010, 29.

[48] Internal BP document (BP-HZN-MBI 128990); Sims, interview, December 14, 2010.

[49] Internal BP document (BP-HZN-MBI 128990).

[50] *Ibid.*